# SUICIDE AMONG THE ARMED FORCES
## Understanding the Cost of Service

Antoon A. Leenaars
*Windsor, Ontario*

Death, Value, and Meaning Series
Series Editor: Darcy L. Harris

Baywood Publishing Company, Inc.
AMITYVILLE, NEW YORK

Copyright © 2013 by Baywood Publishing Company, Inc., Amityville, New York

All rights reserved. No part of this book may be reproduced or utilized in any form or by any means, electronic or mechanical, including photo-copying, recording, or by any information storage or retrieval system, without permission in writing from the publisher. Printed in the United States of America on acid-free recycled paper.

**Baywood Publishing Company, Inc.**
26 Austin Avenue
P.O. Box 337
Amityville, NY 11701
(800) 638-7819
E-mail: baywood@baywood.com
Web site: baywood.com

Library of Congress Catalog Number: 2013017864
ISBN 978-0-89503-873-9 (cloth : alk. paper)
ISBN 978-0-89503-874-6 (paper)
ISBN 978-0-89503-825-8 (e-pub)
ISBN 978-0-89503-826-5 (e-pdf)
http://dx.doi.org/10.2190/SUI

**Library of Congress Cataloging-in-Publication Data**

Leenaars, Antoon A.
  Suicide among the Armed Forces : understanding the cost of service / Antoon A. Leenaars, Windsor, Ontario.
        pages cm -- (Death, value, and meaning series)
  Includes bibliographical references and index.
  ISBN 978-0-89503-873-9 (cloth : alk. paper) -- ISBN 978-0-89503-874-6 (pbk. : alk. paper) -- ISBN 978-0-89503-825-8 (e-pub) -- ISBN 978-0-89503-826-5 (e-pdf) 1. Soldiers--Suicidal behavior--Canada. 2. Soldiers--Mental health services--Canada. 3. Soldiers--Suicidal behavior--United States. 4. Soldiers--Mental health services--United States. 5. Suicide--Canada--Prevention. 6. Suicide--United States--Prevention. 7. Canada--Armed Forces--Medical care. 8. United States--Armed Forces--Medical care. 9. Psychology, Military. I. Title.
  HV6545. 7,L44 2013
  362.28088'35500973--dc23
                                                                                                    2013017864

# Dedication

This book is dedicated to Hospital Corpsman Chris Purcell, General Emory Upton, Admiral Mike Boorda, and all our heroes and veterans who died by suicide and to the survivors of their deaths: the family, friends, and fellow Armed Forces members.

And to my Captain, Dr. Edwin S. Shneidman; he first introduced me to the tragedy decades ago. He also taught me about military suicide's complexity.

# Table of Contents

Preface . . . . . . . . . . . . . . . . . . . . . . . . . . . . . . . vii
Acknowledgments . . . . . . . . . . . . . . . . . . . . . . . . . ix

### PART ONE
### Introduction . . . . . . . . . . . . . . . . 1

**CHAPTER 1**
The Military and Suicide . . . . . . . . . . . . . . . . . . . . . . 3

**CHAPTER 2**
Suicide . . . . . . . . . . . . . . . . . . . . . . . . . . . . . . . 29

**CHAPTER 3**
The Psychological Autopsy . . . . . . . . . . . . . . . . . . . . . 81

### PART TWO
### Historical Study . . . . . . . . . . . . . 103

**CHAPTER 4**
Military Suicide: A Classic Population Study . . . . . . . . . . . 105

**CHAPTER 5**
Military Suicide: A Historical Individual Case Study . . . . . . . 121

### PART THREE
### Current Study . . . . . . . . . . . . 151

**CHAPTER 6**
Suicide among the American Armed Forces . . . . . . . . . . . . . 153

**CHAPTER 7**
Suicide among the Canadian Forces . . . . . . . . . . . . . . . . 165

**CHAPTER 8**
Surveillance and the Reliability of Military Suicide Statistics . . . . . . 173

## PART FOUR
## Beyond Suicide . . . . . . . . . . . . 179

**CHAPTER 9**
The Many Faces of Violence: Homicide, Accidental Deaths,
Self-Harm, and Incarceration . . . . . . . . . . . . . . . . . . . . . 181

## PART FIVE
## Military Efforts . . . . . . . . . . . . . 203

**CHAPTER 10**
The Psychology of Military Suicide . . . . . . . . . . . . . . . . . . . 205

**CHAPTER 11**
Posttraumatic Stress Disorder . . . . . . . . . . . . . . . . . . . . . 217

**CHAPTER 12**
Suicide Prevention in the Military . . . . . . . . . . . . . . . . . . . 243

## PART SIX
## A Case Study . . . . . . . . . . . 259

**CHAPTER 13**
A Soldier's Story Told: A Psychological Autopsy . . . . . . . . . . . 261

## PART SEVEN
## Prevention and Policies . . . . . . . . . . 309

**CHAPTER 14**
Military Suicide: Policies and Prevention . . . . . . . . . . . . . . . 311

**References** . . . . . . . . . . . . . . . . . . . . . . . . . . . . . . . 329
**Index** . . . . . . . . . . . . . . . . . . . . . . . . . . . . . . . . . 351

# Preface

Almost a million people die by suicide each year (World Health Organization estimate). This is staggering; indeed, a report of the WHO (2002), *World Report on Violence and Health*, found that more people die by self-directed violence than terrorism, wars, and homicides combined. The report noted the existence of high-risk or vulnerable groups. A current high-risk vulnerable group is our armed service personnel, our soldiers, and veterans. This may be a surprise to many. The *fact* is that it may well be at near epidemic levels in the military. Not since the *great suicide epidemic of the American Civil War* have we seen so many of our heroes die by their own hand. Today, there is one military personnel who dies by suicide every day.

Questions arise: Why? What do we know about suicides in the military? Among American Armed Forces? Among Canadian Forces? Are rates high? Or alternatively, low? Is suicide the same or different in the United States and Canada? Do we know the causes, patterns, and associations? Is suicide among soldiers different than among civilians? Are there similarities (commonalities) and/or unique causes and patterns?

Of course, indigenous people, African Americans, females, police, people with posttraumatic stress disorder (PTSD) and, apropos of the human life diversity, people in the military and veterans die by suicide, and it is often useful to group suicides under these rubrics. However, it can be made more meaningful if one goes further and also looks at underlying general patterns. In the analysis of suicide, there is only human suicide, and much of it can be understood in terms of the same principles. The psychological pain of the soldier or veteran is, of course, different, but in some ways, by virtue of our human quality, we have a number of important psychological characteristics in common. A basic task urged by John Stuart Mill is to answer the question: What are the common and different (unique) factors in military suicide?

How do we study military suicide? What idiographic (individual) and nomothetic (population) approaches can we use? Could we undertake psychological autopsies? There are multiple legitimate acceptable methods in science. From a systematic literature search, some of the best knowledge on the topic in the Armed

Forces is identified. What we do know today: The rate of suicide is high and increasing in the U.S. Armed Forces. There are no published Canadian studies on reliable military rates. In both countries, there is a relative lack of understanding military suicides. This is significantly due to the very complexity of suicide. Suicide is not like water, wherein all water freezes at 32 degrees Fahrenheit. Suicide is multidetermined. There are at minimum individual, relationship, community, and societal factors (WHO, 2002). There are so many issues and risks in suicide. The aftershocks of war are suicides, but also mental health (psychological) disorders (e.g., PTSD), disabilities, motor vehicle accidents, homicides, homicide-suicides, and acts of terrorism. There is no question that our vulnerable heroes and veterans are at risk for many problems, not only suicide. They pay a high price. There is a cost! To be brief, I believe that the government, the Department of Defense, the military, veteran groups, health providers, and other stakeholders need to develop and support more research, programs, and care for suicidal and disabled armed services personnel and veterans. This is a *reason* for writing this book. Finally, there are many valid perspectives on military suicide; I here offer a sociologist's one. We hope this volume assists and allows us to better understand the needless suicides among our warriors in the armed services.

# Acknowledgments

First and foremost, I need to make explicit my debt to Dr. Edwin Shneidman. Dr. Shneidman has served in the military; he was a Captain in the Army during World War II and was for years on staff at the Veterans Administration. He then asked, "Why does a soldier kill himself or herself?" Dr. Shneidman has been my matchless captain for my studies in military suicide and suicidology in general.

The heart of this book is in the story told: Chris Purcell's life and death. His family, Mike, Helene, and Kristin; and his friend, Derek Ozawa, shared their memories and reflections. Myra Morand (Center for Suicide Prevention, Calgary, Alberta, Canada) and Dr. Lindsey Leenaars, my daughter, helped with the literature searches. Susanne Wenckstern, my wife, and Mario Boudrias helped type the manuscript. There are many more individuals and services. I want to thank Dale Lund and the excellent editorial staff at Baywood, especially Bobbi Olszewski, once more, went beyond the call of duty. The support of my family is always appreciated, despite their occasional question of why I do it—writing books. Much, I think, is like what Virginia Woolf said about her diary when reflecting on her writing. Woolf wrote, "I wonder why I do it. . . . Partly, I think, from my old sense of the race of time 'Time's winged chariot hurrying near.'" An author wants to write "it" down, something that I wanted to communicate to myself and, always, to the reader.

Appreciation is acknowledged to the following with permission, if needed under Copyright Act, to reproduce material that appear in this volume:

## REFERENCES

Shneidman, E. (1977). The psychological autopsy. In L. Gottschalk, F. L. McGuire, E. C. Dinovo, H. Birch, and J. F. Heiser (Eds.), *Guide to the investigation and reporting of drug-abuse deaths* (pp. 179–210). Washington, DC: USDHEW, U.S. Government Printing Office. In the public domain. With permission of David Shneidman.

Shneidman, E. (1993). An example of an equivocal death classified in a court of law. In E. Shneidman (Ed.), *Suicide as psychache* (pp. 211–246). Northvale, NJ: Jason Aronson. With permission of David Shneidman.

Michie, P. (1885). *The life and letters of Emory Upton*. New York, NY: Dr. Appleton & Co. In the public domain.

National Defence (Canada). (2010). *Report of the Canadian Forces Expert Panel on Suicide Prevention*. Ottawa, Canada: Author. In the public domain.

U.S. Department of Defense. (2010). *The challenge and promise: Strengthening the force, preventing suicide and saving lives*. Washington, DC: Author. In the public domain.

U.S. Department of the Army. (2009). *Health promotion, risk reduction, and suicide prevention*. Washington, DC: Author. In the public domain.

# PART ONE

## Introduction

There should be discernable stages in our development of understanding suicide among armed forces. There is a natural progression from conceptualization, to understanding, and then to application and practice. The three chapters in this section reflect some basic conceptualizations on the military and suicide.

Following the WHO's recommendation to understand violence (war-related deaths, suicide, and homicide), I take an ecological perspective: individual, relationship, community, and society factors. It is a systems view. In the first chapter, I examine the military, beyond the individual level, and I explicate the collective military ("green") culture. I answer the specific question on the relation of war and suicide through a unique study on the rates and associations of war and suicide during and after the Yugoslavian War. In the second chapter, I look at and answer, "What is suicide?" Suicide is a multidimensional event. I offer a way to psychologically understand suicide by detailing the most worldwide tested theory of suicide. It answers the question, How can we best understand the soldier's suicide?

The psychological autopsy (PA) has been the gold standard for investigations of suicide; in fact, the PA began in the military after WWII. This is the work of my mentor, Edwin Shneidman. I explicate, in the final chapter in this section, using Captain Edwin Shneidman's own words, what a PA is.

# CHAPTER 1

# The Military and Suicide

> The day soldiers stop bringing you their problems is the day you have stopped leading them. They have either lost confidence that you can help them or concluded that you do not care. Either case is a failure of leadership.
> —Colin Powell, Four-Star General, U.S. Army

War is violence. Suicide is violence. They are lethal violence. War-related violence, suicide, homicide, and other faces of violence have probably always been part of the human experience. There are many possible ways of defining violence. The World Health Organization (WHO, 2002) defines violence as

> the intentional use of physical force or power, threatened or actual, against oneself, another person, or against a group or community, that either results in or has a high likelihood of resulting in injury, death, psychological harm, maldevelopment or deprivation. (p. 5)

Intentionality is central. The matter of intentionality is one of the more complex aspects of the definition of violence. It is core to the definition of war-related deaths. The *Oxford English Dictionary* defines intentional as "done on purpose." *Intentionality* is the noun. It is to have as one's purpose. It is a conscious act. One intends death in war. One has a purpose: to kill, to induce annihilation.

## A SYSTEMS THEORY FOR UNDERSTANDING VIOLENCE AND WAR

> I am an American, fighting in the forces which guard my country and our way of life. I am prepared to give my life in their defense.
> —Article I of the Code of Conduct for Members of the Armed Forces of the United States (Cutler, 2002)

Charles Figley (1983) no stranger to the field of trauma, posttraumatic stress disorder (PTSD), and war, has stated, "The nature of war is destructive." There is intent to kill. It is done on purpose. The military is a profession that is sanctioned to use deadly force. There are different points of view on who is whom—we are the heroes or warriors or martyrs, who must kill; and the other side are the enemies or infidels, who must be killed. Although the military has always served other duties, such as peacekeeping, the intent of armed forces is to inflict combat stress and death on the enemy! Violence is expected, beyond what is normally accepted (Christian, Stivers, & Sammons, 2009). There are physical, mental, emotional, and spiritual stresses of war, some of which are everlasting!

Before I turn to combat stress, first called "friction," I need to turn to a few basic cultural concepts and a comment on military training and indoctrination. Although there are very important differences among the Navy, Army, Air Force, Marine Corps, and so on, the military culture can be described as a *culture of relatedness* (Kagitçibasi, 1996), in which interpersonal relations are of central importance, versus a *culture of separateness* such as the American culture, wherein personal autonomy is valued. This corresponds to *collectivism* and *individualism* (Triandis, 1995). Individualism places the individual at the center of a system of values, behavioral choices, and convictions, and emphasizes personal autonomy, independence, and self-actualization. It has been shown that individualistic societies have higher suicide rates than collectivistic societies. One possible explanation for this difference may be that suicide in these individualistic nations may be seen as a final step in self-determination—taking control of one's own destiny. Collectivistic cultures are compatible with acceptance of traditional authority and adherence to core values and moral traditions, especially in matters of life, death, and war (Kemmelmeier, Wieczorkowska, Erb, & Burnstein, 2002). But also they may not foster acceptance, even consciousness, of individualism, including a person's psychological pain, anguish, dishonor, and even suicide risk.

In a series of studies, I have shown that in collective cultures, such as India and Turkey, like previous cross-cultural studies on suicide in individualistic cultures, such as the United States, the United Kingdom, and Canada, there are great commonalties. However, collective cultures also have very specific cultural differences. There is a great deal more of an overall difference in indirect expressions, which at the very least, calls for greater cross-cultural study of this unique aspect of possible culturally specific risk factors. It may be most relevant to the system of the military. The fact that people in collectivistic cultures expressed more indirectness was unexpected from the previous cross-cultural research, but maybe not from a broader cultural view of the military and other cultures of relatedness. Collective culture encourages soldiers, and thus suicidal warriors, to adhere to core values and moral tradition, including in matters of suicide (Kemmelmeier et al., 2002). There is great stigmatization regarding self-harm and psychopathology in collective cultures. The culture has strong sanctions against suicide and suicide attempts. *One is not a warrior!* People in the military, in fact, take pride in their values and collective style; but maybe for suicidal soldiers, it fosters not communicating the intent, not even being conscious of individual pain. Secrecy is paramount. The indirectness may, thus, add to the lethal mix in a vulnerable and self-harsh person. It may well be that military beliefs foster indirectness of expressing intrapsychic pain. Consciousness of individualism is not fostered; relatedness is. The suicidal state may, thus, be more veiled, clouded, or guarded. This is called dissembling or masking and is a significant factor in suicide risk (Leenaars, 2004), which will be addressed in detail in chapter 2. We can already speculate that the lethal dynamics may be associated to collectivistic processes, but this speculation is only provisional; replication in the military is needed.

Christian et al. (2009) contend that military culture differs significantly from the majority American (and Canadian) culture in that "the military expresses a collective vs. individualistic ethos, has clearly defined and codified social hierarchies, explicitly regulates the expression of emotion in many circumstances, does not use material wealth as an index of social standing or power, and promotes a self-concept rooted in history" (p. 31). A systematic review of studies examining the military cultures of several countries concluded that

> many Western militaries share a separate, more hierarchical and collectivistic culture that often allows individual service members to communicate and cooperate better with one another than with civilians from their own countries. However, military culture is not monolithic . . . each branch of the armed forces armed services instills "core values" . . . The military attempts to do so systematically, and immense pressures are exerted to ensure compliance with these values. (Christian et al., 2009, p. 31)

In the military, collectivism is the norm. The day that an armed services member joins the service, he or she is taught to subordinate the self to the group (Christian et al., 2009). McGurk, Cotting, Britt, and Adler (2006), in fact, question whether individuals from Asian cultures, known for being communities of relatedness, would adjust easier to military life. My own studies on collectivism vs. individualism would support that prediction, but also with further implications. Christian et al. (2009) deduced that

> diminution of the importance of the individual in favor of the organization, team, platoon, or unit is a universal military dynamic. . . . Various internal military documents identify unit cohesion as one of the most important factors in protecting against combat stress and maintaining unit morale in both peace and wartime. (p. 33)

Christian et al. (2009) theorized that "Underlying the core values is an overall commitment to the service of one's country and a recognition that this service could ultimately lead to the giving up of one's own life in that service" (p. 37). These authors go on to state that this core belief (concept) "not only serves to motivate individuals further but may also be a protective factor in maintaining unit morale and thus may be protective against combat stress" (p. 38). In addition, it is a protective factor maybe also against suicide risk in most, but *not* for a few vulnerable soldiers.

Military training has been seen as indoctrination. McGurk et al. (2006) view military training within the framework of indoctrination. It is not simply education and training. It is deeper into the conscious and unconscious mind or psyche. "Indoctrination implies a more intense form of persuading individuals to adopt behaviors that are far outside of their previously held worldview" (Christian et al., 2009, p. 42). Al-Qaeda does the same.

Internalization is a central process. Recruits begin on the very first day to

> more actively integrate the values of the group into their own individual worldviews. Conforming to the group is encouraged through social pressure and the continued emphasis on group norms. Because the individual now has incorporated the group's values, norms, standards, and behaviors into his or her own self-concept, inclusion into the group becomes important. Individuals are now more internally motivated to behave in a way that is acceptable to the group. (Christian et al., 2009, p. 42)

There are additional benefits to the group. McGurk et al. (2006) state that through indoctrination, members are also taught to dehumanize and deindividualize the enemy. A core belief is that the enemy is not human. The indoctrination process, in fact, involves the use of a lot of effective cognitive techniques. Grossman (1996) theorized that the dehumanization created by these processes facilitates the ability to kill, sometimes beyond the context of combat. It allows for violence. It allows one to see the other as a terrorist or infidel. Christian et al. (2009) further argued, "Aggression beyond what would normally be acceptable within society is a fact of life that is encouraged in the military" (p. 43). Violence is acceptable. Is it only in battle? Is it towards oneself, in honor? I do not wish to imply that all this indoctrination results in only negative consequences. It creates, for example, resilience or hardiness. Kelly and Vogt (2009) state that the

> factor that may be implicated in how military personnel respond to stress and trauma exposure is hardiness. . . . Hardiness is defined as consisting of three components: commitment, or sense of purpose and meaning; control, defined by the belief that one can influence the events and challenge, the perception that change is a challenge and a normal part of life. (p. 97)

Hardiness or, as mental health providers call it, resilience, is known to mediate the relationship between combat exposure and stress, PTSD, depression, and suicide risk (Kelly & Vogt, 2009; King, King, Fairbank, Keane, & Adams, 1998). It allows one to persevere despite the war friction. *Resilience is all about Sisyphean perseverance!*

In their insightful paper, Christian et al. (2009) conclude, "The military, with its emphasis on service, collectivism, values, and self-sacrifice for others, gives many young people an opportunity to connect with a higher cause" (p. 45). Yet it also introduces—maybe indoctrinates—the warrior to a great deal of stress outside the range of normal human experience, such as in harm's way, even toward death.

*War stress is unforgiving.* An often-used legacy on war stress is of the Prussian War theorist, General[1] Carl Gottfried von Clausewitz. The general used the word *friction* to describe the stresses of combat. His book, *On War* (von Clausewitz, 1982), is regarded as one of the most influential works on military philosophy. Charles Figley and William Nash in the 2007 editor's foreword to *Combat Stress* wrote,

> In Chapter VII, "The Friction in War," he [von Clausewitz] offers a rather simple lesson, but one that is often not carried forward to subsequent generations, on the concept of friction war. He notes that no amount of training or preparation prepares combatants for the friction or the unexpected and distressing experiences of combat—not even those veterans of other battles. The friction of war occupies the mind and distracts the warfighter from the true mission. Combat friction, or simply combat stress, can easily lead to stress injuries unless the warfighter's leaders are fully aware that these are the consequences of war, having little to do with courage, fear, allegiance to duty, or competence. And it is equally important to understand that attending to the stress injury does not imply a lack of courage, weakness, or competence, and furthermore such attention will decrease the likelihood that the injury will become a stress disorder during and following military service. (p. xv)

Figley and Nash (2007) go on to conclude, "Combat stress is like no other as are the memories it creates, and subsequent consequences of these memories" (p. xv). *This is war!*

William Nash (2007a) asked the obvious question, "How might the conscious, intentional wielding of 'combat stress' as a weapon affect how military personnel perceive their own stress?" He answers, "First and foremost, experienced warriors understand that 'combat stress' is not a by-product or side effect of war that can be sanitized away; war is stress" (p. 13). Carl von Clausewitz (1982), as discussed, called it "friction"—*war-related violence in the military is a friction or a sting.*

Following von Clausewitz, Nash (2007a) stated,

> Since most of the terrors, horrors, and hardships of war are unavoidable, it is imperative that warfighters learn to perform effectively despite the "friction" generated by these stressors. To be most effective, warriors must strive to become tough and resistant to the steady stream of physical and mental stressors that impact on them. Ideally, warriors must even learn to ignore combat and operational stress—to not even allow a conscious awareness of stressors and their impact. . . . The genius of great warriors is to fight as if there is no terror, horror, or hardship. In their minds, there can be none—at least, not until the fight is over. (p. 14)

---

[1] In the tradition of the military, I will refer to ranks in the armed forces in the text.

In the military, war stress is a test of personal competence and hardiness—being a real warrior. Nash (2007a) notes,

> To the extent participation in war is perceived by warriors as a test of their personal strength, courage, and competence, admitting to combat stress "symptoms" may be tantamount to admitting failure. Even if some stress symptoms are understood to be due to unavoidable stress injuries, and not merely personal weakness or cowardice, developing stress symptoms can bring with it considerable shame. Warriors volunteer for, train for, and expect themselves to conquer all the stressors of war, even the worst terrors and horrors of ground combat. Therefore, it is hard for warriors to not perceive stress symptoms of any kind as evidence of personal weakness and failure. (p. 18)

William Nash goes on to state that it is not singly the many stressors of deployment that are overwhelming, but also it is the additive nature of the day after day, year after year, until a breaking point is reached. It "is always the sum of all stressors over time that weighs down the proverbial camel to the point of damage" (Nash, 2007a, p. 18). And suicide is the sum of all the stressors over time for some soldiers and veterans. Hoge, Castro, Messer, McGurk, Cotting, and Koffman (2004) found that 17% of heavily engaged soldiers' self-reported significant stress symptoms 3 to 6 months after returning from Afghanistan or Iraq. Like Figley and others, Nash (2007a) points out, however, that "stigma and fear of negative career consequences prevented many of them from seeking care" (Nash, 2007a, p. 36). Silence prevails. Of course, understanding the stressors themselves and their shadowy presence is not enough. We must also understand the unique military culture. Before we proceed, we want to look at violence and a systems model to understand violence better.

## VIOLENCE, SUICIDE, AND ECOLOGICAL MODEL

It is estimated that 1.6 million people die by violence each year. Almost half (800,000) of these are suicides; one-third are homicides (530,000) and one-fifth (320,000) are war related (WHO, 2002). This is an enormous cost of violence, including for the survivors. No single factor or event explains why so many people are violent. Violence is multidetermined. They are complex. War-related deaths, suicide, and homicide and are not like copper or water, where *all* copper conducts electricity and all water freezes at 32 degrees Fahrenheit. They are multidetermined and need a multiaxial approach to understand them (Meehl, 1986)—this is very true about suicide. War-related deaths, suicide, homicide, and the many faces of violence are the result of interplay of individual, relationship, societal, cultural, and environmental factors. This is sometimes called the *ecological model* (Dahlberg & Krug, 2002). The model takes a systems approach (von Bertalanffy, 1968). The model has been applied to a vast array of behaviors, most recently violence, including war-related deaths and suicide (WHO, 2002) (see Figure 1).

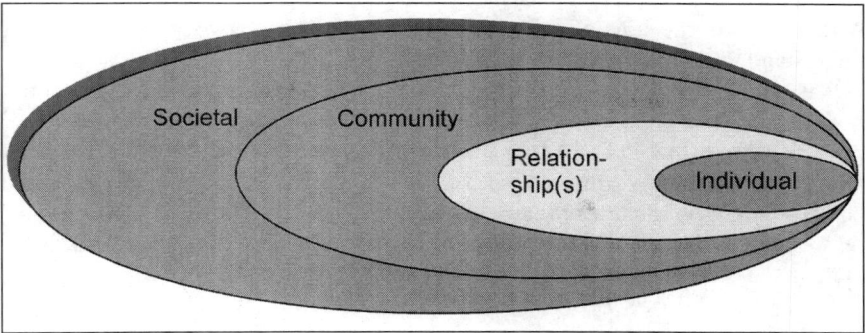

Figure 1. Ecological model for understanding suicide.

The model simply suggests that there are different levels; that is, individual, relationship, community, and societal that influence war-related violence and deaths, suicide and homicide, and thus, by implication, one can address behavior at various levels. Comprehensive (ecological) approaches, thus, would target not only the individual but also the factors beyond the individual. These approaches are also focused on the relationship (e.g., family members, relations with fellow soldiers), community, and societal levels. We will keep this system theory model in mind as we research our topic. First, I want to address some specific issues of systems theory and the military and some specific thoughts on culture.

## THE MILITARY AND WAR: A UNIQUE SYSTEM

Humans live in a "system" of individuals, relationships, community, and society. Ludwig von Bertalanffy (1968), a pioneer in systems theory, noted social phenomena (and military suicide is no exception) must be considered as a "system" (network, organization). The WHO's ecological model is an application of systems theory. Systems theory is a worldview or mindset (Hall, 2011). Probably the most analyzed system is the family system, not the military. The military in general as well as the Navy, Air Force, and Marines specifically need to be studied. In addition, if we are going to address military suicide, a systems approach to the military and its culture becomes necessary. This is true for all collective cultures. Systems theory would state that an interdependent relationship exists between all elements and constituents of the military. The essential factors need always be considered and understood as interdependent factors of a total system. Although no one knows all of the essential elements of a military system, we have enough common ideas to allow us to understand the military better. This is the aim of this section; indeed, I fear anyone who does not understand the system will err in an iatrogenic way.

The attempt to summarize the military system and its culture would not be feasible. It would require a number of volumes, I suspect. Allow me, thus, to offer a few insights into some of what we know. Of course, I am not an expert on military culture and worldview, so I offer only a few essential commonalities derived from the works of Lynn Hall (2008, 2011), a professor and counselor who worked for years for the U.S. Department of Defense Schools (see also Everson and Figley's 2011 edited volume, *Families Under Fire*). However, before I do so, I want to offer some thoughts on culture in general.

## CULTURE:
## A FEW FURTHER THOUGHTS

Before we turn to the topic posed, military suicide, some further questions about culture need to be raised. In the military, the culture is often referred to as a *green* culture (as compared with the *blue* culture of police [Leenaars, 2010a]). Culture is figural in a systems or ecological model. This is true in the military. First, of course, what is it that we are talking about? What is culture? What are we referring to with the term *culture group*? This is a first step. What are we studying in this volume when we state "military or armed forces" in our title? The *Oxford English Dictionary* defines culture as follows:

> 4 The cultivation or development of the mind, manners, etc.; improvement by education and training. $_{E16.}$ 5 Refinement of mind, tastes, and manners; artistic and intellectual development; the artistic and intellectual side of civilization. $_{E19.}$ 6 A particular form, stage, or type of intellectual development or civilization in a society; a society or group characterized by its distinctive customs, achievements, products, outlook, etc. $_{M19.}$ 7 The distinctive customs, achievements, products, outlook, etc., of a society or group; the way of life of a society or group.

Although this explanation falls short, it provides at least a first attempt at definition. It suggests that culture is a cultivation (or read: indoctrination) and this *results* in a community or society having shared values, skills, rules, and knowledge; shared ways of doing things—even in suicide. This would be true for soldiers, in the given service, and veterans. They are a community. It is a unique system. Culture is a collective meaning: to allow people to know who they are, where they come from, and where they are going—again, even in suicide. This is entirely consistent with Leong and Leach's (2008) definition of culture as *worldview*. Soldiers, our heroes, have a worldview. One of the best definitions of culture that I have ever read is from *The Royal Commission on Aboriginal Peoples* (1995) in Canada, which states,

> Culture is the whole complex of relationships, knowledge, languages, social institutions, beliefs, values and ethical rules that bind a people together and give a collective and its individual members a sense of who they are and where they belong. (p. 25)

Culture is rooted in one's people, whether from a society such as the United States or Canada or a group such as a military service. Culture is one's meaning, who you are and where you belong—even in death (Leenaars, 2009a; Leenaars, Maris, & Takahashi, 1997; Leong & Leach, 2008). As I stated some years ago, "The individual—such as John—is among other things a social being. Individuals live in a meaningful world. Culture may well give us meaning in the world. It may well give the world its theories/perspectives" (Leenaars et al., 1997, p. 2).

Although I do not think that culture will ever be clearly and distinctly defined, at least the definitions offered allow us to know better what we are talking about. What are we talking about when we write *green culture*? This means that by the very fact of being from a different group or community or society or even nation, we are different. We are our culture (or subculture). Soldiers are their green culture. Specific examples of the green culture are to adhere to the "macho" outlook or tradition and to take great pride in the achievement of possessing an armed service gun. This raises, thus, a number of questions on our topic of suicide and war-related violence: Is suicide always the same? Alternatively, is it different? Are people in the United States, from, say, a military service, so different from other Americans when they kill themselves? Are there commonalities? Or are there factors or aspects that are different, such as specific relatedness aspects? How can we study suicide in different communities, such as the military? Can we apply our understanding of suicide and military-related violence to all people? To veteran groups? If we need to change our theory with soldiers, what do we need to change? Do we need to change at the individual or relationship(s) or community or societal levels? What are they? Can being a member of an armed service group contribute to suicide? Are there barriers or walls to wellness often called "green" walls? But after we have answered these questions, there are many more, not only in study but in praxis. What military culturally sensitive interventions are needed? What is evidence-based? What policies and procedures in the model are needed to be sensitive to suicidal soldiers, but not enable green walls? Is some of it suicidogenic or even warogenic? We know the green culture fosters deaths. It enables military-related violence. It fosters killing. It trains in lethal use of the gun. It results in an unbearable sting for the survivors. Having raised such questions, we hope the reader will think beyond the individual soldier or veteran and think about the relationship(s), social, cultural, and environmental factors too. You need to know the green culture, among many other factors or aspects, to know the soldier and his/her suicide. I will next turn to the observations of Lynn Hall and many others on the military culture, beginning with the very core identity of a soldier: *the warrior or hero*.

## SYSTEM, CULTURE, AND THE MILITARY

Who is the soldier (again, with apologies, I mean also sailors, airmen, Marines, and an endless list today)? Who is the warrior? What is at the core of the unique military culture? Probably one of the most obvious facts is that being a soldier is an identity—a self. It has its own characteristics, patterns, and associations. I will briefly discuss some core aspects and encourage the reader to turn to Hall's writings for the details.

On the identity of the warrior, Hall (2011) writes,

> On a deeper, perhaps more psychological level, many who join the military feel a need to "merge their identity with that of the warrior." . . . so it is not uncommon for young men to merge their identity with that of a warrior by being a part of something meaningful, which is a motivation to join the military. (p. 34)

On needs satisfied in the search for identity, the military offers escape (so does suicide). On escape, Hall (2011) writes,

> The military also satisfies a need for some young people to get away from whatever they have experienced growing up, "a need for dependence . . . [drawing them] to the predictable, sheltered life". . . [the] attempt to flee from problems at home, for the most part, however, does not end the problems; as in other segments of American society, sometimes the violence, gang mentality, or addiction issues are simply brought with them into the military. (pp. 34–35)

Lynn Hall (2011) isolates some core characteristics of the military system. One of the most obvious is the rigid authoritarian structure. Many individuals thrive in the authoritarian work environment. Hall also described isolation and alienation as characteristics of such structures. She writes, "The warrior society is also characterized by isolation and alienation from both the civilian world and the extended family" (p. 36). The military is a hierarchical system. Christian et al. (2009) write, "A defining feature of military culture is the emphasis on hierarchical relationships with clear power differences. Social status is well defined within the military and determines everything, from peer group identity to obedience to orders" (p 34). The military is a class system; there are the officers and the enlisted ranks. The emphasis in the collective military culture is on maintenance of the rigid boundaries between ranks. Hall (2011) writes, "it is essential for the functioning of the organization to maintain a rigid hierarchical system based on dominance and subordination" (p. 37). The only way that officers and enlisted ranks are equal is in dying in harm's way.

The mission is all-important in the Armed Forces. Martin and McClure (2000) point out that the conditions and demands of "a total commitment to the military—typically a commitment to one's unit, the unit's mission and its members" (p. 4). This is the military's duty. Further, Hall (2011) notes that there is the ever-present preparation for trauma or disaster in the mission. She writes,

> Civilians often seem to blithely overlook a central truth military people can never afford to forget, that at any moment they may be called upon to give their lives—or lose a loved one—to serve the ends of government. Even if it never comes to that, [they] sacrifice a great deal in the course of doing a job that most civilians on some level understand is necessary to the country as a whole. (pp. 40–41)

Hall (2011) observed that the general characteristics of the military system often lead to the existence of strong psychological characteristics. First identified by Mary Wertch (1991), these traits include

1. Secrecy. The military is all about secrecy—the importance of keeping what goes on at work separate from home and vice versa. It creates a culture of secrecy; there is a masking or dissembling. One is ordered to keep silent from spouses, children, and even the doctor, psychologist, and so on.
2. Stoicism. In the military, it is important to keep up appearances. One must be a warrior, a soldier. The appearance of stability and strength is everything. There is a huge stigma to imbalance, instability, mental illness, depression—and especially, suicidality. Despite the constant preparation for change, disaster, or another deployment, thoughts and feelings must not be expressed. Families must also be unaware of the thoughts and feelings.
3. Denial. The soldier must keep all the feelings, fears, or "normal" familial stresses under wraps.

> While in a relatively few number of families this includes domestic violence, marital problems, child abuse, or other "reportable" offenses, what it does mean for most families is that feelings are not expressed, fears are not shared, and the need to ask for help or request assistance goes unnoticed. (Hall, 2011, p. 41)

Suicide must be denied.

Hall concludes, "Secrecy, stoicism, and denial are, in fact, crucial for success of the warrior, success of the mission, and ultimately success of the military" (2008, p. 42). At the same time, these characteristics often influence whether soldiers seek help for the friction. "To the extent that seeking psychological treatment is defined as "weakness," soldiers may be slow to pursue services" (Reger, Etherage, Reger, & Gahm, 2008, p. 30). This reluctance to seek help is a price. It is the military way.

The very purpose of these common factors, of course, is the very aim of the military—violence. Honor and sacrifice is everything in the military. The warrior kills. It is a male psyche. It is about violence. Sam Keen (1991) stated of killing that the soldiers

> have therefore had their bodies and spirits forged into the shape of a weapon. . . . It is all well and good to point out the folly of war and to lament the use of violence. But short of a utopian world from which greed, scarcity, madness, and ill will have vanished, someone must be prepared to take up

arms and do battle with evil. We miss the mark if we do not see that manhood has traditionally required selfless generosity even to the point of sacrifice. (p. 47)

Hall (2011) noted that this is not new to the 20th century. She cites Gary Paulsen, the author of a small book called *A Soldier's Heart* (1998). In his book, Paulsen shares the story of one young Civil War soldier. The soldier, despite experiencing, witnessing, and being confronted with events that involve actual and threatened death or serious injury of his fellow soldiers, was consumed with honor. Paulsen writes.

> But he could not run away. None of the others had and he couldn't. . . . The training must work, he thought. I'm doing all this without meaning to do it. He felt like a stranger to himself, like another person watching his hands move over the rifle. (p. 33)

As noted by Hall (2008), and Nash (2007a), the soldier must go the distance, despite threat to physical integrity of self and others. Honor should be above all else. The concept of honor is, in fact, so central to the military system that every therapist working with soldiers and military families will encounter this (Hall, 2008).

It is a double bind that soldiers are restricted from seeking help. Hall (2008) found that soldiers are "admonished by their chain of command for any domestic matter in which the police are called." This results in problems going underground (read: denied). This is true for asking for help for PTSD, depression, and suicide risk. Yet we know these problems persist and grow. This leads to neglect on the battlefield and at home. The neglect or inattention may endure for 20 or more years of the military career and then continues as a veteran. The career military male, in fact, marries his military service first, not the spouse. Keith and Whitaker (1984) noted that the male-female marriage might then be seen as an extramarital affair. One is first married (identification) to the military. Lyons (2007) and Hall (2008) suggested that the secrecy, not telling the horrible stories, might result in enormous aftershocks—even intentional death.

Shame is powerful. The use of shame is one military toughening-up process. James Gilligan (1996) noted that this results in masking. He observed,

> Behind the mask of cool or self-assurance that many violent men clamp on to their faces—with a desperation born of the certain knowledge that they would "lose face" if they ever let it slip—is a person who feels vulnerable not just to the "loss of face" but to the total loss of honor, prestige, respect, and status—the disintegration of identity, especially their adult, masculine, heterosexual identity; their selfhood, personhood, rationality and sanity. (p. 66)

Gilligan concluded that the "most dangerous men on earth are those who are afraid that they are wimps. Wars have been started for less" (1996, p. 66). Hall (2008, 2011) echoes that view. There is a need for conformity and secrecy. This, Hall (2011) clarified, creates many barriers to treatment, which requires an openness and examination of a person's pain and suicide risk. McKelley (2007) concluded that this results in a warrior's aversion to help-seeking behavior. It is unwarrior-like, unsoldier-like, unmanly. *It is taboo!* Although beyond the scope of this volume, there are, of course, very significant green cultural factors in how you treat the soldier, male or female. There are obvious clinical implications; this is one benefit of a system's view.

In the military, the soldier's values and beliefs play an important role in accomplishing the mission and, it can be argued, in protecting individuals against suicide. The military's collective value system and the importance of group cohesion may be, of course, protective factors against stress and suicide risk. Christian et al. (2009) have argued that

> the relationship between a patient and his unit must be considered in treatment. Considering the bonds that service members develop with one another (buddies) and the ethic of being responsible to the group, service members can often have very intense grief reactions and feelings of guilt when they lose someone in their unit. (p. 45)

This is especially so after a suicide. Soldiers prefer to rely on one another instead of turning to a mental health provider when dealing with stress and violence. Furthermore, "individuals who are sent back home due to injuries sustained in combat will often feel guilty that they left their fellow soldiers and may be very anxious about the safety of their unit" (Christian et al., 2009, p. 46). Christian et al. (2009) further note that "individuals who leave the military, especially due to circumstances beyond their control such as an injury that prevents further service, often experience distress, guilt, and anxiety about returning to the civilian world" (p. 46). Being "separated" from the military is a *deep pain—a pain of pain.* It is a deep loss, a narcissistic injury. Suicide, therefore, can be one escape.

In addition, this system and culture results in spouses and children growing up in a dysfunctional system. Parent absence is common. "This is a society with a great deal of parent absence and, with the changes in the military of the last two decades, sometimes both parents are absent at the same times" (Hall, 2011, p. 38). This affects the family and the soldier deeply, and all too often painfully. There is a lack of trust. "When one is raised unable to trust in the stability, safety, and equity of one's world, one is raised to distrust one's own feelings, perceptions, and worth" (Donaldson-Pressman & Pressman, 1994, p. 18). The military system becomes the family system. The values are the family's values, often for generations. The military taboos, including secrecy, become the family's taboos!

## THE INTEGRATION OF WOMEN INTO THE MODERN MILITARY

> My decision to register women confirms what is already obvious throughout our society—that women are now providing all types of skills in every profession. The military should be no exception.
> —President Jimmy Carter

Carrie Kennedy and Rosemary Malone (2009; see also Sherrow, 2007) noted "Women are no longer serving behind the front line. They are becoming increasingly integrated into combat roles, and the proportion of the military that is female is rising steadily" (p. 67). They go on to state that in America,

> Beginning in the 1980s, women began to be more fully integrated into traditional military roles. The issue of deploying women, particularly for the purpose of ground combat, is contrary to traditional views of women and roles held by women. . . . Some countries have integrated their armed forces, to include placing women in combat roles. In 1989, a Human Rights Review Tribunal in Canada found that excluding women from combat duty was discriminatory, and the tribunal mandated the full integration of women into combat assignments. . . . Israel is often cited as being the most progressive nation with regard to equal treatment of men and women, as both are drafted: however, even though the combat exclusion ended for women in 1995, no women have been assigned to combat units. (pp. 70–73)

The role of U.S. military women continues to evolve today, having served, fought, and died in the front lines, and now in long-prohibited forces: infantry, armor, and special operations. There have been many consequences; some are rape, sexual abuse, and sexual harassment. Combat-related experiences are not the only trauma in wars today. Aside from war friction, the most lasting trauma is rape. Sexual assaults last. In addition, when war trauma and sexual trauma occur together, it is an everlasting nightmare. Unfortunately, sexual assaults, especially of women, occurred/occur frequently in Iraq and Afghanistan. The rate is much higher than among civilian populations. It is compounded by the fact that the aggressor is often military personnel (Montgomery, 2009). It is truly a tragedy of our time. The facts and issues remain hidden, and this also adds to it being a pervasive problem. There is secrecy. That is deadly! The integration of women into the predominantly male green societal structure of the military, in fact, has not been without its suicides. Victor Montgomery, in his 2009 book, *Healing Suicidal Veterans*, presents the case of Jan, a story of rape, pain, and hardiness. Jan called a crisis line, weeping. After establishing rapport, Vic began to explore: "Thank you for calling. How may I help you?" Jan stated,

> "It is hard for people outside the war zone to understand how living in high stress, primitive conditions can affect your ability to make decisions," she declared, her voice quivering. "I didn't report the sexual harassment and attack immediately, because I felt an obligation to continue the mission

and not burden others." She explained that she also wondered how the mostly male upper command would perceive the report. "What would it do to my career and promotions? . . . the Army was my life. . . .

"I was sexually and emotionally abused, harassed, and raped by a fellow Army National Guard soldier while on duty in Iraq. When I deployed home, I felt I could not share this information with my family or anyone else, for that matter. I felt disconnected from my family. I began to isolate myself and am pretty much a loner today. I feel I have lost everything . . . my dignity and honor. And so I decided to leave the service after eight years; I couldn't continue wearing the uniform. I felt dirty . . . I was and still am angry . . . disgusted at myself. I just can't find anything to be happy about . . . I don't smile, Vic. I always had a smile on my face. It's like it is painted over with a brushstroke . . . a blank face, nothing there. I look in the mirror in the morning and don't know who I am anymore. I'm ugly. I was a happy person, Vic. . . .

"I had a lot of friends. I enjoyed being around people. I have no desire to have a relationship. He hurt me, Vic. I hurt all over. I will never be the same again. What's the use of living? I have nothing to look forward to but this same miserable life I live day in and day out. I go to work come home and sit like a zombie. I have difficulty sleeping many nights. . . .

"I feel ashamed and embarrassed beyond belief. . . . I feel like hanging up. I am having a hard time holding this stuff in . . . can't hold this inside any longer. . . . I am going nuts."

Victor Montgomery soon sensed the danger, and he began to develop options and a plan of action.

"Jan, are you thinking about suicide?"

"I have thought about it . . . especially lately," she replied forlornly.

"What have you thought about?" I continued to probe.

"Cutting my wrists with a razor . . . and going to sleep, hopefully escaping from these hopeless feelings. I am tired. Some days are worse than others," Jan responded despondently.

"Have you done anything to hurt yourself?" I asked.

"Oh, it seems I always hurt myself in one way or another, I cut myself on my legs when I am feeling badly. It kind of relieves the tension and takes the edge off. I got therapy when I was a teenager, for cutting. I never have been one for using drugs or alcohol. My father was a drunk and I swore I would never be like him. I hated my mother for living with him and putting up with all his abuse. So I found I would get what relief I needed when I cut myself. The feeling is . . . well, I don't really want to go into it right now."
(Montgomery, 2009, pp. 59–71)

Vic assessed high lethality and developed a rescue plan. The trained crisis center team sprung into action, retrieved information, and began an established rescue protocol, and Jan stayed alive. Yet many such abused soldiers and veterans die by their own hand. These are the wages of war.

Today, soldiers, who also happen to be women, are making history during the current wars. Women are serving in positions never before held by women and, despite casualties, the overall result is positive. I will go further; it has saved many needless deaths.

## SYSTEMS OR PUBLIC HEALTH APPROACH

Before I turn to the topic of war and suicide, I need to make clear for some readers, what is public health. A comprehensive approach calls for both individual and collective care (Leenaars, 2005). Most Americans or Canadians are well aware of individual mental health in the United States and Canada. However, people's knowledge of public health is less so. Thus, I offer a brief perspective on the topic from a larger framework for a comprehensive strategy for suicide prevention in the military.

Public health (or community health or systems health) is not new; indeed advocates of this approach (Satcher, 1998) often point to a mid-19th century example as epitomizing the approach. In 1854, a major cholera outbreak occurred in England. Dr. John Snow started to investigate the epidemic and found a common link: all the victims drank water from London's Broad Street pump. He then made an intervention; he removed the handle from the pump. Soon after that, the cholera epidemic ended. Dr. Snow is a father of public health; his work is a classic in our field.

Suicide is the result of the interplay of individual, relationship, social, cultural and environmental factors This, as we discussed earlier, is called the ecological model (Dahlberg & Krug, 2002; Friedli, Jenkins, McCullock, & Parker, 2002 WHO, 2002). The model simply suggests that there are different levels, that is individual, relationship, community, and societal that influence suicide (or for that matter, war-related death, homicide, etc.), and thus, by implication, one can understand and address (evaluate and treat) behavior at various levels. Public health approaches target the factors beyond the individual. They are primarily focused on the community and societal levels, but may also be at the relationship level (e.g., family members, relations with peers).

Borrowing from David Satcher (1998), former Surgeon General of the United States, there are at least four steps to the public health approach:

1. The first step is geared toward answering the question, "What is the problem?" In this step, we try to define and describe the problem and to measure its magnitude.
2. The second step is to try and answer the question, "What is the cause or causes of the problem?" Here we analyze data, patterns, and associations in order to determine the probable cause(s).

3. The third step is our attempt to intervene in order to prevent or reduce the occurrence of the problem(s). Here we often carry out demonstration projects or clinical trials in an effort to show that something can work to prevent a problem. After we have determined that a prevention strategy works, we move to the next step.
4. The fourth step is to implement the strategy with the broader population and to evaluate the outcome of that intervention (pp. 326–327).

Such an approach can be applied to the problem of suicide in the military—as it can be to cholera, child abuse, and so on. The WHO has, in fact, advocated for the public health approach, and this is within the domain of the armed forces' responsibility. There is a lot that they are doing, and there is much that they can do. Toward the end of this book, I will present some recommendations.

A basic in health care is clearly defining the problem. What is suicide? What are the patterns and associations? Can we determine the probable causes and intervene effectively? Can we be evidence-based?

## WHAT IS EVIDENCE-BASED?

Drs. Snow and Satcher's methods are systematic, comprehensive, and evidence-based. What is evidence-based practice? This is a complex question, not a simple one as many espouse. Evidence-based medicine involves the conscientious, explicit, and judicious use of current best evidence, as delineated in research studies and in individual case studies, to understand and to make decisions about understanding and the care of individual patients or of the larger general population. The practice of evidence-based medicine means integrating individual clinical and/or public health expertise with the best available external evidence from systematic research. The following is one appropriate method for the topic of this volume: systematic reviews.

My intent in this volume is to present the evidence from a systematic review. This is in keeping with the aim of evidence-based practice. I hold to the tenet that evidence-based knowledge is *the* key to effective prevention of suicide in the military. If we are empirically based, we can understand and treat suicidal soldiers more effectively. The purpose of this book is understanding of military suicide; thus, we ask, "What is the problem?" or "What do we know about suicide in the armed services?"

To address these and more questions, we performed a systematic literature review. Mann, Apter, Bertolote, Beautrais, Currier, Haas, Hegerl et al. (2005) present an excellent example on how such a review is done. Electronic literature searches of all articles available until December 12, 2009, were conducted of all literature archived at the Centre for Suicide Prevention, Calgary, Alberta, Canada (M. Morant, personal communication, December 12, 2009). The center houses over 40,000 documents on the topic of suicide, called the SIEC database. We identified reports in "military" and all text fields. We used a broad term for military, such as armed forces, Army, Navy, and Air Force. In addition, we searched for articles targeting systematic reviews. We searched the databases at PubMed/Medline and PsycINFO (with the last search on September 3, 2012). We did not *a priori* specify "relevant" journals; we did not exclude any source. We searched public military documents. We also searched books and chapters in the SIEC database. We attempted to search as far back as possible and wanted to be as broad as possible. From the results, we initially identified hundreds of articles. Abstracts were reviewed, and full text articles that met inclusion criteria were retrieved. Articles and all reports were reviewed in detail, summarized, highlighted, and catalogued, a basis for a better beginning understanding of people who died by suicide in the military. Further, for the review, we also included clinical descriptions and case studies. Studies were included if they reported on the primary topic of interest, namely, military suicide, attempted suicide, suicide ideation, and related types of violence, such as homicide and traffic accidents or, when applicable, intermediate outcomes, including help-seeking behavior, "unit watch," and gun control.

From the search, we identified some of the best knowledge on the topic in the Armed Forces to date. The U.S. sources are very good (Ritchie, 2010). The literature on suicide among the Canadian Forces is very limited, but also there are a few excellent sources (Jetley & Cooper, 2010). The same appears to be true around the world; indeed, many countries do not report data. It is kept a national secret. The U.S. military, unlike many in countries, such as Canada, has, however, identified suicide as a priority and has begun establishing recommendations. The U.S. military offers us one of the best evidence-based sources available.

## WAR AND SUICIDE

What is known about suicide and war? Does war cause suicides? Durkheim (1951), considered a master social theorist on suicide, predicted that the relation between suicide and war will show a decreasing suicide rate during the war, an increase in postwar periods of economic depression, gradually returning to prewar rates when hostilities cease. Additionally, Halbwachs (1978) noted that the decline in suicide rates during the war was not confined to the belligerent countries only. Rather, people felt compassion for the misfortunate nations, promoting help collectively for the suffering people. In fact, the suicide rates declined during World War I and World War II in most nations. Durkheim hypothesized that the drop in suicide rates during wartime is due to the stronger social integration of its participants. War will increase collectivism. The degree of bonding between individuals in support of each other and against a common enemy increases.

Selaković-Bursić, Haramic, and Leenaars (2006) examined war and suicide in the former Yugoslavia. We looked at the suicide rate prewar, during wartime, and postwar. The former Yugoslavia became a communist country after World War II, with the autocratic President Josip Broz (Tito) as the sole governing body of the country. A new nation of communist idealism was formed, and the people believed that they lived in an ideal country, governed by principles of freedom from oppression by the ruling classes, equality, and brotherhood for everybody. However, collectivism was promoted with the belief that enemies lurked on all frontiers. In the years following of Tito's death, chauvinistic-ethnic separatists gained control of the government, blaming each other for the maladies that the country faced. Socioeconomic factors exacerbated conditions, culminating with the inevitable outbreak of civil war in 1991. Since data on suicide during the communist regime in the former Yugoslavia are not available to anyone (read: secrecy), the goal of this research was to focus on suicide rates prewar, during wartime, and postwar from 1989 to 2002. To test Durkheim's (1951) hypothesis *in vivo*, actual wartime situations on two different occasions, the war of 1991–1994, and again in 1999, were examined. Some data were still missing; of course, this is not unusual in communist countries (Leenaars, Lester, Lopatin, Schustov, & Wenckstern, 2002; Wasserman & Varnik, 1998).

During the late 1980s, the political and socioeconomic crisis in Yugoslavia reached new heights. The crisis had its peak influence in Serbia in 1999. The escalating crisis began in Serbia during and after the war in Croatia and Bosnia, reaching its peak during the war in Kosovo (1995) and during the NATO bombing of Serbia (March–June 1999). The greatest inflation ever experienced in the former Yugoslavia occurred in 1993. Following Durkheim's (1951) general social model of suicide during and after war, we predicted that there should be a decrease in suicide rates during the war years of 1991–1994 and once again in 1999 (March–June), followed by a sharp increase in postwar periods in times of economic crisis.

For the method, data on the number of suicides in Serbia and Montenegro were obtained from the Statistical Bureau of the Republic of Serbia (Republicki Statisticki Zavod) in Belgrade for the period of 1989–2003. Data were gathered on the basis of a census conducted in 1991 and 2001 (Selaković-Bursić, 2001). Mean averages were applied to the suicides registered by the census. Other relevant data such as the results of a psychological autopsy, conducted in the Novi Sad area of Serbia in 1999, were included to supplement the surveillance data.

As stated before, suicide is a multidetermined event; war may be one influential factor (Durkheim, 1951). The results of our study demonstrated an increasing trend in acts of suicide in Serbia and Montenegro, from prewar, wartime, to postwar during 1989–2003. Thus, within Durkheim's general social model of suicide and war, a unique pattern of suicide emerged in Serbia and Montenegro. Our findings are most consistent with his specific notions on suicide and unpopular wars. There was a drop in suicide rates prewar in 1990, and that could be explained by the overall economic well-being of the population, since this was a year of economic reforms and a fairly stable standard of living. With the beginning of the war in 1991, suicide rates increased dramatically, reaching a peak in 1992 during the times of the fiercest fighting. These results do not conform to Durkheim's prediction of suicide rate decreases during war times, but other researchers have found this to be not always true. It is likely that it is not simply war or not, but the psychological meaning of the event, as in all suicides whether at an individual level or beyond (Leenaars, 2004). Some wars may bring about integration, but not all. In Serbia, the war was not popular for many Serbians. The suicide rate increase, not surprisingly, was especially found in the male population under the age of 40. This is the population that was most affected.

Our findings indicated a slight drop in suicide rates upon cessation of war in 1995. Subsequently, suicide rates went up again in 1997, just 2 years after the war in Bosnia, reaching unprecedented heights 3 months after the NATO bombing, subsiding once again after 2000 to prewar levels. Thus, Durkheim's (1951) theory seems to hold true for a postwar return to prewar suicide rates. An explanation for these suicide patterns can be further found by imposing the societal factor of economics. The postwar period in Serbia and Montenegro of 1995–1999 saw hyperinflation, with its peak in 1997 (just 2 years after the war in Bosnia) and once again in 1999 (after the NATO bombing). Simultaneously, from 1995 to 1999, suicide rates increased steadily, peaking in 1997 and 1999. War and postwar periods are not sufficient to account for all the fluctuation observed, of course. Suicide is complex, even during war.

There was an increase in male suicide rates during the war period, and this could possibly be explained by the military draft. Hearst, Newman, and Hulley (1986), for example, showed men eligible for the draft lottery in 1970–1972 in California and Pennsylvania had an elevated suicide rate, but the argument is not convincing because the draft existed in other wars, whether there was an increase or decrease in the rate of suicide. Yet our study was the first study of civil war, and thus on native soil (Stack, 2000). This is relevant; following the break-up of former Yugoslavia, there was a war from 1991 to 1994, when the Dayton agreement was signed. Although, officially Serbia was never at war, men were drafted into the army and saw active combat duty as well. In Serbia and Montenegro, military service is still mandatory. It is of note that military researchers in Yugoslavia stated that they reported that the suicide rates in the army decreased over the last 80 years. We wonder if propaganda explains these "facts," because underreporting or not reporting at all is consistent with decades of communist practice (Leenaars, Lester, et al., 2002). There is secrecy. (Is this different about military suicide in the United States and Canada?) There are no published data on the actual suicides in the military during the war years. Of course, many agents of communist countries have stated that there were no suicides in their country; suicide had been cured by communism, although the reality was quite different (Leenaars, Lester, et al., 2002).

Apart from gender and age, the increase in suicide rates during the war years could be explained by the availability of firearms (Leenaars, Cantor, et al., 2002). There is a close relationship between the availability of suicidal means and the method of suicide. In Serbia and Montenegro, historically, hanging seemed to be a predominant method of suicide, constituting more than two-thirds of all suicides. The use of firearms as a preferred method of suicide doubled during the war years and after war. The availability of firearms, of course, increased during and after the war. Furthermore, firearms became a commodity easily obtained on the black market, while restrictions on possession of firearms were loosely imposed. Leenaars and colleagues (2002) have clearly demonstrated that availability of means increases the use of that means for suicide around the world. Additionally, even if there is a restriction on weapons, it is a preferred method for those who do have guns available, such as military personnel. Yet firearms become more available during all wars, so it may not be a key factor here, although the change in method is unusual. Most of the previous studies of war and suicide have been in the United States; firearms are readily available there and have been and still are the most common method. Thus, the previous studies may not apply. The civil war in Serbia was different than all other wars studied. The change to lethal means may well be associated with the increase in suicide rates noted; and we can ask of the current volume, is this true in the American Armed Forces? Canadian Forces?

The second war period of 1999 during NATO bombing was a traumatic experience for the whole population of the country. This war was different. Novi Sad (Vojvodina) suffered heavy bombing and extensive damage, and a very different pattern in the rates appeared in this region. A psychological autopsy of suicides in the Novi Sad area (Selaković-Bursić, 2001) recorded a 50% decrease in the number of suicides as compared with the previous year of 1998. In 1999, unlike the other years, there was a common cause. These results confirm Durkheim's (1951) model of social integration during wartime, when the primary goal of each individual seemed to be the preservation of life and not self-destruction. Individuals were supporting each other. There was collectivism. The suicide rates, as Durkheim predicted, however, increased dramatically just a few months after the war. It is of note that, in the Selaković-Bursić study (2001), respondents (usually family members) cited the trauma in the country *and* war as contributing factors. It was not war alone.

In the psychological autopsy study, stressful or traumatic life events seemed to contribute significantly to suicidal outcomes. The suicidal person was estranged. He or she lost his/her identity (self), relationships, and community. We found depression, loneliness, and physical illness, such as traumatic brain injuries (TBIs). Hopelessness was a determining factor in most suicides. People saw no future, only an endless pain, an enduring suffering. The survivors reported that the suicidal person believed that "*All* was lost, and suicide was the best solution." Thus, the questions asked were, What are the contributing factors? What is the cause(s) of suicide during war? What are the associations and patterns? Regardless of the answers, our study shows that war is stress. The intentional use of force in war resulted in injury, death, psychological harm, adjustment problems, or, in one word—*PAIN*.

In conclusion, in Selaković-Bursić, Haramic, and Leenaars (2006) study, it was found that there were significant changes in suicide rates, especially for male suicides in Serbia and Montenegro during the periods of war. The belief that suicides always decrease in war may well be a myth. There are associations between war and suicide, but they are not simple. War and suicide have an intimate relationship.

What do we know about suicide in the American military? The Canadian military? What are the rates during the Iraq and Afghanistan wars? Prewar? Are they high? Low? What do we know? Are the data reliable? Valid? Will secrecy prevail? Are military leaders in the United States and Canada any different than Tito? What will be the postwar rates among veterans? We will address these topics in a further chapter; however, I want to next make some comments about the survivorship of the soldier's death by suicide.

## SURVIVING THE SUICIDE IN THE ARMED FORCES

> A person's death is not only an ending; it is also a beginning—for the survivors. Indeed, in the case of suicide, the largest public health problem is neither the prevention of suicide . . . nor the management of attempts . . . but the alleviation of the effects of stress in the survivor-victims of suicidal deaths, whose lives are forever changed.
> —Edwin Shneidman, *Deaths of Man*, 1973a, p. 33

How do survivors of suicide adjust to one of the worst possible traumas that occur in a war? On this question, we are fortunate that we have some beginning wisdom. One fact is absolute: The pain of the suicide becomes the pain of the survivor. Anguish, guilt, anger, sadness, shame, and anxiety are a sample of the pains. These responses have a haunting existence. Arnold Toynbee (1968) makes the point that death is a two-party event. He writes,

> The two-sidedness of death is a fundamental feature of death—not only of the premature death of the spirit, but also of death at any age and in any form. There are always two parties to a death; the person who dies and the survivors who are bereaved . . .
> The sting of death is less sharp for the person who dies than it is for the bereaved survivor.
> This, as I see it, is the capital fact about the relation between living and dying. There are two parties to the suffering that death inflicts; and in the apportionment of this suffering, the survivor takes the brunt. (pp. 327–332)

Toynbee here resonates Saint Paul's question, "O death, where is thy sting?" As we see it, the brunt is even more severe for the survivors of suicide in the military. This suffering is even greater than what is expected from war. The loss is more intentional. There are deep feelings of rejection, often perceived as the ultimate one. There is frequently a belief that the sting was malicious. Survivors of soldier's suicide ask, "Did he (or she) want me to hurt so?" The grief can be very complicated.

Suicide is a dyadic event. It occurs in a relational context (system). Shneidman (1972) noted,

> I believe that the person who commits suicide puts his psychological skeletons in the survivor's emotional closet—he sentences the survivor to deal with many negative feelings and, more, to become obsessed with thoughts regarding his own actual or possible role in having precipitated the suicidal act or having failed to abort it. It can be a heavy load. (p. 10)

The load is indeed a heavy burden; the skeletons live. The "extent of the problem" in the general population, however, has only been recognized in the last 15–20 years. There has, in fact, been a rapidly growing effort by many—professionals, psychologists, social workers, psychiatrists, nurses, and the lay public—to understand the "skeletons": the problem. Yet taboos that are present in general society are even more stigmatizing in the military.

Kamerman (1993) has been studying the families of one of the most documented suicide epidemics: the almost 100 suicides among New York police officers in the 1930s. Kamerman notes the survivors have lived with the legacies of their deaths. In a unique study, the families have been followed for over 60 years now. The main conclusion: The suicides of the police officers have affected the lives of wives, sons, and daughters, grandchildren, and great-grandchildren. There is such pain. Indeed, for some, the suicide became the central organizing event in their lives. Yet others learned to survive, requiring painful adjustment; but for some, just living their lives. However, for all there was an *illegacy*. Regrettably, like so much else, there is little, actually one, study of our warriors. There is of late one description of the impact of a suicide of a U.S. Army soldier deployed in Iraq (Carr, 2011):

> A narrative from a treating psychiatrist's perspective describes both acute and long term (four months) effect of the suicide on the members of the soldier's unit, other soldiers at the same base, members of the medical team who attempted to resuscitate him, and the mental health care providers. (p. 95)

Many soldiers were affected, some common in the literature, some unique to a military system; not surprisingly, this was true of the healthcare providers, even the psychiatrist. There was traumatization and suicide risk. The question posed: How will the soldier's or veteran's suicide affect the lives of family, friends, fellow soldiers, health providers, and the public? *We need to know!*

## CONCLUDING INTRODUCTORY THOUGHTS

Military service has unique stresses. The nature of war is traumatic. Trauma is a wound or injury. The intent of war is to create stress, to traumatize, and to kill. This has been known for millennia. It is known that trauma and loss carry on throughout life. Trauma can have an *infinite sting*, and many veterans can tell you that. In the military, our heroes, whether Army, Navy, Air Force, or so on, are confronted with events that are outside the range of normal human experience. The Afghanistan War is a good example. Armed services men and women or soldiers (in the generic sense) are more frequently traumatized than people in the general public, not only by one event but continuously. Many events involve actual and threatened death and serious injury. That is war. There are threats to the physical and psychological integrity of self and others, which even occurs during peaceful times. It would be normal (not crazy or abnormal) to respond with intense fear, helplessness, and horror.

Traumatic events would "horrify, repulse, disgust, and infuriate any sane person" (Rudofossi, 2006) After the Vietnam War, the now obvious question was asked, "Why shouldn't that be true for heroes and veterans?" Regrettably, after traumatic experiences, a common response is "Tragedy in the military happens, learn to deal with it" or "Snap out of it" or "Just get over it" or "This is army life." Silence, of course, is a core part of military existence. Silence is powerful, but it may also become dysfunctional—even lethal. Aftershocks to traumatic events may include not only alcohol abuse, rape, and violence, but also suicide; and all too often, homicide occurs as well. Yet keeping secrets is core to the military system/culture. Forgetting, avoidance, phobias, and inhibition, as was first well documented in survivors of PTSD among Vietnam veterans, only traumatize the person more. Today, we see the same among our Iraq and Afghanistan heroes. The armed services person is "numbed out" or "zoned out." In this book, I will attempt to break down the green wall of silence, which is exemplified by the statement, "Don't talk about it."

Loss or trauma all too often results in the armed services member becoming "disenfranchised"; a core loss in the green culture that many people experience. He/she is then caught in a catch-22 (no-win cycle) of persistently reexperiencing the event (such as recurrent and intrusive distressing recollections, recurrent disturbing dreams, acting as if the event was reoccurring) and persistent avoidance (such as efforts to avoid activities and duties, having an affair, leaving the Armed Forces, restricting affect), with increased symptoms of difficulty falling or staying asleep, irritability, difficulty concentrating, hypervigilance and/or exaggerated startle response. TBIs especially make these processes even more complicated. Yet there is hope; despite the taboos ("You have to be the tough guy"), soldiers can heal; and hopefully this book will offer some beginning paths to healing and a better military. There are abundant existing services in the United States and Canada. However, I believe that we can do more.

This topic presents an arduous challenge. This book will ask some tough questions: What is the trauma? What are the painful short and long-term consequences of a trauma for the armed services members and veterans? Why does suicide occur? What is the unbearable pain? What are life-saving military factors? What are the evidence-based recommendations and interventions that might allow a suicidal soldier or veteran to courageously overcome not only the shocks but also the aftershocks? What are the barriers (green walls) to wellness in the military? What can be done to more effectively educate and help our at-risk for suicide (and homicide, self-harm, motor vehicle accidents, etc.); our warriors and veterans, and even the Admiral? Furthermore, of course, in light of current events in America and Canada, there is the dire system/service question: Will we continue to allow our soldiers to suffer the traumatization and die by their own hands? I will ask key questions; this is called the Socratic method—Socrates knew that this form of exploration empowered the person; being asked the question, not the teacher (or therapist, or the sergeant). There will be many questions throughout the book that will focus on the heart and soul of the soldiers and veterans.

# CHAPTER 2
# Suicide

Suicide is a multidimensional event (Leenaars, 1988, 2004; Shneidman, 1973b, 1985; Zilboorg, 1937). Suicide and suicidal behavior are multifaceted events. We discussed this earlier under the ecological model (WHO, 2002). There are biological, psychological, intrapsychic, interpersonal, sociological, cultural, and philosophical elements in the event. Thus, suicide and suicidal behavior cannot be reduced to a single factor. This complexity of causation indicates the necessity of a parallel complexity of knowledge. Indeed, because suicide is a multifaceted problem, it needs to be understood on several different levels at once. This is equally true of homicide (Allen, 1980) and what presents to us too often today, war-related deaths. We will explore these topics, but we will first discuss suicide in some detail. Unequivocally, one has to know about suicide to understand a suicidal soldier or veteran.

Most frequently, people identify external causes (e.g., combat stress, ill health, being abandoned by the spouse, a demotion) as to why the person killed him/herself. This view is too simplistic, although often the suicidal person holds that perspective. This is not to suggest that a recent traumatic event (e.g., a divorce, a traumatic brain injury [TBI]) cannot be identified in many suicides. However, although there are always situational aspects in every suicidal act, they are only one aspect of the complexity, which I hope to demonstrate in this book.

Suicide is a multidimensional malaise (Shneidman, 1985). Within the larger frame of the ecological model, at the individual level, suicide is seen as a state of being, a human malaise or a condition or discomfort. Any element of the malaise is a legitimate avenue to understanding suicide. Studies of serotonin have a place. Studies of military culture have a place. Studies of the effect of gun control have a place. In fact, I oppose any reductionistic model in understating suicide. Suicide is a multifaceted event and is open to study by multiple disciplines; in this chapter, I offer a psychological/psychiatric perspective, primarily at the individual and relationship levels. Let us begin with Shneidman's (1985) arboreal image to understand suicide:

An individual's biochemical states, for instance, are the roots. An individual's method of suicide, the contents of the suicide note, the calculated effects on the survivors and so on, are the branching limbs, the flawed fruit, and the camouflaging leaves. But the psychological component, the problem solving choice, the best solution of the perceived problem, is the main "trunk." (pp. 202–203)

From a psychological point of view, I would like to offer a few observations on the question, "Why do soldiers die by suicide?"

## DEFINITION OF SUICIDE

Understanding begins with definition. Briefly defined, suicide is the human act of self-inflicted, self-intentioned cessation (Shneidman, 1973b). Suicide is not a disease (although there are many who think so); it is not a biological anomaly (although biological factors may play a role in some suicides); it is not an immorality (although it has often been treated as such); and it is not a crime in the United States, Canada, and most countries around the world (although it was for centuries). Yet, as a footnote, in armed services, there may be exceptions to "it is not a crime." For example, an exception may occur in the height of battle if a soldier attempts suicide, resulting in needless deaths among his fellow warriors. This can be called *suicide-by-battle*, for lack of a better term.

Suicide may today be defined differently depending on the purpose of the definition—medical, legal, administrative, and so on. In the United States and Canada (and most of the countries reporting to the World Health Organization) suicide is defined (by a medical examiner or coroner) as one of the four possible modes of death. An acronym for the four modes of death is NASH: natural, accident, suicide, and homicide. This fourfold classification of all deaths has its problems. The major deficiency is that it treats the human being in a Cartesian fashion, namely, as a biological machine rather than appropriately treating him or her as a motivated, intentional, biopsychosocial organism; that is, it obscures the individual's intentions in relation to his or her own cessation, and further completely neglects the contemporary concepts of psychodynamic and cognitive psychology regarding intention, including unconscious motivation.

Shneidman's entire book, *Definition of Suicide* (1985), can be seen as a necessary step to more effective understanding and prevention of suicide. In the book, Shneidman argued that we desperately need a clarification of the definitions of suicide—definitions that can be applied to needful persons. In his definition "suicide is a conscious act of self-induced annihilation, best understood as a multidimensional malaise in a needful individual who defines an issue for which the suicide is perceived as the best solution" (p. 203).

I will be using Shneidman's definition in this volume.

The two giants in the field of suicidal theorizing at the turn of the 20th century were Emile Durkheim and Sigmund Freud. Durkheim, in *Suicide* (1951), focused on society's inimical effects on the individual, while Freud, eschewing the popular notions of either sin or crime, gave suicide back to intentional man but put the locus of action in man's unconscious. Since around 1900, a host of psychological theories aside from Freud's have focused on the individual; for example, those of Alfred Adler, Ludwig Binswanger, Carl G. Jung, George Kelly, Karl Menninger, Henry A. Murray, Edwin Shneidman, Harry Stack Sullivan, and Gregory Zilboorg, to name some of the best known (Leenaars, 1988). This list was made with Edwin Shneidman in the mid-1970s to early 1980s. From a 2012 perspective, there would be obvious additions. I would add, at least, Aaron T. Beck (Beck, 1963, 1967, 1976; Beck & Greenberg, 1971; Beck, Resnick, & Lettiere, 1974; Beck, Rush, Shaw, & Emery, 1979) and David Lester (1987); I will discuss their theories in more detail below. These suicidologists have given us a rich history to understand suicide; thus, we will use the ten suicidologists' theories to understand suicide among armed services personnel and veterans better.

Intentional, Subintentional, and Unintentional

Freud (1974a), Shneidman (1963), Murray (1967), and others have speculated that beyond intentional suicides, there is a vast array of subintentional inimical behaviors. The very lifestyle of some soldiers seems to truncate and demean their lives so that they are as good as dead; for example, a soldier who runs recklessly into a field of fire in battle. Self-destruction is not rare. Often, alcoholism, drug addiction, mismanagement of armed service guns, unnecessary high speed flying of a jet fighter, and masochistic behavior can be seen in this light (Farberow, 1980; Freud, 1974a; Murphy, 1992). We, in this volume, will explore in a latter chapter the many faces of violence and suicide.

A related concept is "subintentioned death" (Shneidman, 1963). This concept asserts that there are many deaths that are neither clearly suicidal nor clearly accidental or natural. These are deaths in which the decedent played some covert or unconscious role in "permitting" his or her death to occur, either "accidentally" or by "inviting" homicide, or by unconsciously disregarding what could be a life-extending medical regimen and thus dying sooner than "necessary." An important aspect of the issue at hand is that suicide is an intentional act. As Litman (1984) noted,

> The concept, which defines a death as suicide rather than an accident, is intention. For example, we assume that when a man shoots himself in the head with a gun, he intended to die. Therefore the death was a suicide. However, if in fact, he intended to survive, for example, if he thought the gun was not loaded, the death was accidental. (p. 88)

Thus, the concept that defines a suicide, including among the armed services, is intention. Suicidal intent can be defined as "understanding the physical nature and consequences of the act of self-destruction" (Nolan, 1988, p. 53). There is an understanding of the finality of the self-directed violence. What is the soldier's intention relating to his or her being dead? The verb "to intend" means to contemplate, to plan, to purpose. The noun, "intention" indicates a psychological exertion for a purpose, to an end (Litman, 1988). Intention was critical; in suicide, there must be intention to die. The person has, in his/her mind, the intent to be lethal. Litman (1988) stated, "The concept of suicide requires that the self-destructive action has, for at least one of its purposes or goals, the death of the person" (p. 71). Litman (1988) further stated,

> The concept "intentional" signifies to me that the individual in question understood, to some degree, his or her life situation and also understood, to some degree, the nature and quality of the self-destructive action (the proposed action representing to some degree, in the person's mind, killing one's self as a solution to the life situational problem). (p. 72)

Litman (1988) further noted,

> In suicide, a person has it in mind to end a distressing life situation by a self-destructive act, which carries a known predictability for causing death. Death is understood, in the mind of the person, as an end to his or her earthly existence. *When one's own death is being used instrumentally to solve life's problems, we are talking about suicide.* (p. 71)

Thus, the person who intends to die, by Litman's (1988) definition, would have to understand the finality of the act. A common question, often asked by people is, "How is suicide intent determined?" Nolan (1988) stated, "Since the suicide victim is dead and unavailable for direct inquiry as to his intention, professionals charged with making such determinations have developed the standard investigatory technique now known as a psychological autopsy" (p. 56).

Litman (1988) makes the further obvious point about determination of intention even in living people. He stated,

> Unfortunately, absolute certainty about human intentions is seldom achieved even with the living, including our patients, colleagues, and families. We constantly act upon our own evaluations of others' intentions based upon their verbal communications, their behaviors, their previous track records and the social context. (p. 78)

This is what a psychological retrospective study, called a psychological autopsy, does; it is "an excellent window for viewing and understanding intention" (Litman, 1988, pp. 78–79). I will discuss this topic in detail in the next chapter. Briefly, the psychological autopsy is a window into what the suicidal person had in mind.

There are exclusions in any argument of intentionality. The most frequent legal exclusion is the question, as defined by law, of "sane or insane." Suicide is an intentional, self-directed act and many "insane" people, by law (not all), cannot commit suicide since that person lacks "the requisite mental capacity to form intent to commit suicide" (Nolan, 1988, p. 52). There are people who do not understand the finality of the act; some soldiers with a TBI may be a good example. Litman (1988) offers the example of a man who climbed into a lion's area of a zoo under the delusion that he was a biblical prophet with lion-taming powers, but the lion killed him. There are further exclusions, such as whether the soldier's blood alcohol level is at a significant level that would impair that person's ability to understand his or her act. Drug intoxication can be the same. Yet again, these exceptions rarely apply.

Therefore, we are well advised to keep the concept of intentional, subintentional, and unintentional in mind, especially in forensic military cases (Litman, 1988). For example, on November 5, 2009, Major Nidal Malik Hasan killed 13 people and wounded 30 more at the Soldier Readiness Center at Fort Hood, Texas. No one predicted that trauma (and the deadly aftershocks). What was his intent? Did he intend homicide-suicide? Did he intend to be a martyr? Why would a Major in the military service act that way?

## Attempted Suicide

A previous attempt is one of the best clues to future attempts (Beck, Kovacs, & Weissman, 1975). However, not all previous attempters go on to attempt again (or kill themselves). Approximately 15% of individuals who previously attempted do so, versus 1.5% for the general population (Leenaars & Lester, 1994). Although it is obvious that one has to "attempt" suicide in order to die from suicide, it is equally clear that the event of "attempting suicide" does not always have death (cessation) as its objective. It is acknowledged that often the goal of "attempted suicide," sometimes referred to as the parasuicide (which includes self-harm, such as self-cutting or ingesting large amounts of alcohol), is to change one's life (or to change the behavior of the "significant others" around one) rather than to end it (Shneidman, 1985), such as wanting to escape from the front lines. However, I wish to stress, as have others (e.g., Stengel, 1964), that it is useful to think of the "attempter" and the "completer" as two sets of overlapping populations: (a) a group of those who attempt suicide, a few of whom go on to die by suicide, and (b) a group of those who die by suicide, many of whom previously attempted it. A great deal has to do with the *perturbation* and *lethality* associated with the event.

Perturbation refers to how upset (disturbed, agitated, "insane") the individual is—rated as low, medium, and high—and may be measured by various means (e.g., self-reports, biological markers, psychological tests, observations). Lethality is roughly synonymous with the "deathfulness" of the act and is an important dimension in understanding any potentially suicidal individual. Lethality can also be rated as low, moderate, and high. An example for measuring lethality is the following assessment item, derived from Shneidman (1967): "During the last 24 hours, I felt the chances of my actually killing myself (committing suicide and ending my life) were: absent, very low, low-medium, fifty-fifty, high medium, very high, extra high (came very close to actually killing myself)." A critical distinction in suicide (and often, for that matter, homicide) is that lethality, no perturbation, kills. All sorts of people are highly perturbed but are not suicidal. The ratio between suicide attempts and completions ranges from 4 to 1 to about 8 to 1—one completed suicide for every four to eight attempts; however, in young people, some reports have the ratio at 50 to 1, even 100 to 1. The ratio, in fact, appears to vary a lot between nations and across specific groups; that is, gender, age, and cultural group. We do not know the rates in the military.

## CONTEXTUAL OBSERVATIONS

From a vast potential list (see Hawton & van Heeringen, 2000), I need to offer a few contextual clues, focused on clarifying military suicide.

### Biological Roots

"No brain, no mind" is one of Henry Murray's reminders to his student Edwin Shneidman. Shneidman's position is best presented in his 1985 book, *Definition of Suicide*, to which I have previously referred. Therein, he states his principles (see his Chapter V, "A Formal Definition, With Explication"). He writes, "Suicide is a multifaceted event and that biological, cultural, sociological, interpersonal intrapsychic, logical, conscious and unconscious, and philosophical elements are present, in various degrees, in each suicidal event" (p. 202).

Stoff and Mann's (1997) edited volume begins to outline the understanding of the neurobiology of suicide. On a critical note, however, one could see Stoff and Mann's view as being too reductionistic. They reduce suicide to only biological roots. They write,

> Efforts aimed at identifying the potentially suicidal individual using demographic, social developmental and psychological factors offer too weak a prediction to be of substantial clinical utility. It is believed that the biological perspective, which has grown out of the expanding research on the biological basis of mood disorders, is a predominant approach to suicide research. It can assist in the investigation of risk factors that predispose a person to suicidal behavior and that increase understanding of etiology, treatment, and ultimately, prevention. (p. 1)

Utilizing only this view will lead us astray. A person is not simply a biological anomaly. Once more, suicide is complex; more complex than most people, even experts, are aware. It is not only the suicidal brain. It is not only the suicidal mind.

Asberg, Traskman, and Thorien (1976) undertook the classic study in suicidology of the biological roots. They identified a substance known as 5-H1 AA in cerebrospinal fluid as a biochemical marker in some suicides. Yet, although the Asberg study is now over 35 years old, there is relatively little well-documented, verified knowledge regarding the neurobiology of suicide today (Hawton & van

Heeringen, 2000). Be that as it may, Maltsberger (2001) and others are optimistic about our biological understanding of suicide in the future, especially in the relation of suicide to affective disorders (e.g., bipolar disorder/manic-depressive disorder) and their neurobiological correlates. At least, in some suicides, biological correlates may be strikingly relevant.

Brain Dysfunction and Traumatic Brain Injuries

The importance of brain dysfunction and its relation to cognitive deficits, learning disabilities, aggression (violence) regulation, impulsivity, and other abilities are well documented. The relation of brain dysfunction and socioemotional problems in people is, however, a more neglected topic in the literature. Not only does a cognitive disorder (e.g., TBI, learning disability, Alzheimer's) render a person at risk for socioemotional problems including suicide, but also there are particular subtypes of impairments/disabilities that may result in different levels of suicide risk. As a way of understanding this clue, I cite a brief literature review on acute brain injuries, such as TBIs.

Specific cerebral deficits associated with brain dysfunction render people at risk for very specific problems such as planning and sequencing social events and such; and sometimes there are problems in executive functioning such as mood regulation, thinking before acting, and so on. Attention deficit/hyperactivity disorder (ADHD), in all its subtypes, is an example. There are soldiers, of course, with ADHD, some with very high IQs. There are diverse brain dysfunctions, some that result in chronic difficulties, while some are acute and more temporal in nature. The acute problems may be due to strokes or to recovery after being hit by a bullet in the head (or in a self-inflicted shooting). There are, in fact, many possible war-related injuries and other risk factors in the Army. Although further empirical studies need to be conducted in the neuropsychology of suicides, these observations clearly warrant attention. Indeed, Rourke, Young, and Leenaars (1989) have shown that one possible outcome of brain (central processing) deficiencies, especially in the right hemisphere, is suicidal behavior, as well as other socioemotional problems, such as depression.

Special Reference to Traumatic Brain Injuries

I must make special reference to traumatic brain injuries (TBIs). This is because it would be accurate to state that TBIs in the military today would be at epidemic levels if the same rate occurred in the general population. It may, in fact, be a huge suicide risk factor among armed forces.

Too little is known about traumatic brain injury. TBI is the disruption of normal brain functioning that occurs secondary to any number of traumas, including a bump, blow, or jolt to the head or a penetrating head injury (Centers for Disease Control and Prevention [CDC], 2009). Breshears, Brenner, Harwood, and Gutierrez (2010) note,

> Secondary to injuries sustained by U.S. military personnel in Iraq and Afghanistan, there has been increased focus regarding the assessment and treatment of deployment-acquired TBI, especially mild injuries caused by explosions/blasts (Warden, 2006). Estimates of deployment-acquired TBI range from 15% to 23% (Hoge, Auchterlonie, & Milliken, 2006). Furthermore, there has been growing concern and debate regarding the psychiatric/emotional sequelae that is often associated with TBI in the military population. (p. 349)

Only recently has the military brought the issues of TBI and suicide independently into the public eye. They are rarely addressed together. TBIs have been identified, however, as a very significant factor in soldiers' suicides (Ritchie 2010). Ritchie (2010) noted that *there is a relationship between TBI and suicide* This may answer one of the main questions in a public health approach, "What is the cause(s) of the problem?" Very little is known about TBIs and suicide risk. Teasdale and Engberg (2001) presented the seminal population-based study on TBI and suicide. They found that the suicide rate among individuals with TBI was 2.7 to 4.1 times higher than that of the general population. Of course it depended on the nature of the injury, as suggested earlier. Not all brain injuries or dysfunctions are alike; some present greater risk. Teasdale and Engberg further determined that the clear majority of individuals who attempted suicide postinjury had no history of suicide behavior, such as attempts predating TBI onset. The findings suggest that for at least a subset of TBI patients, injury status has the potential to independently pose increased risk for suicidal behavior (Breshears et al., 2010).

Lisa Brenner (2010) has stated that the most recent military data, indeed, suggest that TBIs are a huge risk factor for suicide among soldiers and veterans. Despite difficulties in the study, she suggested that the risk of TBI and the psychological/emotional issues that results, and often the alcohol/substance abuse is *21 times higher*. Imagine a risk factor 21 times higher; there are very few such risk factors! *TBIs are a significant suicide risk factor!* Further, Brenner also noted comorbid problems. Many injured soldiers with TBIs can also be diagnosed with PTSD. She stated that one needed to discern between posttraumatic stress disorder and TBIs. One must discriminate. They may be overlapping, but they are also different. Veterans with TBIs report more headaches and dizziness. Symptoms related to TBIs are constant but not necessarily those related to PTSD. Of course, they can overlap. It is complex. Depression and alcohol abuse are also common subsequent reactions. Brenner, like Ritchie (2010) concluded that we needed to know more.

In their study, "Predicting Suicidal Behavior in Veterans with Traumatic Brain Injury: The Utility of the Personality Assessment Inventory," published in the *Journal of Personality Assessment*, Breshears et al. (2010) presented one of the best studies to date. The question: What measures of psychopathology and suicidality can have functional utility and construct validity with respect to the individuals with a history of TBI and suicide risk? What evidence-based method can be used? Breshears et al. used an archival method. They utilized archival data from 154 veterans of the U.S. military who sustained TBIs. They note,

> Of the 154 subjects . . . a total of 11 (7.1%) subjects engaged in suicidal behavior within the 2-year follow-up period, and five (45.5%) of these had more than one documented incident, resulting in a mean number of suicidal behavior incidents of 1.8 ($SD = 1.1$). The severity of suicidal behavior among the 11 subjects encompassed highly lethal suicide attempts with injuries and explicit or implicit intent to die as well as suicide attempts without injury and one incident of self-harm behavior with injury (e.g., taking a medically serious overdose of medications reportedly in an attempt to incite hospitalization). (p. 350)

Of course, as stated in this chapter elsewhere, the prediction of suicide and related behaviors is inherently problematic (Pokorny, 1983). Yet there are certain risk factors that do suggest increased risk. Probably one of the best risk factors is a previous suicide attempt (Mann, Waternaux, Haas, & Malone, 1999; Shneidman, 1985). In Breshears et al.'s study (2010), the medical archives of 29.25% of subjects included histories of prior suicidal behavior, and of these, 24.4% engaged in suicidal behavior during the 2-year period. Most important, no soldiers who lacked such histories engaged in suicidal behavior during the 2-year period studied. They concluded that when a patient with a TBI had a history of suicidal behavior, the presence of high levels of psychological distress was associated with increased risk for future suicidal behavior. The risk was greater than that suggested by suicidal behavior alone. The presence of a TBI, thus, increased the risk. The common timeline risk may range anywhere from 3 to 18 months. They further concluded that TBIs are an independent risk factor for suicide. Furthermore, a TBI, especially if associated with PTSD, is a very significant risk factor, both acute and chronic. This risk, in fact, continues throughout life.

## Physical Disabilities and Illness

I would be remiss if I did not, as a further contextual observation, at least note the importance of physical problems in suicidal behavior in some people (Barraclough, 1986); and this is especially true in war, given the large number of war-related injuries. Physical injury/illness interacts with an individual's emotional functioning; indeed, some illnesses directly affect one's emotions. Empirical study regarding illness and suicide is urgently needed; there is little in the military (see Koren, Hilel, Idar, Hemel, & Klein, 2007). Currently, research suggests that physical illnesses associated with suicidal behavior include anorexia, bulimia, diabetes, epilepsy, traumatic brain injury (as discussed above), and muscular dystrophia (Barraclough, 1986). Some individuals with physical disabilities who are at risk are those with limb amputations or spinal injuries resulting in quadriplegia; regrettably, these types of injuries are common in war. However, it is important to realize that not all such people are suicidal.

## BEHAVIORAL CLUES

In understanding suicide, we need to be aware of behaviors that are potentially predictive of suicide. However, there is no single definitive predictive behavior. The two helpful concepts, which have already been discussed, are lethality and perturbation. The clues below are applicable to all age groups, although the mode of expression may differ depending on age and numerous other factors. I present here a few insights.

### Previous Attempts

A previous attempt is a good clue to future attempts, especially if no assistance is obtained after the first attempt. However, not all previous attempters go on to attempt again (or kill themselves). Regrettably, despite the lethality of these risk factors, too frequently such behavior is not taken seriously, or minimized, even covered up by fellow soldiers. "It was only an accident," a soldier said after his fellow soldier took his service revolver, placed the gun to his head, and shot himself. The soldier said, "I saw the trigger pulled"; yet he could not believe that it was a suicide. It was too traumatizing.

### Verbal Statements

As with behavior, the attitude toward individuals making verbal threats is too frequently negative. These statements can be seen as manipulative or "just for attention." This attitude results in ignoring the behavior of a person who is genuinely perturbed and potentially suicidal. This is especially true in the military, due to the previously discussed military culture. Examples of verbal warnings are the following: "I'm going to kill myself" or "If they take my gun, I'll kill myself," both being very direct. Other more indirect examples are the following: "I am going to see my (deceased) mother" or "I give up" or "I can't cope with this war stress."

### Cognitive Clues

The single most frequent state of mind of the suicidal person is cognitive constriction. There is a tunnel vision, a narrowing of the mind's eye. There is a narrowing of the range of perception or opinions or options that occur to the mind. Frequently, the person uses words like "only," "always," "never," and "forever." The following are examples: "No one will ever love me. Only my wife loved me"; "Sam was the only one who loved me"; "The antigay attitude will always be that way"; and "Either I'll kill the sergeant or myself."

## Emotional Clues

The person who is suicidal is often highly perturbed; he or she is disturbed, anxious, and perhaps agitated. Depression is frequently evident. Suicidal people may feel boxed in, rejected, harassed, and unsuccessful. Some frequent feelings reported by patients are the following: anger, anxiety, emptiness, loneliness, a sense of loss and sadness. A common emotional state in most suicidal people is hopelessness/helplessness. Statements like the following may signal hopelessness: "Nothing will change. Even the Captain rejected me. It will always be this way." Whereas helplessness may be verbalized as "There is nothing I can do. There is nothing the Secretary of Defense can do to make a difference." Shame and disgrace may also be especially unbearable pains in a military culture.

## Sudden Behavioral Changes

Changes in behavior are also suspect. Both the outgoing individual who suddenly becomes depressed and isolated and the normally reserved individual who starts being outgoing and drinking more may be at risk. Such changes are of particular concern when a precipitating painful event is apparent. Being less careful and prudent in battle may be an important clue. Reckless behavior, such as inappropriate displays of the gun or drinking in battle or flying at high speeds in the Navy jet, may be clues. Making final arrangements such as giving away unusual personal belongings such as a service medal or a favorite firearm may be ominous (but is by no means a definitive clue). A preoccupation with death, such as reading and talking about death after the death of a friend in battle, may be a clue. There are so many possible behavioral cues and changes, and these may not all be only about suicide or suicidal thoughts.

## Life-Threatening Behavior

The following are examples of life-threatening behavior found in suicide. A veteran who had killed himself had previously been seen leaning out of an open window at the VA hospital, and at another time, playing recklessly with a gun. Following the rejection by his girlfriend, a 24-year-old soldier died in a single-car accident on an isolated road after having had several similar accidents. A 70-year-old female veteran, a nurse, with detailed knowledge of drugs, died from drug "mismanagement," despite the nurse in her residence controlling her medication. Self-destruction is not rare. A verbal statement might be something like, "Who cares if I drink too much; everyone in the Army does." Often alcoholism, noncompliance with treatment, and inappropriate use of firearms can be seen in this light, as previously discussed.

## Suicide Notes

Like previous attempts and verbal statements, suicide notes are important clues; however, they are often read but not listened to by the reader. About 18% to 37% of adults leave notes (samples have varied greatly in the percentages). There is no verifiable estimate available for soldiers or veterans.

Here is a sample:

1. Adolf Hitler killed himself on April 30, 1945. Hitler is probably one of the great perpetrators of war in history worldwide—World War II. He wrote a last will on the 29th of April. He is believed to have died from a gunshot wound. There is no question in our mind that it is a genuine suicide note:

   As I did not consider that I could take responsibility, during the years of struggle, of contracting a marriage, I have now decided, before the closing of my earthly career, to take as my wife that girl who, after many years of faithful friendship, entered, of her own free will the practically besieged town in order to share her destiny with me. At her own desire she goes as my wife with me into death. It will compensate us for what we both lost through my work in the service of my people.

   What I possess belongs—in so far as it has any value—to the Party. Should this no longer exist, to the State; should the State also be destroyed, no further decision of mine is necessary.

   My paintings, in the collections which I have bought in the course of years, have never been collected for private purposes, but only for the extension of a gallery in my home town of Linz on Donau.

   It is my sincere wish that this bequest may be duly executed.

   I nominate as my Executor my most faithful Party comrade, Martin Bormann.

   He is given full legal authority to make all decisions.

   He is permitted to take out everything that has a sentimental value or is necessary for the maintenance of a modest simple life, for my brothers and sisters, also above all for the mother of my wife and my faithful co-workers who are well known to him, principally my old Secretaries Frau Winter etc. who have for many years aided me by their work.

   I myself and my wife—in order to escape the disgrace of deposition or capitulation—choose death. It is our wish to be burnt immediately on the spot where I have carried out the greatest part of my daily work in the course of a twelve years' service to my people.

   Given in Berlin, 29th April 1945, 4:00 A.M.
   [Signed] A. Hitler

2. Kurt Cobain was a very famous, exceptionally talented rock star; his band, Nirvana, invented the "grunge" sound. Yet he painfully struggled; he had an emotional disturbance (psychopathology). On the 5th of April 1995, he shot himself. Here is his note:

To Boddah

Speaking from the tongue of an experienced simpleton who obviously would rather be an emasculated, infantile complain-ee. This note should be pretty easy to understand.

All the warnings from the punk rock 101 courses over the years, since my first introduction to the, shall we say, ethics involved with independence and the embracement of your community has proven to be very true. I haven't felt the excitement of listening to as well as creating music along with reading and writing for too many years now. I feel guilty beyond words about these things.

For example when we're back stage and the lights go out and the manic roar of the crowds begins, it doesn't affect me the way in which it did for Freddie Mercury, who seemed to love, relish in the love and adoration from the crowd which is something I totally admire and envy. The fact is, I can't fool you, any one of you. It simply isn't fair to you or me. The worst crime I can think of would be to rip people off by faking it and pretending as if I'm having 100% fun. Sometimes I feel as if I should have a punch-in time clock before I walk out on stage. I've tried everything within my power to appreciate it (and I do, God, believe me I do, but it's not enough). I appreciate the fact that I and we have affected and entertained a lot of people. I must be one of those narcissists who only appreciate things when they're gone. I'm too sensitive. I need to be slightly numb in order to regain the enthusiasms I once had as a child.

On our last 3 tours, I've had a much better appreciation for all the people I've known personally, and as fans of our music, but I still can't get over the frustration, the guilt and empathy I have for everyone. There's good in all of us and I think I simply love people too much, so much that it makes me feel too fucking sad. The sad little, sensitive, unappreciative, Pisces, Jesus man. Why don't you just enjoy it? I don't know!

I have a goddess of a wife who sweats ambition and empathy and a daughter who reminds me too much of what I used to be, full of love and joy, kissing every person she meets because everyone is good and will do her no harm. And that terrifies me to the point to where I can barely function. I can't stand the thought of Frances becoming the miserable, self-destructive, death rocker that I've become.

I have it good, very good, and I'm grateful, but since the age of seven, I've become hateful towards all humans in general. Only because it seems so easy for people to get along that have empathy. Only because I love and feel sorry for people too much I guess.

Thank you all from the pit of my burning, nauseous stomach for your letters and concern during the past years. I'm too much of an erratic, moody baby! I don't have the passion anymore, and so remember, it's better to burn out than to fade away.

    Peace, love, empathy.
    Kurt Cobain

Frances and Courtney, I'll be at your altar.
Please keep going Courtney, for Frances.
For her life, which will be so much happier without me.
I LOVE YOU, I LOVE YOU!

3. A 21-year-old male soldier had traveled to Detroit to bury his mother, who died by overdose on pills. He wrote,

Mary
I'm sorry I had to do this, but I had no choice. After Mom died, I no longer had any reason to go on. I've just realized it.

I have no future and life is entirely pointless. I've been deceiving myself all along and its time it stopped.

On the train on the way back from Detroit to Fort Bragg, I thought of this. Returning to the stinking Army made me realize all I was doing was wasting another 3 years. And then what was I going to do? Well, there isn't any positive answer.

I love you all very much and was glad you and Sharon and Mike and the kids and I got together.
 Love,
 Bill

4. A 20-year-old male soldier, who had a physical disability resulting from an automobile accident, was distraught over the recent death of a close friend and told his girlfriend that he was going to kill himself. Shortly thereafter, she broke up with him. He shot himself in the head with a .38 revolver. He wrote,

I loved her so much. I'm so sorry but this is how its gotta be. I hate life. Life's done me wrong. Please God forgive me but I'm weak. My lifes finished.

5. A 38-year-old married male, who was a Reserve Officers' Training Corps (ROTC) instructor who was depressed over his marriage and stressed in his work. He hanged himself in his garage. He wrote, "Call the police. I've killed myself in the Garage."

Based on the discussion thus far, especially behavioral clues, how can these factors help us in predicting risk in people, especially so in battle. However, before we begin to answer these questions, we need to discuss evaluation of risk in general.

## EVALUATION OF SUICIDE RISK

No one really knows how to assess or study suicide risk perfectly. One of the most frequent questions asked about suicide risk is, "How do you predict suicide risk?" More specifically, the question in front of us is, "How do you study/assess a soldier's suicide risk?" Indeed, suicide risk evaluation may well be the most complex clinical task that mental health professionals and researchers face. This is true in the military. For example, there are great differences in a response that may be elicited if you ask a person, "Did you try to kill yourself?" vs. "Did you attempt suicide?" Furthermore, risk cannot be reduced to one or two questions. What is essential is to learn that assessment and prediction are interwoven with understanding.

In the 1960s and 1970s, there was a focus on the prediction of suicide, and suicidologists believed that it would eventually be possible to predict which individuals out of a population would ultimately complete suicide (Beck et al., 1974). However, it was soon realized that the statistical rarity of suicide; that is, low base rate (Pokorny, 1983) and the imperfection of the prediction instruments led to an enormously large number of false positives; so many, in fact, that the prediction instruments were of little use to clinicians or to those planning to study suicide.

In the 1980s and 1990s, the focus shifted to assessment (Maris, Berman, Maltsberger, & Yufit, 1992; Yufit & Lester, 2005). That is, rather than predicting the future occurrence of suicide in people, the intent was to assess potentially suicidal people in a more general sense, taking into account all of their life experiences; psychological characteristics; and specific culture, such as the military's, which are relevant to future suicidal behavior (Maltsberger, 1986; Shneidman, 1985). Indeed, it is a common belief that prediction and assessment are mutual processes, and any separation is an artificial one. They are not separate categories.

Assessment *a priori* requires clear definition. What is *it* that we are assessing? Studying? This, as we learned from Shneidman (1985), is labyrinthine. In assessing suicide risk in people, whether for research or intervention, we need to be aware of behaviors that are potentially predictive of suicide. However, there is no such definitive behavior. Suicide is a multidimensional event, and a multiaxal approach would be needed (Leenaars, 2004; Meehl, 1986; Millon, Krueger & Simonsen, 2010). Two concepts (introduced earlier) that are essential in prediction are *lethality* and *perturbation* (Shneidman, 1973b, 1985, 1993).

There have been numerous attempts to construct tests to study and predict suicide and related phenomena. In response to awareness of the inherent difficulties in predicting suicide, Maris et al. (1992) and Yufit and Lester (2005) reviewed available assessment instruments used to study suicidal behavior. Their conclusion: *Few, if any, are useful.* Each question, each clue, each test, *by itself*, has little utility. They must be placed into the context of a larger array of facts. That is the basis of sound science (see below). Pompili and Tatarelli (2010) offer the same conclusion in an updated evidence-based review. Would one test or question have predicted a soldier's suicide? Homicide-suicide? (See section "How Can We Best Understand Suicide?")

Based on the discussion to this point, I need to bring to your attention that one can show very few behavioral clues, and even say "no" to a question of risk, and still be highly suicidal (see page 75, "Dissembling: Clues to Suicide Reconsidered"). Often soldiers show no clues, due to their indoctrination into a military culture. Nonetheless, clues or questions may assist in many cases, but not all. We, however, do not know what percentage.

All this pessimism should *not* be read to mean that no question, no risk questionnaire, no lethality evaluation, and so on, could be used. There are useful measures, such as, Beck's Suicide Intent Scale (BSIS; Beck et al., 1975), Smith's Lethality of Suicide Attempt Rating Scale (Smith, Conroy, & Ehler, [LSARS] 1984), Shneidman's list of predictive clues among the gifted Terman subjects who died by suicide (Shneidman, 1971), and my own, the Thematic Guide for Suicide Prediction (Leenaars, 2004). These procedures are designed to assess a person's own narratives (stories), what the person says or writes in a suicide note, or interview, or Internet writing (e-mails). It is likely that no one behavior or one question, including a test score, will provide all of the information needed to assess and predict suicide or for that matter, study military suicidal behavior. A soldier's suicide risk could not and cannot be assessed at one time by one or two questions, one interview, or such. There is no single "bump on the head" that will tell us whether a veteran is suicidal or not, much less how suicidal that person is. This is a major problem in military study to date. It is important to ascertain what tool would help us to assess suicide risk in the military. What would a warrior's risk look like? To answer these questions, we need a detailed look at theory and suicide.

## THEORY AND SUICIDE: A FOREWORD

As a crucial point of introduction, there are, of course, views that theory should not play a role in understanding suicide. Suicidology should only be tabular and statistical. However, I believe that theory, explicit and implicit, plays a key role in understanding any behavior. Theory is the foundation in science (Kuhn, 1962). Newton, Einstein, Beck, and all great scientists are great because they were theorists. It is only through theory, Shneidman (1985) once noted, that we will sort out the booming buzzing mess of experience. In fact, it can be argued that "sciences have achieved their deepest and most far reaching insights by descending below the level of familiar empirical phenomena" (Hempel, 1966, p. 77). Theory may well be in the eye of the beholder (Kuhn, 1962), but it is pivotal in scientific understanding whether one is a researcher, armed forces officer, or clinician. There is nothing as useful as good theory. People must make formulations about things in order to understand them (Husserl, 1973). This does not mean that John Stuart Mill's set of basic rules for science (and, for that matter, for military investigation) have to be abandoned. In his *System of Logic*, Mill (1984) reported a set of canons for inductively establishing causality. These are the Methods of Difference, of Agreement and Difference, of Residues, and of

Concomitant Variation. In our study of military suicide, for example, Mill's Method of Difference does not need to be relinquished; as we will see, in fact, it will be essential. John Stuart Mill has answered many questions. Thus, it would be wise to borrow the ideas of our leading theorists to answer empirically the question, "Why did the soldier kill herself? Why did the veteran kill his spouse and then himself?" *What does psychological theory tell us?*

Psychological Theories of Suicide

Shneidman (1985) suggests that a psychological theory regarding suicide begins with the question, "What are the interesting common psychological dimensions of suicide?"—not what kind of people die by suicide. This question, according to Shneidman, is critical, for they (the common dimensions) are what suicide is. Not necessarily the universal, but certainly the most frequent or common characteristics provide us with a meaningful conceptualization regarding suicide.

The modern era of the psychological study of suicide began around the turn of the 20th century with the investigations of Sigmund Freud (1974a, 1974b, 1974c, 1974d, 1974e, 1974f, 1974i). Freud placed the focus of blame on the person; specifically, in the person's unconscious. Even if the person communicated that he or she had intentionally planned the suicide, the focus of the act will be found in the unconscious. Since around 1900, there has been a host of psychological theories that have attempted to define suicide. Indeed, a (if not the) major advance in the psychology of suicide in the last century was the development of various models that have attempted to understand this complicated human act. Besides Freud, they at least include Adler, Binswanger, Jung, Kelly, Menninger, Murray, Shneidman, Sullivan, and Zilboorg. What we have discovered so far is that suicide can be defined differently from various psychological theories. I do not mean to suggest that all these views are mutually exclusive or equally accurate or helpful. I hope that these alternative constructions in some way help to clarify the central issue—understanding why soldiers died by suicide.

I will next turn to theory in line with the discussion of what science is. It follows, among other individuals, Paul Meehl's ideas. We must have an evidence-based theory. However, what is science? What does the reader need to know before we can discern a good theory?

How Can We Best Understand Suicide?

Humans need to make meaning of things. We are not only *homo erectus*, but we are also *homo-meaning makers*. This is true for the topic that presents itself, suicide. In fact, the search for meaning is not new. Humankind developed reliable and useful theories and classifications long before modern science. Only science has done it better; the thought and logic is evidence-based. There has been enormous value in developing explicit definition and methods (Meehl,

1978; Millon, 2010; Millon, Krueger, et al., 2010). This systematic effort takes enormous time, not a few minutes of speculation. It has, however, been worth the effort, and this is true in suicidology.

Let me ask the fundamental question: *Is it possible to understand suicide?* Is there logic that we can use? Is there evidence? Facts? Are there useful clinical experiences (e.g., pain levels, cognitions, behaviors, affects) that cluster together? If they occur together, are there commonalities? Do they make sense? Do they make sense as clusters or patterns? Are they consistencies over time? Gender? Age? Culture? Armed forces? To answer these questions calls for careful study.

The main problem at hand is that suicidology at the individual level is intrapsychic (i.e., existing or taking place within the mind or psyche). These mental processes are hidden and sometimes masked. Yet these introspective aspects have been for millennia central in defining and categorizing suicide. For example, if a clinical psychologist interviews a suicidal soldier in Kabul, and the soldier, during the assessment, makes some statements such as "I'm depressed all the time," "The pain is unbearable," or "I want to end it all," how would the psychologist classify the content. She/he needs to make meaning of the statements or protocols. The therapist wants to understand, predict, and control. The question is, "Is the statement a particular or specific instance of a classification, say, 'depression' or 'unbearable psychological pain'?" The psychologist will want to make observable observations, both subjective and objective. Of course, the trick in this difficult business is the question of how the specific instance fits into a more abstract classification. The psychologist, like all scientists, must make judgments about whether the content of a protocol is understandable within a given classification. The trick in suicide risk evaluation is to conclude accurately "yes" or "no." This, in fact, may well be the most difficult task that a clinician faces.

In science, intrapsychic concepts are known as "open concept" (Millon, 2010; Pap, 1953). Like any concept, they must, however, be defined by some reliable and meaningful, albeit diverse, empirical facts or events. For example, the concept of "unbearable psychological pain" may be assessed by the soldier's own story, his penultimate note, a score on a psychology test, a history of previous attempts, a TBI, reckless behavior on the battlefield, or so on. Although the term *unbearable pain* of a person cannot be observed directly, it can be inferred and verified. One can conclude intentionality (see earlier discussion of Litman's most insightful definition and classification on intention and suicide).

Of course, there are some people who object to terms like *unbearable pain*. It cannot be touched and is not objectively measurable. It is not like water. Therefore, it does not exist; and if it did, it is background noise or surplus meaning. These researchers argue that our science of suicidology is muddled. Of course, there are confusions, and problems; yet this is not a reason for eschewing a better empirical effort. One will, however, not be able to do it perfectly.

Open concepts have meanings with reference to theory—a theoretical network of variables and clusters. Absolute anchoring or verifying is not possible. At our current stage of knowledge on suicide, we still need creative methods. The very nature of our topic makes it so; we need open concepts (Millon, 2010; Shneidman, 1985).

What data from the stream of events and processes ought to be studied? "Is everything grist for the taxonomic mills" (Millon, 2010, p. 153)? What structural schema should we use to organize the elements of the event? Should we use formal statistical techniques, such as empirically based clusters? Or should we just use clinical speculation?

What elements must be chosen? How must they be classified? There are different formulations; yet I believe that multidetermined or multiaxal models offer the best fit. Suicide cannot be classified like SARS or cholera. In the infectious-disease (medical) model, we just observe, identify cases, disentangle confounders, and identify the cause. If it were only that simple! The disease model is too limited. Suicide is a multiaxal event. Millon (2010) stated that "The multiaxal structure aligns many of the potential relevant factors that can illuminate the nature of a clinical condition, and produce a means of registering their distinguishing attributes" (p. 160). Thus, it should not be a surprise that defining suicide and assessing risk on the battlefield is demanding and often difficult. Schemas of unitary elements are a wishful fantasy. Complex systems or models are needed. This makes it more demanding. It is not simply taking a sample of bodily fluids.

The question about suicide is how good are our schemas, classifications, and generalizations? Suicide is not a homogenous group. It is not like water (All water freezes at 32 degrees Fahrenheit). It is not like copper (All copper conducts electricity). It is a complex phenomena to classify "yes" or "no"; however, it is also not as heterogeneous as "white things," wherein few generalizations can be made. How then would you classify it? People with similar expressions of suicide risk are neither like water nor white things. Regrettably, there are some who have attempted to classify suicide with robust generalizations, as if it were water or copper. There are even manuals and cookbooks to do so. Kendler and Zachar (2008) attributed this to the core belief that these scientists, despite other scientists' good efforts before, have now finally classified suicide successfully.

And a further point on this issue: Good classification systems require more than discussing facts or evidence; it also requires that we make decisions based on our culture and values (Zachar & Kendler, 2010). However, these are issues beyond the scope of this book; it will need a whole volume itself, even for the military. Finally, Shneidman (1985) and Meehl (1993) have both cautioned us to adopt the ethic of being cautious—*don't jump to conclusions* and *don't put things in a nutshell.*

## How Can We Be More Valid?

One approach to validation, introduced by Paul Meehl in the 1950s, is called "construct validation" (Cronbach & Meehl, 1955). This method is based on the rigor of empiricism (Ayer, 1959; Carnap, 1959). People, like A. J. Ayer (1959) espoused strict operationalism. That is, operationalists believe that abstract or open concepts had to be explicitly defined in terms of observable data. The methodology had to be empirically explicit (100%) and replicable (100%). For example, water freezes at 32 degrees Fahrenheit. Paul Meehl's genius lay in questioning the *absolute* commitment to operationalism. (Is it ever possible with suicide?) It is not that we should throw out Ayer's or Carnap's operationalism; rather, it has to be developed for complex open concepts, such as intrapsychic phenomena, intention, and suicide. As we have learned more about associations and patterns to suicide, we discovered that unbearable pain is associated with other elements, say, such as depression or vulnerability; and the experience of pain extends into the interpersonal realm. Suicide and any construction must be explicitly defined, said Meehl, by observable events, but implicitly defined in terms of other constructs. Beyond the observable, there are theoretical constructs, such as "unbearable pain." In Meehl's evolved empiricism, pain is construed realistically, not instrumentalistically—in the practical, it is not simply, "Please fill out this suicide questionnaire" or "Give me some bodily fluids."

Ayer (1959) and Carnap (1959) would argue that a theoretical term like *unbearable pain* is a verbal label to what is being measured (Zachar & Kendler, 2010). For Meehl, it is more. Take Zachar and Kendler's (2010) example of "depression"; one can simply insert "unbearable pain" for every mention of "depression." They write,

> The goal is to use the measure to infer observable consequences; for Meehl the primary objective of scientific interest is the latent variable called "depression," and our measure is considered to be a fallible indicator of that latent construct. Different measures of depression are conceptualized to be assessing the same thing, but each one is at best a partial measure. Observable evidence is used to triangulate on the construct and better understand its full nature.
>
> Applied to psychiatric classification, the diagnostic criteria for mood disorders are also fallible indicators. Each indicator samples only a part of the domain of depression, and they all work better together to represent the whole domain. These indicators can be assessed for diagnostic validity, but the construct of depression cannot be reduced to its indicators. (p. 138)

Although some characteristics or elements in suicide may be considered poor criteria, say, vulnerable ego or mental constriction, because they are muddled, they are essential. One cannot be a literalist; it is all too multifarious.

The construct validation approach is an evolution of empiricism, notably of the operationalists. It is a revision. Zachar and Kendler (2010) write,

> One of the things the revised empiricism focused on was laws (Feigl, 1970; Hempel, 1966). Laws describe patterns or regularities in nature that must occur. Knowing about the regularities allows one to predict what will occur, and also to explain why events did occur.
>
> An example of a law is "All copper conducts electricity." If the law is valid, then whenever presented with an individual piece of copper, we know that it has to conduct electricity. This law-like or "nomological" approach to science would seek to confer on "Bill will respond to treatment X because he has schizophrenia" the same degree of certainty that we attribute to statements such as "This metal will conduct electricity because it is copper." Conducting electricity and responding to treatment X would be lawful consequences of the physical nature of copper and schizophrenia.
>
> In this tradition, a theory is considered to be an interconnected network of theoretical terms and observations. Laws and hypotheses relate observations to theoretical terms. Laws also relate theoretical terms to each other. The whole edifice is called a "nomological network." (p. 139)

For example on our topic, written expressions, such as in a suicide note, might be associated with theoretical constructs such as unbearable pain, aggression, and helplessness/hopelessness, which are then associated to an even more abstract open concept, namely, suicidality or suicidal risk. Thus, the question: "Can we find our laws?" I believe that we can. Only through patient research can we find some of the similarities or patterns or what Shneidman called "commonalities." We can then use the constructs to understand and predict imminent risk. For example, we might predict imminent risk if a soldier wrote in his note, "It will always be this way. The sergeant will never change. He will never accept that I'm gay. There is nothing I can do, except kill the sergeant and myself."

It is all about verification—finding evidence that supports a theory or schema, the various associations or patterns (Cronbach & Meehl, 1955). Of course, there are different kinds of validity, such as predictive and concurrent, which are important. In addition, there are further strategies for testing theories in a logical empirical way (see Cronbach & Meehl, 1955; Meehl, 1986). One further point: Meehl (1986, 1990) thinks that the relationship between observation/evidence and a theory is complicated. In addition, this has implications for not only understanding and predicting suicide, but also controlling it. There will always be *alternative constructions*, thus, it may be wise to start with different points of view on our topic.

How have different and diverse psychological theories come into being? Traditionally, the main classifications have been a product of clinical experience (Menninger, 1963; Shneidman, 1985). Some respected clinicians and scholars, such as Karl Menninger, Sigmund Freud, Aaron Beck, and Edwin Shneidman, have developed theories of suicide. Until recently, suicidological taxonomies were based solely on clinical observation—this was true of Freud, Menninger, and Shneidman—but not all, such as Beck's. This is not to say that, for example, Freud's ideas are not useful, only that they need to be put to the test. Verification, within Meehl's method, needs to be undertaken. This will result in a better evidence-based theory.

Presenting suicidal patients similarities or common factors is useful, and I believe that making them empirical has an even greater advantage. Of course 60 years ago, Shneidman did not have the advantage of the abilities of today's computers. Computers have made a huge difference. Empirical approaches today can "combine cases in more subtle ways than a clinician's; combinations of features too complex to grasp intuitively may yield better classifications than simple combinations" (Andreason & Grove, 1982, p. 45). There have been rapid developments in mathematical techniques; we can analyze and synthesize vast bodies of data, say, for example, thousands of suicide notes. There is a growing body of methods. Millon (2010) writes,

> This growing and diverse body of quantitative methods can be put to many uses, of which only a small number are relevant to the goal of taxonomic construction. Some statistical techniques relate to the validation of existent nosologies (e.g., discriminant analysis) rather than to their creation. Among those used for taxonomic development, some focus on clinical attributes as their basic units, whereas patients themselves are the point of attention for others. For example, factor analysis condenses initially diverse sets of clinical attributes and organizes them into potential syndromic taxa. Cluster analysis, by contrast, is most suitable for sorting patient similarities into personological taxa. (p. 166)

I have strongly argued for almost 4 decades that to classify suicide, *cluster analysis is most suitable*. Of course, questions remain. Do our cluster algorithms mirror the natural structure of suicidality? Are they better than the traditional clinical categories? Is the evidence that we have more accurate and replicable? Is it useful in our predictions? Do they have applicability and implications for direct response? If so, what are they?

Allow me to quote Theodore Millon (2010) again:

> In the early stages of knowledge, the categories of a classification rely invariably on observed similarities among phenomena. . . . As knowledge advances, overt similarities are discovered to be an insufficient, if not false, basis for cohering categories and imbuing them with scientific meaning. . . . As Hempel (1966) . . . pointed out, theory provides the glue that holds a classification together and gives it both its scientific and its clinical relevance. (p. 139)

In Hempel's (1966) discussion of classificatory concepts is most important, he wrote that

> the development of a scientific discipline may often be said to proceed from an initial "natural history" stage . . . to subsequent more and more "theoretical" stages. . . . The vocabulary required in the early stages of this development will be largely observational. . . . The shift toward theoretical

systematization is marked by the introduction of new, "theoretical" terms . . . more or less removed from the level of directly observable things and events.

These terms have a distinct meaning and function only in the context of a corresponding theory. (pp. 139–140)

As Hempel (1966) stated, Millon (2010) believes that "mature sciences progress from an observationally based stage to one that is characterized by abstract concepts and theoretical systemizations" (p. 167).

A theoretically anchored taxonomy or schema allows one to generate greater insight. One cannot be free of theory. Those who claim to have eschewed theory are naïve. Probably unknowingly, they have given primacy to direct observation, a test score, or a biological anomaly. No one today defends strict operationalism—well, maybe a few.

Allow me one more question and to quote Theodore Millon (2010) once more:

What distinguishes a true theoretically based taxonomy from one that merely provides an explanatory summary of known observations and inferences? Essentially, the answer lies in its power to generate new attributes, relations, or taxa—that is, ones other than those used to construct it. This generative power is what Hempel (1966) termed the "systematic import" (p. 6) . . . of a scientific classification. (p. 168)

We will allow you to judge the theoretically-based taxonomy to be presented. Does it help you understand? Do you know better why a soldier kills himself? Does it help you to predict her attempt? And, most importantly, does it help you to control that lethal event in a sea of lethal violence—war?

## AN EVIDENCE-BASED THEORY OF SUICIDE

Theory must begin with definition (Shneidman, 1985). Thus, I refer the reader back to the formal definition of suicide, provided earlier in this chapter on page 30. Suicide is not simply a psychopathological entity in the DSM-IV (American Psychiatric Association, 1994). (It is actually only cited twice as a behavior; once under Depression and once under Borderline Personality Disorder.) We do not agree with those who point to an external stress as the sole cause of suicide. We also do not agree that it is only pain. We tend to place the emphasis on the multideterminant nature of suicide. Suicide is *intrapsychic*. It is stress and pain, but not simply the stress or even the pain, but the person's inability to cope with the event or pain. The issue of any schema about human personality is one that makes an individual an individual (Murray, 1938). It should be the study of the whole organism, not only the stress or pain. People do not simply commit suicide because of pain, but because it's unbearable; they are mentally constricted; they have a mental/emotional disorder; they cannot cope; and so on.

However, from a psychological view, suicide is not only intrapsychic, it is also *interpersonal* (or stated differently, it is beyond the individual level in the ecological model, it is also relationship(s), community, societal). The suicidal individual is not only depressed, mentally constricted, and so on, but he or she is also cut off from loved ones, ideals, and/or even the community. The suicidal person is estranged. People live in a world (a society). Individuals are interwoven; the suicidal person is painfully so. We disagree with those who point to only some intrapsychic aspects such as anger turned inward or primitive narcissism to explain suicide. Suicide occurs in a person *and* between people (or some other ideal; e.g., being loved by someone, being in the armed forces). It is a dynamic interactional system. Yet the intrapsychic world is figural. Suicide occurs as a solution in a mind. The mental processes are the foreground (such as the pain, depression). This is an important difference. It is in the inner world that a person makes the decision to jump, shoot, or such. It is here that he or she decides, "This is the best solution." To put it simply, no drama, no stage. It is the intersection between the two phenomenologies that is essential to understand in suicide. It is, for example, not simply loss of the gun on the stage, but how the person's drama unfolds this very personal, individual stage. Metaphorically speaking, suicide is an intrapsychic drama on an interpersonal stage.

To begin, keeping the concepts of perturbation and lethality in mind, suicide can be clinically understood from at least the following templates or patterns or clusters (Leenaars, 1988, 1989a, 1995a, 2004), each will be followed by the specific concepts or protocol sentences under the cluster. Protocol sentences are testable hypotheses. It is a short statement stating a truth or concept. It is a concept expressed tersely in a few telling words. Although protocol sentences may be an "open concept," they must be testable (although some form of specificity is implied for the sentence to be testable). One must be able to determine the truth or falsity of the statement. They should also be subject to the possibility of verification (falsifiability). The protocol procedure was first introduced by Carnap (1959) and applied in my own research in suicide for over 35 years. Protocol sentences (or concepts) are one means of classifying an event.

## INTRAPSYCHIC DRAMA

### I. Unbearable Psychological Pain

The common stimulus in suicide is *unbearable* psychological pain (Shneidman, 1985, 1993). The enemy of life is pain. The suicidal person is in a heightened state of perturbation, an intense mental anguish. Author William Styron (1990) called it, "The howling tempest of the brain." I call it the tortured mind. It is the pain of feeling pain. Although, as Menninger (1938) noted, other motives (elements, wishes) are evident, the person primarily wants to flee from pain experienced in a trauma, a catastrophe. The fear is that the trauma—the crisis—is bottomless: an eternal suffering. The person may feel any number of emotions such as boxed in, rejected, deprived, forlorn, distressed, and especially hopeless and helpless. It is the emotion of impotence, the feeling of being hopeless/helpless that is so painful for many suicidal people. The situation is unbearable, and the person desperately wants a way out of it. The suicide, as Murray (1967) noted, is functional because it abolishes painful tension for the individual. It provides escape from intolerable suffering. Suicide is escape.

The specific protocol sentences (concepts) in the cluster are

1. Suicide has adjustive value and is functional because it stops painful tension and provides relief from intolerable psychological pain.
2. In suicide, the psychological and/or environmental traumas among many other factors may include incurable disease, threat of senility, fear of becoming hopelessly dependent, feelings of inadequacy, humiliation. Although the solution of suicide is not caused by one thing, or motive, suicide is a flight from these specters.
3. In the suicidal drama, certain emotional states are present, including pitiful forlornness, emotional deprivation, distress and/or grief.
4. The person appears to have arrived at the end of an interest to endure and sees suicide as a solution for some urgent problem(s) and/or injustices of life.
5. There is a conflict between life's demands for adaptation and the person's inability or unwillingness to meet the challenge.
6. The person is in a state of heightened disturbance (perturbation) and feels boxed in, harassed, especially hopeless and helpless.

## II. Cognitive Constriction

The common cognitive state in suicide is mental constriction (Shneidman, 1985). Constriction, that is, rigidity in thinking, narrowing of focus, tunnel vision, concreteness, and such, is the major component of the cognitive state in suicide. The person is figuratively "intoxicated" or "drugged" by the constriction; the intoxication can be seen in emotions, logic, and perception. The suicidal person exhibits at the moment before his/her death only permutations and combinations of a trauma (e.g., defeated in a battle, military scandal, a TBI, rejection by the spouse back home). The suicidal mind is in a special state of relatively fixed purpose and of relative constriction. In the face of the painful trauma, a possible solution became *the* solution. This constriction is one of the most dangerous aspects of the suicidal mind. (I highlight this more below.)

The specific protocol sentences (concepts) in the cluster are

7. The person reports a history of trauma (e.g., poor health, rejection by a significant other, a competitive spouse).

8. Figuratively speaking, the person appears to be "intoxicated" by overpowering emotions. Concomitantly, there is a constricted logic and perception.
9. There is poverty of thought, exhibited by focusing only on permutations and combinations of grief and grief-provoking topics.

## III. Indirect Expressions

Ambivalence, complications, redirected aggression, unconscious implications, and related indirect expressions (or phenomena) are often evident in suicide. The suicidal person is ambivalent. There are complications, concomitant contradictory feelings, attitudes, and/or thrusts, often toward a person and even toward life (see dissembling, page 75). Not only is it love and hate, but it may also be a conflict between survival and unbearable pain. The person experiences humility, submission, devotion, subordination, flagellation, and sometimes even masochism. Yet there is much more. What the person is conscious of is only a fragment of the suicidal mind (Freud, 1974b). There are more reasons to the act than the suicidal person is consciously aware of when making the final decision (Freud, 1974b, 1974c; Leenaars, 1988, 1993). The driving force may well be unconscious process.

The specific protocol sentences (concepts) in the cluster are

10. The person reports ambivalence; e.g., complications, concomitant contradictory feelings, attitudes, and/or thrusts.
11. The person's aggression has been turned inward; for example, humility, submission and devotion, subordination, flagellation, masochism, are evident.
12. Unconscious dynamics can be concluded. There are likely more reasons for the suicide than the person is consciously aware of.

## IV. Inability to Adjust/Psychopathology

People with all types of pains, problems, and such are at risk for suicide. Psychological autopsy studies suggest that 40% to 90% of people who kill themselves have some symptoms of psychopathology and/or problems in adjustment (Hawton & van Heeringen, 2000). It has been claimed that as high as 60% appear to be related to mood disorders. Although the majority of suicides may thus fit best into mood nosological classifications (e.g., depressive disorders, bipolar disorders [manic-depressive disorders], adjustment disorder with mixed anxiety and depressed mood), other emotional/mental disorders have been identified. For example, anxiety disorders (PTSD is especially prevalent in soldiers and veterans), schizophrenic disorders (especially paranoid type), panic disorders, borderline personality disorders, and antisocial personality disorders have been related to suicides (Leenaars, 1988; Sullivan, 1962, 1964).

Anxiety may well be an equally important pain, next to depression (Fawcett, 1997). People with schizophrenia have a very high rate (about 5%; not the often-cited 10%) (Palmer, Pankratz, & Bostwick, 2005). Yet there are other disorders not specified that may result in risk. From the psychological autopsy studies, it is learned that many may have no disorder identifiable in the DSM-IV (or some other classification scheme, such as ICD). The person may be simply paralyzed by pain that life, a future, and such are colorless and unattractive.

However, in its varieties, depression is the most frequent disorder; however, it must be understood by any reader that not all suicidal people are depressed and that not all depressed people are suicidal. It is often cited that 15% of people who develop depression ultimately kill themselves. Bostwick (2000) has, however, clearly demonstrated in a meta-analysis of the research that this is a myth. It may well, in fact, be as low as 2%. Most importantly, it is important to remember that suicidal people experience unbearable pain, not always depression and even if they do experience depression, the critical stimulus is the "unbearable" nature of the depression (as in some other mood). Suicidal people see themselves as in unendurable pain and unable to adjust. His/her state of mind is, however, incompatible with accurate discernment of what is going on. Having the belief that they are too weak to overcome difficulties, these people reject everything except death—they do not survive life's difficulties.

The specific protocol sentences (concepts) in the cluster are

13. The person considers him//herself too weak to overcome personal difficulties and therefore rejects everything, wanting to escape painful life events.
14. Although the person passionately argues that there is no justification for living on, the person's state of mind is incompatible with an accurate assessment/perception of what is going on.
15. The (suicidal) person (S) exhibits a serious disorder in adjustment
    a. S's reports are consistent with a manic-depressive disorder such as the down-phase; e.g., all-embracing negative statements, severe mood disturbances causing marked impairment.
    b. S's reports are consistent with schizophrenia; e.g., delusional thought, paranoid ideation.
    c. S's reports are consistent with anxiety disorder (such as obsessive-compulsive, posttraumatic stress); e.g., feeling of losing control; recurrent and persistent thoughts, impulses, or images.
    d. S's reports are consistent with antisocial personality (or conduct) disorder; e.g., deceitfulness, conning others.
    e. S's reports are consistent with borderline personality disorder; e.g., frantic efforts to avoid real or imagined abandonment, unstable relationships.

f. S's reports are consistent with depression; e.g., depressed mood, diminished interest, insomnia.
g. S's reports are consistent with a disorder not otherwise specified. S is so paralyzed by pain that life, future, etc., is colorless and unattractive.

## V. Ego (Vulnerable Ego)

The ego, with its enormous complexity (Murray, 1938), is an essential factor in the suicidal scenario. The OED defines ego as "The part of the mind that reacts to reality and has a sense of individuality." Ego strength is a protective factor against suicide. The biological perspective has equally argued this conclusion; van Praag (1997) has, for example, clearly documented a biological aspect to suicidal people, such as increased susceptibility to stressors, labile anxiety, and aggression regulation. Suicidal people frequently exhibit a relative weakness in their capacity to develop constructive tendencies and to overcome their personal difficulties (Zilboorg, 1936). The person's ego has likely been weakened by a steady toll of traumatic life events (e.g., loss, rejection, abuse, demotion, transfer to a perceived lesser unit [a "promotion"]). This implies that a history of traumatic disruptions—*pain*—placed the person at risk for suicide; it likely mentally and/or emotionally handicapped the person's ability to develop mechanisms (or ego functions) to cope. There is, to put it in one simple word, vulnerability. There is a lack of resilience. A weakened ego correlates positively with suicide risk.

The specific protocol sentences (concepts) in the cluster are

16. There is a relative weakness in the person's capacity for developing constructive tendencies (e.g., attachment, love).
17. There are unresolved problems (a "complex" or weakened ego) in the individual; e.g., symptoms or ideas that are discordant, unassimilated, and/or antagonistic.
18. The person reports that the suicide is related to a harsh conscience; i.e., a fulfillment of punishment (or self-punishment).

## INTERPERSONAL STAGE

## VI. Interpersonal Relations

The suicidal person has problems in establishing or maintaining relationships. There frequently is a disturbed, unbearable interpersonal situation. A calamity prevailed; one of the most common is a marital break-up. A positive development in those same disturbed relationships may have been seen as the only possible way to go on living, but such development was seen as not forthcoming. The person's psychological needs are frustrated. Suicide appears to be related to an unsatisfied or frustrated attachment need, although other needs, often more intrapsychic, may be equally evident, for example, achievement, autonomy, dominance, honor. Suicide is committed because of thwarted or unfulfilled needs, needs that are often frustrated interpersonally. The possible needs that are frustrated or blocked are expansive. Table 2.1 presents a partial list of needs, adopted from Henry A. Murray's *Explorations in Personality* (1938). What were the needs of the soldier who died by suicide?

The specific protocol sentences (concepts) in the cluster are

19. The person's problem(s) appears to be determined by the individual's history and the present interpersonal situation.
20. The person reports being weakened and/or defeated by unresolved problems in the interpersonal field (or some other ideal such as health, perfection).
21. The person's suicide appears related to unsatisfied or frustrated needs; e.g., attachment, perfection, achievement, autonomy, control.
22. The person's frustration in the interpersonal field is exceedingly stressful and persisting to a traumatic degree.
23. A positive development in the disturbed relationship was seen as the only possible way to go on living, but such development was seen as not forthcoming.
24. The person's relationships (attachments) were too unhealthy and/or too intimate (regressive, "primitive"), keeping him/her under constant strain of stimulation and frustration.

## VII. Rejection-Aggression

Wilhelm Stekel first documented the rejection-aggression hypothesis at the famous 1910 meeting of the Psychoanalytic Society in Freud's home in Vienna. Adler, Jung, Freud, Sullivan, and Zilboorg have all expounded variations of this hypothesis. Loss is central to suicide; it is, in fact, often a rejection that is experienced as abandonment. It is an unbearable narcissistic injury. This injury is part of a traumatic event that leads to pain and in some, self-directed aggression. In the first controlled study of suicide notes, Shneidman and Farberow (1957) reported, for example, that both hate directed toward others and self-blame are evident in notes. The suicidal person is deeply ambivalent and, within the context of this ambivalence, suicide may become the turning back upon oneself of murderous impulses (wishes, needs) that had previously been directed against a traumatic event, most frequently someone who had rejected that individual. Biological research in the field has demonstrated a neurobiological link between aggression and suicide. Despite a minimizing of this fact by some (e.g., Shneidman, 1985), aggression, whether other or self-directed, has for example, an association to serotonin dysfunction (Asberg et al., 1976). Freud's hypothesis appears to have a biological basis within the biopsychosocial view of suicide. Aggression is, in fact, a common emotional state in suicide, and of course, the

Table 2.1  A Partial List of Murray's Psychological Needs

---

Abasement. To submit passively to external force; accept injury, criticism, punishment; to surrender; become resigned to fate; blame or belittle self.

Achievement. To accomplish something difficult; master, manipulate, organize physical objects, human beings or ideas; to overcome; to excel oneself.

Affiliation. To enjoyably cooperate or reciprocate with an allied other; to please and win affection; to adhere or remain loyal to a friend or group.

Aggression. To overcome opposition forcefully; to fight; to attack or injure another to oppose forcefully or punish other.

Autonomy. To get free, shake off restraint; break out of social confinement avoid or quit activities of domineering authorities; be independent and free.

Counteraction. To make up for failure by restriving; overcome weakness or repress fear; to maintain self-respect and pride on a high level; overcome.

Defendance. To defend or vindicate the self against assault, criticism, blame conceal or justify a misdeed, failure, or humiliation.

Deference. To admire and support a superior; praise, honor, or eulogize; yield eagerly to influence of another; emulate an exemplar.

Dominance. To control other humans; influence or direct others by command suggestion, or persuasion; or to dissuade, restrain, or prohibit others.

Exhibition. To make an impression; be seen and heard; to excite, amaze, fascinate entertain, shock, intrigue, amuse, or entice others.

Harm avoidance. To avoid pain, physical injury, illness, and death; escape from a dangerous situation; to take precautionary measures.

Infavoidance. To avoid humiliation; avoid or quit conditions that lead to scorn derision, indifference, or embarrassment.

Inviolacy. To protect the self; remain separate; maintain distance; to resist others' intrusion on one's own psychological space; isolated.

Nurturance. To gratify the needs of another person, especially one who is weaker to feed, help, support console, protect, comfort; to nurture.

Order. To put things or ideas in order; to achieve arrangement, balance, organization, tidiness, and precision among things and ideas.

Play. To act for "fun." To enjoy relaxation of stress; to laugh and make jokes to seek pleasurable activities for their own sake.

Rejection. To exclude, abandon, expel, separate oneself or remain indifferent to a negatively seen person; to snub or jilt another.

Sentience. To seek and enjoy sensuous experience; to give an important place to creature comforts and satisfaction of the senses—taste, touch.

Succorance. To have one's needs gratified by the sympathetic aid of another be supported, sustained, guided, consoled, taken care of; protected.

Understanding. To ask questions; be interested in theory; speculate, analyze, generalize; to want to know the answers.

green culture fosters such. Suicide may be veiled aggression—it may be murder in the 180th degree (Shneidman, 1985).

The specific protocol sentences (concepts) in the cluster are

25. The person reports a traumatic event or hurt or injury (e.g., unmet love, a failing marriage, disgust with one's work).
26. The person, whose personality (ego) is not adequately developed (weakened), appears to have suffered a narcissistic injury.
27. The person is preoccupied with an event or injury, namely, a person who has been lost or rejecting (i.e., abandonment).
28. The person feels quite ambivalent, i.e., both affectionate and hostile toward the same (lost or rejecting) person.
29. The person reports feelings and/or ideas of aggression and vengefulness toward him/herself although the person appears to be actually angry at someone else.
30. The person turns upon the self, murderous impulses that had previously been directed against someone else.
31. Although maybe not reported directly, the person may have calculated the self-destructiveness to have a negative effect on someone else (e.g., a lost or rejecting person).
32. The person's self-destructiveness appears to be an act of aggression, attack, and/or revenge toward someone else who has hurt or injured him/her.

VIII. Identification-Egression

Freud (1974b, 1974d) hypothesized that intense identification with a lost or rejecting person or, as Zilboorg (1936) showed, with any lost ideal (e.g., health, promotion, employment, gun ownership) is crucial in understanding the suicidal person. Identification is defined as an attachment (bond) based upon an important emotional tie with another person (object) (Freud, 1974d) or any ideal. If this emotional need is not met, the suicidal person experiences a deep pain (discomfort). There is an intense desperation and the person wants to egress, that is, to escape. Something must be done to stop the anguish. The suicidal person wants to leave, to exit, to get out, to get away, to be gone, to be elsewhere—not to be—to be dead. Suicide becomes the only solution and the person plunges into the abyss. *Suicide is escape*.

The specific protocol sentences (concepts) in the cluster are

33. The person reports in some direct or indirect fashion an identification (i.e., attachment) with a lost or rejecting person (or with any lost ideal [e.g., health, freedom, employment, all A's]).
34. An unwillingness to accept the pain of losing an ideal (e.g., abandonment, sickness, old age), allows the person to choose, even seek to escape from life and accept death.

35. The person wants to egress (i.e., to escape, to depart, to flee, to be gone), to relieve the unbearable psychological pain.

In concluding, the theory outlined is only one point of view. Yet the elements have utility in understanding suicide. Indeed, to begin to address the question, "Why do people kill themselves?" or more specifically, "Why did the veteran die by suicide?" we need a psychology of suicide. We must answer the question, What are the important common psychological dimensions of suicide? rather than What kind of people die by suicide? The question is critical, for these common dimensions (or "sameness") are what suicide is. Not necessarily the universal, but certainly the most frequent or common characteristics provide us with a meaningful conceptualization and classification of suicide.

## RESEARCH ON THEORY: AN EMPIRICALLY BASED UNDERSTANDING

Understanding the act of suicide and motives behind suicidal behavior seems extremely important worldwide (WHO, 2002), and in order to do so empirically, many researchers from around the world have used different methods. Shneidman and Farberow (1957), Maris (1981), and others have suggested the following avenues: national mortality statistics, retrospective psychological investigations (often called psychological autopsies), the study of nonfatal suicide attempts, and the analysis of documents (such as suicide notes). All of them have their strengths; however, there are problems in obtaining them in many countries or cultures (such as the military), including in subpopulations (or subcultures), such as soldiers, sailors, airman, and Marines. Yet each of these methods has been shown to extend our understanding of suicide and suicidal behavior (Hawton & van Heeringen, 2000; Leenaars et al., 1997). I will next examine the theory in the *prima facia* evidence in cases of suicide—suicide notes (Leenaars, 1999).

Early research on suicide notes largely used an anecdotal approach that incorporated descriptive information (Frederick, 1969). Subsequent methods, using Frederick's (1969) scheme for methods of analysis have used content analysis, classification analysis, and theoretical-conceptual analysis. Each of these approaches has had utility, although Frederick suggested that simple content analysis has limitations. Traditional classification schemes use data such as age, gender, marital status, educational level, employment status, and mental disorder (see, for example, Ho, Yip, Chiu, & Halliday, 1998). Ho and his colleagues (1998) developed the most widely used classification scheme; they studied suicides notes in Hong Kong. Ho et al.'s scheme is largely based on forensic data that are gathered at postmortem investigation. They found that suicide notes written by young people were longer and richer in emotions than those written by older people. A similar classification scheme has been used in the United States, India, Mexico, Japan, and Turkey. However, there are limitations in these studies. The data are not entirely consistent, and differences in collection occur between researchers in different countries. There are also limitations in the generalizability of the findings. For example, unlike Ho et al.'s finding, the team from Turkey (Leenaars, Sayin, Candansayer, Akar, Demirel, & Leenaars, 2010) found the notes from the elderly longer. Differences of demographic features like age and gender are also inconsistent. *The main finding to date is that suicide note writers are essentially similar to suicides who did not leave notes.* The studies have also supported the value of the data that showed that one can validly study suicide by studying suicide notes, and the notion that suicide is complex (Ho et al., 1998; Leenaars, 1988; Shneidman & Farberow, 1957), warranting, among other things, more in-depth study of suicide notes.

Only a very few studies have utilized a theoretical-conceptual analysis (Frederick, 1969), despite the assertion in the first formal study of suicide notes (Shneidman & Farberow, 1957) and in ongoing discussion (Millon, 2010) that such an approach offers much promise. To address this lack, almost 40 years ago, the author applied a logical, empirical analysis to suicide notes. The method permits a theoretical analysis of suicide notes, augments the effectiveness of controls, provides construct validation, and allows us to develop some theoretical insight into the vexing problem of suicide that may have cross-cultural application.

The method has been previously described in detail (Leenaars, 1988). It treats suicide notes as an archival source. This source is subjected to the scrutiny of control hypotheses, following an *ex post facto* research design (Kerlinger, 1964). The major problem with the current type of research is the lack of control over extraneous variables and the large number of potentially important antecedent variables; thus, the danger of misinterpreting relationships. Kerlinger (1964) suggested that these problems could be largely overcome by explicitly formulating not just a single hypothesis, but several "control" hypotheses as well. This would call for suicide protocol such as notes to be recast in different theoretical contexts (hypotheses, theories, models) for which lines of evidence of each of these positions can then be pursued in the data. Carnap's (1959) logical and empirical procedures can be utilized for such investigations. To date, the theories of 10 suicidologists, as noted earlier, have been investigated: Adler, Binswanger, Freud, Jung, Menninger, Kelly, Murray, Shneidman, Sullivan, and Zilboorg. In order to test the formulations, Carnap's positivistic procedure calls for the translating of theoretical formulations into observable (specific) protocol sentences (Ayer, 1959), within the context of construct validation approach (Millon, 2010).

To summarize from the world's largest array of empirical studies on suicide notes (e.g., age, gender, method used, nation) of the theories of the 10 suicidologists, a number of theoretical propositions/implications (or protocol sentences) have been identified to be observable in various samples of notes, the very words of a suicidal person. In the method, Leenaars isolated 100 protocol sentences or concepts from each of the ten theorists (10 for each theorist) and reduced them to 35 sentences; 23 protocol sentences were found to be highly predictive (described) for the content of suicide notes (i.e., one standard deviation above the mean of observations), and 17 protocol sentences significantly discriminated genuine suicide notes from simulated notes (i.e., control data) (with 5 sentences being both) (Leenaars, 1989a). Thus, there was not only construct validation, but also predictive and discriminative validity. One unique finding of these studies is that there are a few significant age differences in the suicide notes, but not gender (Leenaars, 2004; Leenaars, De Wilde, Wenckstern, & Kral, 2001). Using cluster analysis, the protocols were reduced to the eight clusters discussed previously, classified into 5 intrapsychic and 3 interpersonal aspects: (a) unbearable pain, (b) cognitive constriction, (c) indirect expressions, (d) inability to adjust (psychopathology), (e) ego (vulnerability), (f) interpersonal relationships, (g) rejection-aggression, and (h) identification-egression (Leenaars, 1996, 2004; Leenaars et al., 2001). Further, from a series of studies, Leenaars (1988, 1996) proposed a metaframe to organize the clusters into *intrapsychic* and *interpersonal* elements. Suicide can be, as discussed, seen as an intrapsychic drama on an interpersonal stage. Suicide can thus be theoretically understood from the proposed theory (templates, constructs, and frames), outlined earlier in detail. There is, without a doubt, evidence for the theory, having great utility, not only for research but also for preventing suicide in the military.

Table 2.2 presents the actual protocol sentences within each cluster, presented as a Thematic Guide for Suicide Prediction (TGSP).

To apply, the TGSP to each of the suicide notes presented earlier, I present the actual scores. (The reader may wish to do so.):

1. Adolf Hitler's note: 1, 2, 3, 4, 6, 7, 9, 12, 13, 15g, 17, 19, 20, 21, 23, 25, 26, 27, 29, 30, 31, 32, 33, 34, 35.
2. Kurt Cobain's note: 1, 2, 3, 4, 5, 6, 7, 8, 9, 10, 11, 12, 13, 14, 15f, 16, 17, 18, 19, 20, 21, 22, 23, 24, 25, 26, 27, 28, 33, 34, 35.
3. Bill, the 21-year-old soldier's note: 1, 2, 3, 4, 5, 6, 7, 9, 13, 14, 15f, 16, 19, 20, 21, 22, 24, 25, 33, 34, 35.
4. The 20-year-old soldier's note: 1, 2, 3, 4, 5, 6, 7, 8, 9, 12, 13, 15g, 16, 17, 19, 20, 21, 22, 25, 26, 29, 33, 34, 35.
5. The 38-year-old ROTC instructor's note: 12, 35.

Like the ROTC instructor's note, some are banal. Poet Sylvia Plath's note read, "Please call Dr. Horder."

Independent research on suicide notes (O'Connor, Sheeby, & O'Connor, 1999), investigation of suicidal Internet writing (Barak & Miran, 2005), and biographical studies of suicides (Lester, 1994) have supported the utility of the approach to note, or any narrative, analysis, adding construct validity. Barak and Miran (2005) found that unbearable pain, cognitive constriction, and problematic interpersonal relationships were especially evident in writings on the Internet about suicide

Lester (1994) found evidence of all aspects of Leenaars' model in the biographies of people who died by suicide. In-depth studies of interjudge reliability (e.g., O'Connor et al., 1999) and over 3 decades of study by the author and colleagues show that, indeed, the percentage of interjudge agreement has been satisfactory (> 85%). Reliability and validity have also been established in different countries.

International studies are not only rare in the study of suicide notes but suicide in general. There are only a few studies, for example, on suicide notes from different countries. It is well known that Canada has a higher rate of suicide than the United States (Leenaars & Lester, 1994), and Leenaars (1992) examined 56 suicide notes from Canada and the United States, whose writers were matched for age and gender (this was the first cross-cultural study of suicide notes). None of the intrapsychic or interpersonal aspects differed. Subsequently, studies from Germany (Leenaars, Lester, Wenckstern, & Heim, 1994), the United Kingdom (O'Connor & Leenaars, 2004), Hungary (Leenaars, Fekete, Wenckstern, & Osvath, 1998), Russia (Leenaars et al., 2002), Australia (Leenaars, Haines, Wenckstern, Williams, & Lester, 2003), India (Leenaars, Girdhar, Dogra, Wenckstern, & Leenaars, 2010), Mexico (Chavez-Hernandez, Leenaars, Chavez-de Sanchez, & Leenaars, 2009), and Turkey (Leenaars, Sayin, et al., 2010) supported this observation. A few differences, however, were observed in some people: some cultural/social groups showed more indirect expressions, such as unconscious processes, ambivalence, and dissembling; and some more interpersonal elements of aggression, murderous impulses, and revenge (Leenaars, 2007). And this, by implication, may well be true in some soldiers, for green cultural or many other reasons, discussed and to be discussed. There are differences by age: younger soldiers will probably show some significant differences from the older group (see below). Yet the theory has been empirically applicable to all people. Thus, the model has significant cross-cultural or social application; be that as it may, questions remain. Does it apply to a military culture such as suicide of soldiers? Probably. Regrettably, militaries do not release suicide notes to allow for nomothetic study. What does the soldier's note say?

The history of suicidology (psychiatry, psychology, sociology, and so on) gives us the ideas, concepts, formulations, and so on, and science gives us the observable "valid" ones. This is the best of all possible psychologies—an empirically supported one. The theory outlined is an attempt to do that. It has a great deal of construct validity. There are few such theories in suicidology. It provides an answer to the question posed; it presents some common elements to answer, Why did the soldier kill himself?

These elements, common to suicide, highlight that suicide is not only due to external "stress" or pain or even unattachment. The common consistency in suicide is, in fact, lifelong adjustment patterns (Shneidman, 1985). Suicidal people have experienced a steady toll of life events, that is, threat, stress, failure, loss and challenge—or in one word, pain—that has undermined their ability to cope. Suicide has a history. The soldier's suicide has a history.

Table 2.2 Thematic Guide for Suicide Prediction
Antoon A. Leenaars, Ph.D., C.Psych.

## I PATIENT DATA

Date: _____

Name _____  Age _____  Sex _____

Date of Birth _____  Marital Status ___ Divorced___

Education Status _____       _____
            (years)                              (degrees)

Current Employment _____

## II SUICIDAL EXPERIENCE

1) Has the patient ever seriously contemplated suicide? (If yes, note particulars)
_____

2) Has the patient ever attempted suicide? (If yes, note particulars)
_____

3) Does the patient know anyone who attempted suicide? (If yes, indicate family, acquaintance, etc.)
_____

4) Does the patient know anyone who committed suicide? (If yes, indicate family, acquaintance, etc.)
_____

## III REFERRAL DATA

1) Purpose _____ Postvention_____

2) What is the referral question? _____ See file_____

3) What is the presenting problem(s)? _____ See file_____
_____

Copyright © 1998 by Antoon A. Leenaars

Table 2.2 (Cont'd.)

## IV INTERVIEW SITUATION
1) Observations _____
   _____
   _____

2) Other procedures (e.g., tests, interviews) _____
   _____

## V INTERPRETATIONS

| | | Low | Medium | High |
|---|---|---|---|---|
| 1) | Perturbation rating: | | | |
| | scale equivalent | 1 2 3 | 4 5 6 | 7 8 9 |
| 2) | Lethality rating: | Low | Medium | High |
| | scale equivalent | 1 2 3 | 4 5 6 | 7 8 9 |

3) Guide summary:
   scores:  I - 1, 2, 3, 4, 5, 6;  II - 7, 8, 9;  III - 10, 11, 12;
   IV - 13, 14, 15;  V - 16, 17, 18;  VI - 19, 20, 21, 22, 23, 24;
   VII - 25, 26, 27, 28, 29, 30, 31, 32;  VIII - 33, 34, 35

Conclusions: _____
_____
_____
_____

## VI REMARKS
_____
_____
_____
_____
_____
_____

Include on back any other relevant data.

Table 2.2 (Cont'd.)

## INSTRUCTIONS

Your task will be to verify whether the statements provided below correspond or compare to the contents of the person's protocols (e.g., interview, written reports). The statements provided below are a classification of the possible content. You are to determine whether the contents in the person's protocols are a particular or specific instance of the classification or not. Your comparison should be observable; however, the classification may be more abstract than the specific instances. Thus, you will have to make judgments about whether particular contents of a protocol are included in a given classification or not. Your task is to conclude, yes or no.

INTRAPSYCHIC                                                                 Circle One

I – Unbearable Psychological Pain

1) Suicide has adjustive value and is functional because it stops painful tension and provides relief from intolerable psychological pain. (P[2])    Yes   No

2) In suicide, the psychological and/or environmental traumas among many other factors may include: incurable disease, threat of senility, fear of becoming hopelessly dependent, feelings of inadequacy, humiliation. Although the solution of suicide is not caused by one thing, or motive, suicide is a flight from these specters. (F & D)    Yes   No

3) In the suicidal drama, certain emotional states are present, including pitiful forlornness, emotional deprivation, distress and/or grief. (P & D)    Yes   No

4) The suicidal person (S) appears to have arrived at the end of an interest to endure and sees suicide as a solution for some urgent problem(s), and/or injustices of life. (P)    Yes   No

5) There is a conflict between life's demands for adaptation and the S's inability or unwillingness to meet the challenge. (P)    Yes   No

6) S is in a state of heightened disturbance (perturbation) and feels boxed in, harassed, especially hopeless and helpless. (P)    Yes   No

II – Cognitive Constriction

7) reports a history of trauma (e.g., poor health, rejection by significant other, a competitive spouse). (P & D)    Yes   No

8) Figuratively speaking, S appears to be "intoxicated" by overpowering emotions. Concomitantly, there is a constricted logic and perception. (D)    Yes   No

9) There is poverty of thought, exhibited by focusing only on permutations and combinations of grief and grief-provoking topics. (D)    Yes   No

---

[2] The letter P refers to a specific highly predictive concept, whereas the letter D refers to a specific differentiating concept of the suicidal mind.

Table 2.2 (Cont'd.)

## III – Indirect Expressions

10) S reports ambivalence; e.g., complications, concomitant contradictory feelings, attitudes, and/or thrusts. (P & D)  Yes  No

11) S's aggression has been turned inward; for example, humility, submission and devotion, subordination, flagellation, masochism are evident. (P)  Yes  No

12) Unconscious dynamics can be concluded. There are likely more reasons to the suicide than the person is consciously aware. (D)  Yes  No

## IV – Inability to Adjust

13) S considers him/herself too weak to overcome personal difficulties and therefore rejects everything, wanting to escape painful life events. (P)  Yes  No

14) Although S passionately argues that there is no justification for living on, S's state of mind is incompatible with an accurate assessment/perception of what is going on. (P)  Yes  No

15) S exhibits a serious disorder in adjustment. (P)  Yes  No

  a) S's reports are consistent with a manic-depressive disorder such as the down-phase; e.g., all-embracing negative statements, severe mood disturbances causing marked impairment.  Yes  No

  b) S's reports are consistent with schizophrenia; e.g., delusional thought, paranoid ideation.  Yes  No

  c) S's reports are consistent with anxiety disorder (such as obsessive-compulsive, posttraumatic stress); e.g., feeling of losing control; recurrent and persistent thoughts, impulses or images.  Yes  No

  d) S's reports are consistent with antisocial personality (or conduct) disorder; e.g., deceitfulness, conning others.  Yes  No

  e) S's reports are consistent with borderline personality disorder; e.g., frantic efforts to avoid real or imagined abandonment, unstable relationships.  Yes  No

  f) S's reports are consistent with depression; e.g., depressed mood, diminished interest, insomnia.  Yes  No

  g) S's reports are consistent with a disorder not otherwise specified. S is so paralyzed by pain that life, future, etc. is colorless and unattractive.  Yes  No

Table 2.2 (Cont'd.)

## V – Ego

16) There is a relative weakness in S's capacity for developing constructive tendencies (e.g., attachment, love). (D) — Yes No

17) There are unresolved problems (a "complex" or weakened ego) in the individual; e.g., symptoms or ideas that are discordant, unassimilated, and/or antagonistic. (P) — Yes No

18) S reports that the suicide is related to a harsh conscience; i.e., a fulfillment of punishment (or self-punishment). (D) — Yes No

## INTERPERSONAL

## VI – Interpersonal Relations

19) S's problem(s) appears to be determined by the individual's history and the present interpersonal situation. (P) — Yes No

20) S reports being weakened and/or defeated by unresolved problems in the interpersonal field (or some other ideal such as health, perfection). (P) — Yes No

21) S's suicide appears related to unsatisfied or frustrated needs; e.g., attachment, perfection, achievement, autonomy, control. (P) — Yes No

22) S's frustration in the interpersonal field is exceedingly stressful and persisting to a traumatic degree. (P) — Yes No

23) A positive development in the disturbed relationship was seen as the only possible way to go on living, but such development was seen as not forthcoming. (P) — Yes No

24) S's relationships (attachments) were too unhealthy and/or too intimate (regressive, "primitive"), keeping him/her under constant strain of stimulation and frustration. (D) — Yes No

## VII – Rejection-Aggression

25) S reports a traumatic event or hurt or injury (e.g., unmet love, a failing marriage, disgust with one's work). (P) — Yes No

26) S, whose personality (ego) is not adequately developed (weakened), appears to have suffered a narcissistic injury. (P & D) — Yes No

27) S is preoccupied with an event or injury, namely, a person who has been lost or rejecting (i.e., abandonment). (D) — Yes No

28) S feels quite ambivalent, i.e., both affectionate and hostile toward the same (lost or rejecting) person. (D) — Yes No

29) S reports feelings and/or ideas of aggression and vengefulness toward him/herself, although S appears to be actually angry at someone else. (D) — Yes No

Table 2.2 (Cont'd.)

| | |
|---|---|
| 30) S turns upon the self, murderous impulses that had previously been directed against someone else. (D) | Yes No |
| 31) Although maybe not reported directly, S may have calculated the self-destructiveness to have a negative effect on someone else (e.g., a lost or rejecting person). (P) | Yes No |
| 32) S's self-destructiveness appears to be an act of aggression, attack, and/or revenge toward someone else who has hurt or injured him/her. (P) | Yes No |
| **VIII – Identification-Egression** | |
| 33) S reports in some direct or indirect fashion an identification (i.e., attachment) with a lost or rejecting person (or with any lost ideal [e.g., health, freedom, employment, all A's]). (D) | Yes No |
| 34) An unwillingness to accept the pain of losing an ideal (e.g., abandonment, sickness, old age), allows S to choose, even seek to escape from life and accept death. (D) | Yes No |
| 35) S wants to egress (i.e., to escape, to depart, to flee, to be gone), to relieve the unbearable psychological pain. (P) | Yes No |

Hopefully, the synthesis presented here will provide a useful clinical and forensic perspective for many on the question, "Why did the veteran commit suicide?" and we will use this theory in understanding the suicide of specific individual within the military service.

Of course, we do not know whether the theory applies to all suicides. Suicide bombers, homicide-suicides, suicide martyrs, altruistic suicides, for example, may differ (Leenaars & Wenckstern, 2004). Today, we have Islamic extremists who, cognitively constricted to the point of being delusional, for the glory of Allah, kill "infidels" and themselves—a homicide-suicide. There are many different kinds of homicide-suicides. Not so long ago, we had the Korean and Vietnamese monks who burned themselves for democratic freedom and independence from American tyranny. There is a long history, the Christian martyrs being one example. Yet we do not know whether these suicides and homicide-suicides are the same or different. We recently published the first study in the world on the writings of suicide martyrs. We (Leenaars, Park, Collins, Wenckstern, & Leenaars, 2010) examined 33 letters of "altruistic suicides" and compared them blindly to a matched sample of the more "common" suicides (as outlined in J. S. Mill's [1984] methods of difference and of agreement or similarity). I will answer in a later chapter some key questions: Are suicide martyrs the same and/or different than suicide? What intrapsychic elements are critical for predicting a to-be martyr to kill him/herself? To kill you? Are interpersonal protocols relevant? What elements? Is ambivalence salient? There are many unique insights offered by the personal documents (e.g., martyrs' last letters, Internet writings) of suicide bombers. As we will learn in a following chapter, this is the topic of the variety of suicides.

On the topic of the suicide terrorist, Maltsberger (2001) insightfully speculated,

> One day a messianic leader, inflamed with malignant narcissistic fury, will become so intoxicated with the desire for revenge and destruction of his enemies that he will no longer be deterred by any self-preservative instinct or compassion for the welfare of his own people. Herman Melville imagined such a leader in Captain Ahab, who so nursed his grievance against Moby Dick, the white whale that bit off his leg, that he destroyed himself and his entire crew in his quest for revenge. (p. 145)

Is the soldier who died by suicide a Captain Ahab? Did he have grandiose fury?

## GENDER DIFFERENCES

The basic gender difference in suicide is that males kill themselves more often than females (although this is not so in China). This is true in the military. In contrast, females attempt suicide more often than males. This gender difference has been found in almost all nations (Canetto & Lester, 1995). The male-female ratio of suicide has remained fairly stable over time. The generally accepted male-female ratio of completed suicides is three or four males to one female, but there is great variation around the world and probably within military cultures.

It is beyond the scope here to examine the vast literature on this topic. Briefly, explanations in the literature (Leenaars, 1988) have varied. Females use different and less lethal methods (drugs vs. shooting). Is this so for female soldiers? We do not know. Individuals with severe emotional disturbances (psychiatric disorders) have higher rates of suicide, and men are more likely to be diagnosed with such disorders. There are also alternative social expectations for men and women in trauma such that males act more catastrophically. Is this true for the female soldier who has been indoctrinated into a collective green culture? Yet Shneidman (1985) has argued that genotypic similarities may be more prevalent than differences. Indeed, the author's research (Leenaars, 1988) on suicide notes of males and females in the general population confirms this. Pain is pain. Frustrated needs are frustrated needs. Constriction is constriction. Maybe there are phenotypic differences (e.g., method, diagnostic label) in suicide but not genotypic ones across gender. Thus, we can use the theory presented to evaluate female and male soldiers.

Further reflections on this topic are most relevant but are beyond space allowed in this volume. For example, if the soldier who died by suicide was a female, and women have significantly lower rates of suicide than men, maybe one should ask

"Is the rate of suicide in female soldiers higher than the general population? Why did the female soldier kill herself? Are the usual protective factors for females absent in female veterans? Is the military culture a factor? Despite the research to date, are there unique factors in female suicide? Are there unique factors in female soldiers' suicide? Are the critical factors the same as male soldiers? What factors are common and unique in female and male soldiers who die by suicide? There are no adequate studies to answer these questions.

## YOUNG ADULTHOOD: A COMMENT

Most soldiers are young adults. Young adulthood is a discrete timeline in development (Frager & Fadiman, 1984; Kimmel, 1974). It has its unique biological, psychological, cultural, and sociological issues, and this is so in the military. I tentatively support the position in limiting this timeline from ages 18 to 25 (Kimmel, 1974). Of course, no developmental period can be rigidly defined chronologically, and at best the 18 to 25 range approximates what can only be defined developmentally, that is, some people mature earlier, others later than the mean.

Adolescence is a stage marked by the development of a sense of identity. The young adult continues to develop this sense of identity, evolving a finer and more discrete sense of who he/she is in relation to others. This is so in the armed services. Not distinct from this process, the demand to master the challenge of intimacy emerges as the central issue of young adulthood. Erik Erikson (1963, 1980) was one of the first to pioneer work on intimacy (intimacy vs. isolation) in young adulthood, noting that one must have a sense of who one is before one can appreciate the uniqueness of another. Although the capacity to relate to others emerges earlier, "the individual does not become capable of a fully intimate relationship until the identity crisis is fairly well resolved" (Kimmel, 1974, p. 23). (Does the military delay the resolution of the identity crisis? Or does it offer an alternative collective warrior identity?) Often, before such development, the individual can only avoid genuine closeness or engage in narcissistic relationships. As Frager and Fadiman (1984) note,

> Without a sense of intimacy and commitment, one may become isolated and be unable to sustain intimate relationships. If one's sense of identity is weak and threatened by intimacy, the individual may turn away from or attack whatever encroaches. (p. 152)

In suicide, this is true, even of oneself!

Research regarding any psychological area of young adulthood is scarce. Even the old volumes on suicide that are apparently directed to include this age group (e.g., Klerman, 1986), disappoint us, for they take note only of our lack of knowledge about this group of individuals. Often young adults are classed together with adolescents, a taxonomic maneuver both theoretically and empirically unsound. There have been lately exceptions to the overlapping taxonomy in the field (e.g., Leenaars, 1989a, 1989b, 1991; Lomas & Lester, 2011); however, clearer definitions are needed in the research of young adults, which needs to be integrated into life span perspectives (Leenaars, 1991).

A question: "Are young adult suicides psychologically different from those of other adults?" Based on the analysis of hundreds of suicide notes across the adult life span, there are enormous commonalities or patterns. Yet a life-span developmental perspective (Leenaars, 1991; Leenaars et al., 2001) is still essential. Despite greater similarities, there are differences. This is an issue of more or less, not presence or absence. Significantly more often, young adults' notes reflect the probable presence of psychopathology or the inability to adjust. More often than during any other age group, we can discern a mental disorder associated with a very vulnerable ego. This likely suggests that young adults' suicides may have a high relation to the crystallization of psychopathology imbalance, or whatever you want to call it, at this age. Psychopathology, of course, can be seen across the life-span, but it also may be most relevant to soldiers in this age group—the vast majority of soldiers.

Thus, the clinician working with young adult soldiers is cautioned about using current research, needing to be clear and distinct about age, cultural issues and a host of other factors in their clinical applications, and this is very much so in the military.

## A NOTE ON COGNITIVE STYLE

The best known faulty syllogism may well be:

All men are immortal
Socrates is a man
Therefore Socrates is immortal.

As a syllogism, the above is valid; but it is faulty because it begins with a first premise that is false. All men are not immortal. It is a false universal inductive generalization. It is basic to realize that a valid syllogism can have a false conclusion. This can happen if one or more of the premises are false. In addition, in the suicidal person, the first premise (sometimes called a core belief) is not only false, but it is also lethal.

Edwin Shneidman (Leenaars, 1993, 1999) has been keenly interested in making explicit the latent logical (cognitive) components of everyday thought. He realized how useful it is to examine the cognitive styles exhibited in each suicidal person. For example, in a terse but insightful paper, "On 'Therefore I Must Kill Myself,'" Shneidman (1999) shows how vitally important it is for a clinician to understand the patient's idiosyncratic logical style and then not agree with that patient's major premise when the premise is the keystone to the patient's lethal (suicidal) syllogistic conclusion. For example, "People who have committed a certain sin ought to be dead; I am a person who has committed that sin; *therefore*, I ought to be dead." For another example, "Soldiers who are rejected by the commander ought to be dead; I am a soldier who has been rejected; *therefore*, I ought to be dead." For another example, "People who kill ought to be dead. I am a soldier who killed; *therefore*, I ought to be dead." The homicidal mind is often the same. "People who disgrace me ought to be dead; I am a person who has been disgraced by the sergeant; therefore, he ought to be dead." This means that people need to know the suicidal person's mind. Therefore, you have to know what suicide is. Further if you do, you will not agree with the cognition or the major premise, "People who cannot live up the demands of the military ought to be dead," or some other *lethal* premise (or, if you prefer, core belief).

Yet, I, akin to Shneidman (1999), think that the logic is wrong. It all depends on the word *therefore*. It is perhaps the most important word in life—and, it seems, death. When we invoke the word *therefore*, we have come to a decision or resolution. Thus, *therefore* is the pivotal word in logic, specifically, syllogisms. Shneidman (1999) writes,

> We all know that there are bridge-words between thought and action. Words like "might," "ought," "should," "must"—which convey *various* amounts of psychological *push*. The point is that *not all "therefores" are psychologically identical*, or to put it another way: not all equal signs are equally equal. Thus the word "therefore" cannot be taken for granted as a word which means "always" or "under all circumstances" as in the syllogism about Socrates. In *that* example, Socrates is *always* mortal, but that example simply illustrates the confinements of traditional Aristotelian syllogisms. (p. 74)

Yet there is an even more important fact: the mind does much more than simply logic. The sinner is an example. The sinner makes the classical error; the major premise is false. All humans are not immortal. In addition, all people who have committed a certain sin ought not to be dead. In the case of the veteran who died by suicide, what were his cognitions, premises, beliefs, and so on?

Once you know what suicide is, I believe that it is easy to conclude the following: The suicidal person, at the moment of taking his/her life, is figuratively intoxicated with his/her overpowering emotions and constricted logic. Are these suicidal people "rational"? Was the soldier "rational"?

Cognitive constriction is an essential element of the suicidal mind. It is imperative that the reader be reminded that the suicidal person defines the trauma. Trauma is a perception, not a thing in itself. An educational example of Shneidman's is the man who always wanted the perfect car. He saved his money and bought a Ferrari. Then the car got a scratch; his response was to write a suicide note, stating, "There is nothing to live for," and therefore shot himself. For almost all, the scratch appears, to use a popular expression, to be small stuff, but not for him. His belief was, "People with a scratch on a Ferrari ought to be dead." His logic can be symbolized by the following:

> People with a scratch on a Ferrari ought to be dead.
> I am a person who has a scratch on his Ferrari.
> *Therefore*, I ought to be dead.

Another Shneidman example is the elderly woman who killed herself after her canary died. Her cognition was, "People with a dead canary ought to be dead." She had her *"therefore."* Thus, we must remember that trauma for both these individuals—and all suicidal people—was/is a perception, not "objective" reality. They had *the* trauma and major premise; he, about a scratch on a car, and she about a dead pet. What was the soldier's trauma(s)? It is, however, so much more. It is not only simply about sinning or having a scratch on a car or suffering from a TBI. It is not only simply about a demotion or a dead pet. It is no simply about losing a loved one. ("People rejected by Mary ought to be dead.")

Figuratively speaking, the suicidal person is "intoxicated" by overpowering emotions. Concomitantly, there is a constricted logic and perception. Although the person defines the trauma (situation) subjectively, one cannot buy into the overpowering emotions and constricted logic presented by the individual, even in a retrospective investigation, a psychological autopsy, or military investigation. Yet many do so after events like suicide.

One of the basics of a suicidal mind is the following: There is poverty of thought, exhibited by focusing only on permutations and combinations of grief and grief-provoking topics. A common premise is that something is either black or white. Something is either "A or non-A." To presume that the suicidal individual either wants to kill him/herself or not is an extremely limited point of view, even if the person generates *only* such permutations and combinations in his or her premises. It is not necessary to require a view of the world as "A or not-A," (e.g., "victory or no victory," "canary or no canary") that is, a view (or belief) of the world exhibiting only permutations and combinations of one content. The individual may only think of grief-provoking content but he/she can also have other fantasies, or cognitions. To suggest such a "A and not-A" world is a limited and harsh view of life, one that neither the suicidal person nor we have to accept. Thus, the task is to not accept the first premise or core belief, and resulting "therefore." What any suicidal person says, writes, communicates, believes, and so on do not limit us. We, in fact, need to not buy into the suicidal person's beliefs/cognitions; some of them may well be intentioned to mask the truth, reality, and so on.

Theory, explicit and implicit, like that presented, plays a key role in understanding any behavior (and for suicidal behavior, it is essential) (Leenaars, 2004). It is only through theory that we will sort out the booming buzzing confusion (mess) of the suicide experience. We do not have to accept the mess. Perception is in the eye of the beholder (Kuhn, 1962). Yet, do we accept the theory of the soldier, suffering from mild cognitive disorder from a TBI? The abandoned person? The man with the scratched Ferrari? The soldier who received a demotion? The soldier who committed a crime and jailed? Thus, it would be wise to reflect on the dead soldier's conclusion, his/her "therefore." We can be relatively sure—no, absolutely sure—that at least the scratch on the Ferrari is not cause to therefore kill oneself. It is also not cause to therefore kill someone else too, such as the sergeant who demoted you.

## DISSEMBLING:
## CLUES TO SUICIDE RECONSIDERED

"Suicide happens without warning." This is a myth that Shneidman (1985) challenged. He stated that the fact was, "Studies reveal that the suicidal person gives many clues and warnings regarding suicidal intentions." Another fact, according to Shneidman, was "Of 10 persons who kill themselves, 8 have given definite warnings of their suicidal intentions." However, is "Suicide happens without warning" really a fable? People, in fact, today believe that "Suicide happens without warning" is a myth. Yet, in the military and in clinical settings, the concern is if 8 out of 10 people give warnings, what about the other 2? Armed services are more concerned about this group too, such as the suicidal warrior in the field of battle who dissembles. He/she may not only risk himself, but also his/her fellow soldiers in battle. Goldblatt (1992) and Litman (1995) have separately noted that a small but noted percentage of completed suicides are seen as having left no clues. A minority of these people is most perplexing to even the most veteran forensic suicidologist (Leenaars, 1997). How do we understand and predict their suicides?

The classical case, albeit a literary one, is Edwin Arlington Robinson's *Richard Cory*. The poem describes Richard Cory as a "gentleman from sole to crown." He was "human," "rich," "favored," "schooled," and in fact, people "thought that he was everything." Then, as the poem ends, Richard Cory unexpectedly puts a "bullet through his head." That suicide happened without warning. Did the veteran's? The Richard Cory-type of patient is, in fact, a person who many of us experience in our career. The odds are greater than 50% of psychiatrists and greater than 20% of psychologists losing a patient to suicide over the course of their careers. (What are the percentages in the military and VA?) A patient dying by suicide is thus not a rare event, although the Richard Cory patient does not compose all of them. The fact is that over 80% of the people who killed themselves did leave clues. The other 20% raise our anxiety, with the Richard Cory-type being even more infrequent—but maybe not in the military. The scenario may occur more frequently in the armed forces, given the climate of secrecy. We need to understand the Richard Cory-type soldier better. Was this Admiral Boorda?

Admiral Mike Boorda was the 25th Chief of Naval Operations. Boorda died by suicide on May 16, 1996. I remember that day well. I was then President of the

American Association of Suicidology (AAS). In those years, the Survivor's After Suicide Conference was a distinct meeting of the AAS. I was at the conference, acting presidential, and I met many guests, some of whom were officers in the Navy. They were distinctive in their uniforms. We had gotten to know each other. After news of Boorda's death was released, fortune would have it that I met with his sailors, who intentionally sought me out. We talked. They were shocked. They could not believe that their Admiral killed himself. He was a great hero, they said. They were in denial and understandably, anxious and agitated. They asked, "Did you hear anything? Have they told you anything?" They *needed to understand* (i.e., the needs of understanding and deference). Was there a contagion? Aftershocks? PTSD? Based on that May 16th, I would predict so. However, we do not know. Secrecy was the order of that day.

Shneidman (1994) had reconsidered his perspective on clues to suicide. He asked "How it is that some people who are on the verge of suicide . . . can hide or mask their secretly held intentions?" Shneidman suggests that many clues are veiled, clouded, and guarded, some even misleading. He argues that there are individuals who live secret lives. These people do not communicate. They often lie. These people do not process and/or mediate the stimuli in the usual way, having a defensive intent, conscious and/or unconscious, to avoid the situation. There are walls, whether green, white (i.e., the health providers' barriers), or otherwise. Often these walls reflect a basic coping style, with conscious and unconscious elements in the process (Leenaars & Lester, 1996). This is pervasive in the military. Shneidman calls it *dissembling*.

To dissemble means to conceal one's motives. It is to disguise or conceal one's feelings, intention, or even suicide risk. These people wear "masks." In the military, it is a green masking. The story that they tell is that they *do not* tell their story; indeed, they themselves may be unaware of the dissembling or masking. There may be self-deception, and the masking may be completely unconscious. There is frequently an enduring personality pattern. It is an enduring pattern of inner experience and behavior, and the green culture may reinforce it. For example, from an early age, they lie about everything, both omissions and commissions. The behavior often deviates markedly from the expectations of family, friends, and even of one's culture—or the culture of a group (or subculture), such as being a soldier (or is it?). Often, unconscious processes are involved. Kurt Cobain dissembled; he stated in his note that he was "faking it and pretending as if I'm having 100% fun." Most clinicians encounter such people, not only suicidal people. Their stories are invalid; sometimes they even intentionally produce or feign a behavior (or symptom).

Of course, secrets are universal; people keep secrets. This is very true in the military. It is probably true that secrets must be kept in the military—some may be life saving. Secrecy is a core value, one's honor, as we discussed in the first chapter. Yet there are several types of secrets that have varying effects on wellness, for the soldier, the military family, and the troops. Berg-Cross (2000) offered the following list: supportive, protective, manipulative, and avoidant. It is probably true that supportive and protective secrets are often positive and may even have healing value for the wounded soldier (Everson & Camp, 2011). They may well be essential in war. Yet the manipulative and avoidant secret—the Richard Cory types—are a source of anxiety, PTSD, depression, and suicide risk. They are deadly, even in the military. They are *never* helpful in the therapeutic sense (Everson & Camp, 2011; Leenaars, 2004). Lethal secrets in the military— and the soldier's family—fall into the latter two categories. They are one of the most lethal (if not the most) aspects of the scenario in military suicide. It is dissembling that kills so many of our heroes; and the military system/culture reinforces the deaths. Of course, this is not intentionally so, but reinforcing nonetheless. It all too often creates needless deaths. What can we do?

On healing, getting past the secrecy, Shneidman (1994) stated,

> We suicidologists who deal with potentially suicidal people must . . . understand that in the ambivalent flow and flux of life, some desperately suicidal people . . . can dissemble and hide their true lethal feelings from the world. (p. 395)

I strongly believe that this is especially important to remember about soldiers who die by suicide. They often dissemble with fellow soldiers, parents, spouse, therapist, the sergeant, everyone. No one often predicts risk. The question raised: "Why?" Is it the military culture?

## PREVENTION/INTERVENTION/POSTVENTION

Finally, in this chapter, I need to offer a note on *prevention*. The classical approach to prevention, whether mental health or public health, is that of Caplan (1964), who differentiated between primary, secondary, and tertiary prevention. The more commonly used terms today for the primary, secondary, and tertiary modes of *prevention* in suicidology are prevention, intervention, and postvention, respectively, and has applicability from the individual to the societal levels. Briefly, the three modes of a comprehensive response to a warrior at risk are as follows:

1. *Prevention* relates to the principle of good mental hygiene in general. It consists of strategies to ameliorate the conditions that lead to suicide— to do something before the event occurs. Preventing suicide is best accomplished through primary prevention, mainly by education. The military must be educated about suicide. Such education—given that suicide is a multi-dimensional event—is enormously complicated. Be that as it may, there are numerous suicide (primary) prevention programs available in military services in the United States, Canada, and around the world.

2. *Intervention* relates to the treatment and care of a suicidal crisis or suicidal problem. Secondary prevention is doing something during the event. Suicide is a multidimensional event and thus is not solely a medical problem, and many people can serve as lifesaving agents. Soldiers especially can be life saving (e.g., "buddy to buddy"). Nonetheless, professionally trained people, psychologists, psychiatrists, social workers, psychiatric nurses, crisis workers, and so on continue to play significant roles in intervention, and the psychologists/psychiatrists with military services belong to this list. Thus, although equally true for prevention and postvention, intervention will call for the development of community and social linkages, a hallmark of an evidence-based response to intervention. There is a lot that any armed services can do.
3. *Postvention*, a term introduced by Shneidman (1973a, 1973b, 1975), refers to those things done after the event has occurred. Postvention deals with the traumatic aftereffects in the survivors of a person who has committed suicide (or in those close to someone who has attempted suicide). There were/are many survivors. Postvention in suicide involves offering mental health and public health services to the bereaved survivors. Family, friends, fellow soldiers, local service providers, and so on, need help in the loss and grief. Postvention includes working with all survivors who are in need. Of course, as is well established (Shneidman, 1985), psychological autopsies are effective postvention and often help in the healing.

## CONCLUDING REMARKS

People in general have considerable difficulty appreciating significant characteristics of suicidal individuals. Most fortunately, a host of suicidologists have given us a rich history of theory for understanding suicide. These suicidologists point out that the suicidal history is understandable, and thus, may be predictable and controllable. This is also true in the military. Our military personnel do not have to die needless deaths.

The main conclusion is that there may be more commonalities (agreement or sameness) in suicide than differences. By virtue of our human quality, whether male or female, whether soldier or not, whether veteran or not, greater similarities than differences are to be expected in all suicidal events (Shneidman, 1991). Maybe there are classifications, common constructs, commonalities, or whatever we wish to call *it*. We presented an array of evidence-based commonalities. Despite clear and distinct commonalities, each individual must still be, however, understood idiosyncratically (uniquely), even if dissembling. This is also true for the soldier in the military.

We need to continue to develop multidimensional models (Goffman, 1974; Kuhn, 1962) to understand suicide; and the suicide of that individual, whether

Adolf Hitler or any soldier, can be researched, and most importantly, can be made useful and life saving. What would have helped with Adolf Hitler? What intervention (psychotherapy, healing circle, medication, hospitalization, and so on) will effectively work with that particular type of individual? What empirically supported recommendations can be made that are effective with the suicidal soldier? Suicidal veterans? We need to be person centered (or oriented)—what kind of person that person is. The goal of our book is simple: to understand the suicidal soldier or veteran. What can we learn from soldiers who died by suicide?

There was, I predict, so much pain and anguish, a psychache in the soldier's mind, who died by suicide. In addition, why did his fellow soldier survive the pain, and she not? What were the fellow soldier's protective factors? Why did *that soldier* die and his father, her ex-husband, his friends, her fellow soldiers, and so on, survive? These are the most important (individual) questions. These are some of the questions one asks in psychological autopsies. Why did the 21-year-old soldier kill himself after his mother's death or the 38-year-old reserve officer jump into death?

Herman Melville, who was a troubled soul, wrote a monumental study of people's anguish and suicide, *Moby-Dick*. In Chapter 36, "The Quarter Deck," Captain Ahab speaks:

> Hark ye yet again—the little lower layer. All visible objects, man, are but as pasteboard masks. But in each event—in the living act, the undoubted deed—there, some unknown but still reasoning thing puts forth the mouldings of its features from behind the unreasoning mask. If man will strike, strike through the mask! How can the prisoner reach outside except by thrusting through the wall?

We—soldiers, commanders, clinicians, survivors, and so on—must understand and reach through *the green suicidal mask*.

# CHAPTER 3
# The Psychological Autopsy

In Edwin Shneidman's own reflections, he does not know whether suicide was looking for him or he was looking for suicide. Part of his motivation came on a day in 1949 while working for the Veterans Administration (VA), when he discovered hundreds of "genuine suicide notes" in the vaults of the Los Angeles Coroner's Office. He was restless and looking for some niche in psychology. In religion, we talk about epiphanies and epiphany moments. His realization that the notes were "genuine" was an epiphanic moment for Shneidman. He had an autonomic reaction with the feeling, without verbalizing, that it was important to say "genuine" suicide note. Within a couple months of saying "genuine" and then "simulated" and then eliciting notes, and then calling Norman Farberow, he was beginning a career—and a discipline. He said, "Oh boy," suicide notes, the golden road to the unconscious of suicide, and suicidology began.

Suicide is a human malaise, and suicide notes are the penultimate act giving a voice to this malaise. In *Voices of Death*, Shneidman (1980a) wrote that such "documents contain special revelations of the human mind and that there is much one can learn from them." Suicide notes allow us to learn about a person, to advance the nomothetic (population) and idiographic (individual) approaches in science, and to aid in the aims of science in general—understanding, prediction, and control.

## A CORE BELIEF STATED BEFORE THE PRESENTATION

A core belief in science is that the idiographic or case study approach allows us to do the main business of psychology: the intensive study of the human person and thus, the suicidal individuals in the military. Case studies, through personal documents, biography, and so on, allow us to understand the suicidal mind better (Shneidman, 1996). Windelband (1904) noted at the turn of the last century, a core belief in science is that we need idiographic and nomothetic pathways in science, and this is also true in military science. Later, I will discuss a historical military case of suicide: General Emory Upton. *It is a soldier's story told.*

## INTRODUCTION TO THE STUDY OF SUICIDE

Within the context of the need for multidisciplinary perspectives, in psychiatry, psychology, how do we study suicide in the military? What evidence or data can we legitimately use? After all, unlike most other areas of investigation regarding complicated human acts, we cannot ask questions of the dead suicidal person. We cannot ask, "Why did you kill yourself in harm's way?" This forces us to look elsewhere. As discussed earlier, the following ways can help find answers to these questions: statistics, third-party interviews, the study of nonfatal suicide attempters and documents (including biographies, suicide notes, letters and other personal documents). Despite strengths in each of the methods, all of these have their limitations. Statistics reflect by themselves only numbers and are, at best, only a fraction of the true figures. (It is the nomothetic story.) Third-party interviews—called psychological autopsy studies—can provide only a point of view that is not the suicide's (it is the survivor's story). Nonfatal attempters are different from completers (it is the attempter's story). Documents may provide only a snapshot of an event that requires a full-length movie (it is, however, the person's story).

Of course, there is the problem of obtaining any of these data, statistics interviews, reports by attempters, and personal documents (this as we learned is especially so in the military). Maris (1981) has brilliantly discussed at length some of these problems. He states, "One of the major problems in understanding self-destructive behaviors is that the data base for such potential explanations is conspicuously absent." However, such a formulation or core belief should not deter us in our study of suicide. Rather, it should make us, as Shneidman (1980a) suggested, rethink the problem (and the specific problems with each source of data) anew (i.e., cognitive problem solving). In the study of violence whether war related or suicide, there may be no single datum or evidence. The issue of *the* data does not only exist in suicidology; it is ubiquitous in science. In science, data are samples; some areas of science have greater accessibility to data, whereas others have less (e.g., astronomy). Equally, as in all science, all sources of data in suicidology have limitations; yet equally they have strengths. For example, personal documents may have limitations, such as providing only a snapshot; but sometimes they provide a vignette of sufficient length so that some essential essences of the entire movie can be reasonably inferred. Examples of such documents are case histories, psychological test results, death certificates, suicide notes, and a host of others. I here offer a few insights on the psychological autopsy, with special reference to the use of documents in the autopsy. It allows us to tell the person who died by suicide and the survivor's story

The psychological autopsy (PA) is the work of Edwin Shneidman (see my 1999 edited volume of his edited works, *Lives and Deaths*. Quotes in this chapter are from that volume). Shneidman's key papers, according to his decision (Leenaars, 1999), on the psychological autopsy are "The Psychological Autopsy" (1977), a paper on what a PA is; "Comment: The Psychological Autopsy" (1994), a very terse statement on what a PA is not; and Shneidman's paper "An Example of an Equivocal Death Clarified in a Court of Law" (1993a), an example of a psychological autopsy. I will present this chapter essentially as Shneidman taught, *res ipsa loquitor*, with a preamble about the main function:

> The main function of the psychological autopsy is to clarify an equivocal death and to arrive at the correct or accurate mode of that death. In essence, the psychological autopsy is nothing less than a thorough retrospective investigation of the *intention* of the decedent—that is, the decedent's intention relating to his being dead—where the information is obtained by interviewing individuals who knew the decedent's actions, behavior, and character well enough to report on them. (Leenaars, 1999, p. 388)

## THE PSYCHOLOGICAL AUTOPSY

The paper, "The Psychological Autopsy" (1977; see Leenaars, 1999) provides, according to Shneidman's opinion, his best paper on the topic. It is an overview of the PA procedure. Here are his words:

> It is probably best to begin by defining a psychological autopsy and its purposes, then to discuss some related theoretical background and ways of actually performing psychological autopsies. The words psychological autopsy themselves tell us that the procedure has to do with clarifying the nature of a death and that it focuses on the psychological aspects of the death . . . ideas, important to understanding the psychological autopsy, need to be discussed. The first is what I have called the NASH classification of deaths.
>
> From the beginning . . . the certification and recordkeeping relating to deaths have implied that there are four modes of death. It needs to be said right away that the four modes of death have to be distinguished from many causes of death listed in the current International Classification of Diseases and Causes of Death (World Health Organization, 1957; National Center for Health Statistics, 1967). The four modes of death are natural, accidental, suicide, and homicide; the initial letters of each make up the acronym NASH. Thus, to speak of the NASH classification of death is to refer to these four traditional modes in which death is currently reported. Contemporary death certificates have a category which reads "Accident, suicide, homicide, or undetermined"; if none of these is checked, then a "natural" mode of death, as occurs in most cases is implied.
>
> It should be apparent that the cause of death stated on the certificate does not necessarily carry with it information as to the specific mode of death. For

example, asphyxiation due to drowning in a swimming pool does not automatically communicate whether the decedent struggled and drowned (accident), entered the pool with the intention of drowning himself (suicide), or was held under water until he was drowned (homicide).

## Purpose of the Psychological Autopsy

As long as deaths are classified solely in terms of the four NASH categories, it is immediately apparent that some deaths will, so to speak, fall between the cracks, and our familiar problem of equivocal death will continue to place obstacles in our path to understanding human beings and their dying. Many of these obstacles can be cleared away by reconstructing, primarily through interviews with the survivors, the role that the deceased played in hastening or effecting his own death. This procedure is called "psychological autopsy," and initially its main purpose was to clarify situations in which the mode of death was not immediately clear.

The origin of the psychological autopsy grew out of the frustration of the Los Angeles County Chief Medical Examiner-Coroner, Theodore J. Curphey, MD, at the time of the reorganization of that office in 1958. . . . As a result, he asked the Los Angeles Suicide Prevention Center to assist him in a joint study of those equivocal cases, and it was this effort—a multidisciplinary approach involving behavioral scientists—which let to my coining the term "psychological autopsy.". . .

At present, there are at least three distinct questions that the psychological autopsy can help to answer:

1. *Why did the individual do it?* When the mode of death is, by all reasonable measures, clear and unequivocal—suicide, for example—the psychological autopsy can serve to account for the reasons for the act or to discover what led to it. . . .
2. *How did the individual die, and when—that is, why at that particular time?* When a death, usually a natural death, is protracted, the individual dying gradually over a period of time, the psychological autopsy helps to illumine the sociopsychological reasons why he died at that time. . . .
3. *What is the most probable mode of death?* This was the question to which the psychological autopsy was initially addressed. When cause of death can be clearly established but mode of death is equivocal, the purpose of the psychological autopsy is to establish the mode of death with as great a degree of accuracy as possible. . . .

## Conducting the Psychological Autopsy

How is a psychological autopsy performed? Talking to some key persons—spouse, lover, parent, grown child, friend, colleague, physician, supervisor, and co-worker—who knew the decedent, does it. The talking to is done gently, a mixture of conversation, interview, emotional support, general questions, and a good deal of listening. I always telephone and then go out to the home. After rapport is established, a good general opening question might be: "Please tell me, what was he (she) like?" Sometimes clothes and material possessions are looked at, photographs shown, and even diaries and correspondence shared. (On one occasion, the widow showed me her late husband's suicide note—which she had hidden from the police!—rather changing the equivocal nature of the death.)

In general, I do not have a fixed outline in mind while conducting a psychological autopsy, but inasmuch as outlines have been requested from time to time, one is presented below with the dual cautions that it should not be followed slavishly and that the investigator should be ever mindful that he may be asking questions that are very painful to people in an obvious grief-laden situation. The person who conducts a psychological autopsy should participate, as far as he is genuinely able, in the anguish of the bereaved person and should always do his work with the mental health of the survivors in mind.

Here, then, are some categories that might be included in a psychological autopsy (Shneidman, 1977):

1. Information identifying victim (name, age, address, marital status, religious practices, occupation, and other details)
2. Details of the death (including the cause or method and other pertinent details)
3. Brief outline of victim's history (siblings, marriage, medical illness, medical treatment, psychotherapy, suicide attempts)
4. Death history of victim's family (suicides, cancer, other fatal illnesses, ages of death, and other details)
5. Description of the personality and life-style of the victim
6. Victim's typical patterns of reaction to stress, emotional upsets, and periods of disequilibrium
7. Any recent—from last few days to last twelve months—upsets, pressures, tensions, or anticipations of trouble
8. Role of alcohol or drugs in (a) overall life-style of victim, and (b) his death. [I would add questions, given the soldier cases, about guns and weapons—AL]
9. Nature of victim's interpersonal relationships (including those with physicians)
10. Fantasies, dreams, thoughts, premonitions, or fear of victim relating to death, accident, or suicide
11. Changes in the victim before death (of habits, hobbies, eating, sexual patterns, and other life routines)
12. Information relating to the "life side" of victim (upswings, successes, plans)
13. Assessment of intention, that is, role of the victim in his own demise
14. Rating of lethality
15. Reaction of informants to victim's death
16. Comments, special features, and so on.

In conducting the interviews during a psychological autopsy, it is often best to ask open-ended questions that permit the respondent to associate to relevant details without being made painfully aware of the specific interests of the questioner. As an example: I might be very interested in knowing whether or not there was a change (specifically, a recent sharp decline) in the decedent's eating habits. Rather than ask directly, "Did his appetite drop recently?" a question almost calculated to elicit a defensive response, I have asked a more general question such as, "Did he have any favorite foods?" Obviously, my interest is not to learn what foods he preferred. Not atypically, the respondent will tell me what the decedent's favorite foods were and then go on to talk about recent changes in his eating habits—"Nothing. I fixed for him seemed to please him"—and even proceed to relate other recent changes, such as changing patterns in social or sexual or recreational habits, changes which diagnostically would seem to be related to a dysphoric person, not inconsistent with a suicidal or subintentioned death. (Leenaars, 1999, pp 387–410)

*This is quite a task.* Before I proceed, there is one very terse, but most important comment needed in answering, "What is a psychological autopsy?" Keeping Mills in mind, "How is the PA different from other retrospective death investigations?" I will next present Captain Shneidman's core beliefs.

## COMMENT ON THE PSYCHOLOGICAL AUTOPSY

Shneidman thought that there were different legitimate approaches of retrospective death investigations. He clarified this point often and presented his position in "Comment: The Psychological Autopsy" (1994; see Leenaars, 1999). It is a one pager for *American Psychologist* that says that the psychological autopsy for Shneidman is about intention. There are different kinds of procedures; some are not psychological autopsies. This did not mean that they had no value, only that these studies were a different method to understand an event. There are many legitimated means; they simply are different (and often each can provide a perspective on the death). For example, if you do a ballistics test and take blood samples and so on, that is not a psychological autopsy, that is a forensic autopsy or a clinical autopsy. If you look at bullet markings, blood type and DNA profiling, that is not intention. The psychological autopsy is about the person's intention vis-à-vis the death.

Captain Shneidman identified minimally four kinds of death investigations: the medical autopsy, the forensic investigation, the statistical or demographic report, and the psychological autopsy. I next present verbatim the commander's brief description of the other types of death-follow-up procedure:

1. **Autopsy**. The autopsy involves inspection and partial dissection of a dead body to learn the cause of death, the nature and extent of disease, and where possible, the mode of death. It is an examination by a physician-pathologist (and ancillary personnel). It is objective; it reports the facts—the

weight of the brain, alcohol content of the blood, appearance of the liver, and so forth. The pathologist acts as an amicus curiae, specifically as a friend to the state, reporting the findings concerning a particular dead person for public and archival record....

2. **Forensic investigation**. The forensic investigation relates to the physical evidence surrounding the death. It may include a plethora of relevant details: windows open, doors locked, trajectory of bullets, powder marks, fingerprints, handwriting analyses, personal documents (suicide notes, threatening letters, cashed checks, etc.). Although these facts can be centrally relevant in either criminal or civil cases and need to be done as thoroughly as possible, a report of forensic details is, of course, not an autopsy at all—and certainly not a psychological autopsy-but an investigation....

3. **Statistical or demographic reports**. If one is interested in, for example, the prevention of suicide, then it is obvious that knowledge of past patterns of behavior of individuals who have committed suicide can be a useful tool. These patterns might be called prodromal indices or premonitory signs. They make up the now well-known "clues to suicide" (Shneidman & Farberow, 1957). What may not be so obvious is that the statistical truths about a large number of committed suicides do not necessarily tell us anything about any particular case. The frequent error in this field is to confuse statistics with individual events and then to argue that because this individual does (or does not) have certain desiderata characteristics of a group, suicide must have (or must not have) occurred. Statistics are made up of individual cases; an individual case is not controlled by statistics. To argue from statistics to an individual case is a tyro's error. . . . One must be on the alert for this kind of reverse reasoning. In any event, the citation of statistical and demographic data, even when cogent and sensible, is clearly not a psychological autopsy. . . . (Leenaars, 1999, pp. 411–412)

What is meant by a *psychological autopsy* is all very different from these procedures. The clarification of the mode (NASH) of some deaths devolves on the intention of the decedent in relation to the death. Suicide, as we read in Chapter 2, is an intentioned death. The PA was devised to clarify what the mode of death was, why the individual did it, and when the person did it (Curphey, 1961; Leenaars, 1999, 2004; Litman, Curphey, Shneidman, Farberow, & Tabachnick, 1963; Shneidman, 1969, 1973b, 1977, 1985). A psychological autopsy is an objective procedure that seeks to make a reasonable determination of what was in the mind of the decedent vis-à-vis his or her own death. It does this by looking at the history of the decedent, the lifestyle, the intrapsychic and interpersonal characteristics, the cognitive style, the psychopathology, and so on, as discussed in this chapter in detail.

> It legitimately conducts interviews (with a variety of people who knew the decedent) and examines personal documents (suicide notes, diaries, and letters) and other materials (including the autopsy and police reports) that

are relevant to the psychological assessment of the dead individual's role in the death. (Leenaars, 1999, p. 414)

Of course, as Shneidman often reminded me over the hundreds of hours of discussions, the results cannot be stated with complete certainty.

I will next discuss the value of documents in the PA.

## PERSONAL DOCUMENTS

Suicide notes are not the only valuable personal document. Indeed, there is a host of personal documents worthy of psychological digs—diaries, poems, letters, biographies, psychological reports, medical files, police records, newspaper reports, and so on. A most fascinating kind of personal document left by suicides is the suicide diary. A suicidal diary is a suicide note with a history. It meets the following criteria: It is a lengthy, literate document kept over a fairly long period of time, often years; the diarist writes explicitly about suicide, including his or her suicidal thoughts, impulses, reflections, and resistances; and, the diarist commits suicide. Rosenblatt (1983) notes that diaries "allow one to see . . . clearly day-to-day changes, long-term trends, and the effects of specific events." A diary is a document about how life is lived. Most of us are as curious about diaries as we are about suicide notes. To cite a diarist, Evelyn Waugh "The routine of their day properly recorded is always interesting." The personality of its keeper is richly found within a diary—a document not only of events but the very writer him/herself.

Why do people write such a document? Mallon (1984) cites Virginia Woolf who later drowned herself, reflecting on her diary, "I wonder why I do it. . . Partly I, think, from my old sense of the race of time 'Time's winged chario hurrying near.' Does it stay it?" Anais Nin notes that the diary is a persona document, an exploration, a growth, a meaningful personal relationship anc much more. Both Mallon (1984) and Rosenblatt (1983) point out that there is a large variety of diaries that are written for various reasons—all diarist: wanting to write "it" down on pages, like the writer of the suicide note. Both the diary and the suicide note are rich personal documents left by their writers something they wanted to communicate to themselves and, almost always, to specific and/or unknown others.

Diaries, like suicide notes, sometimes show us the wretchedness and unbearable pain as it unfolds day to day for the keeper. Franz Kafka, as he worked on *The Castle* and *The Trial*, contemplated and wrote about suicide in his diary Kafka's writings, in fact, are piercing insights into the unbearable suicidal pain In *The Trial*, we read about Mr. K, who each morning rises to the same routine the landlady has his breakfast ready, with always an egg. One morning, there is no egg; there is no landlady. He waits, and the police arrive, arresting him He asks, "What have I done?" He is led to jail at the courthouse. There, he waits becoming evermore anxious, depressed, and forlorn. One day, he is led to the judges in a dreamlike courtroom, and he asks his question again. The head judge says, "You know what you did"; his anxiety becomes unendurable. The judges pronounce a verdict, "Guilty," and Mr. K. is led away. Detached, he is led to his death! Others diarists write about their own suicide attempts. Mallon (2010), for example, cites Lee Harvey Oswald, who attempted suicide and later wrote, "I decide to end it. Soak wrist in cold water to numb the pain, then slash my left wrist . . . I watch my life whirl away." A few diarists write about their death, a fascinating example is the Arthur Inman diaries, probably now the most studied diary (Aaron, 1985; Leenaars & Maltsberger, 1994). There are some 50 references in his diary to death and suicide. He is known to have had three serious suicide episodes and died by suicide at age 68. Here are a few raw passages, taken from Aaron (1985):

> Not growing in stature, I was painfully aware of how offsized for my age I looked and how artificial and sissy I seemed. . . . It stamped me as outside my class, a scorned thing to myself and to others, a blasted tree. I felt then a failure, a person confined within himself. (1913, age 17, p. 98)

> I lay awake at night and thought. I concluded, although not decisively enough to please me, that were my back to fall to pieces in an agonizing way, I would kill myself. (December 1931, age 36, p. 464)

There are few such documents available. Mallon (1984) provides a detailed account of suicide diaries—of Sylvia Plath, Dora Carrington, and others, all worthy of intense study. In 2004, David Lester published a book on a suicide dairy called *Katie's Diary*. It expresses so many of the suicidal protocols, as discussed in Chapter 2. One does not, in fact, have to read far—she says it almost all in the very first entry. Katie writes,

> I am so depressed and suicidal. My body feels restless and tired. I don't know who to turn to for help. I don't want to bother anyone with my battle. I've been acting out in all sorts of ways. I just feel like crying. I presume that all has to do with the fact that I love Mark so deeply. I think he cares for me but, however, I don't know if it's true love for him. It definitely is for me. I really want to marry him so badly. I don't care if he reads my journal at all. I am just so stressed out. I haven't done any of my work. I have a hard time getting along with people. I really hate my body. I really hate my life where it is, with everything I am. I decided to start exercising today. I need to do it every day. I leave all my life's frustrations there out on the track.
> God give me strength. Help me today. I feel so unbelievably lonely and battered. (Unpublished manuscript, not numbered)

To the best of my knowledge, there is no suicidal diary of a soldier, as defined above. It would, I predict, be most enlightening of the soldier's suicidal mind.

Of course, there are a wide array of insightful documents, beyond diaries, notes, and poems. A most insightful account of the suicidal mind, much because of its literary style, is the exegesis by William Styron, in his book, *Darkness Visible* (1990). It is an autobiography. He writes, "For some time now I have sensed in my work a growing psychosis that is doubtless a reflection of the psychotic strain tainting my life." William Styron is no stranger to writing; he is a most respected author. He is author of *Lie Down in Darkness, The Long March, Set This House on Fire, The Confessions of Nat Turner, Sophie's Choice,* and *This Quiet Dust*—books that gave expression to the desolation of melancholia and suicide. He has been given numerous awards, including the Prix Mandial Cinco del Duca, which played a special significance in William Styron's own suicidal malaise. It is his literary skill that makes this document rich. One can learn much from his success at painfully making darkness more visible. After his survival in balancing the life and death scale, Styron advises those who suffer depression, "Chin up." He quickly adds that although this is "tantamount to insult" that "one can nearly always be saved." This is necessary advice to a suicidal soldier; they have had the strength before, and still know how. "Chin up."

A related document—and in many ways a personal one—is the poem. Poems have always been seen as a looking glass, being open like the diary and suicide note to understanding the human mind. This is especially true about the poems left by people who died by suicide, although, of course, they are likely written to be more public than the note but not more public than many diaries. There is a host of poets who killed themselves, including John Berryman, Hart Crane, Cesare Pavese, Anne Sexton, Sara Teasdale and, of course, Sylvia Plath (Leenaars, 2004). These poets' lives, as evident in their poems, diaries, and other documents, were full of pain, depression, and suicidal behavior.

Cesare Pavese is an illustrative example of the poet's inimical life; his life was characterlogically suicidal (Shneidman, 1982). Pavese's childhood was, by his own words, a desolate existence (Pavese, 1961; Shneidman, 1982b). His father died when Pavese was age 6, and his mother was described as "spun steel, harsh and austere." He was haunted by pain all his life. Even in adolescence, he wrote about suicide ("You should know that I am thinking about suicide.") He did so deeply in his life and at the age of 42 on August 27, 1950, he died by suicide. His writings are reflective of his pain as one can read in the last words in his diary:

> August 18th, 1950. The things most feared in secret always happens.
> All it takes is a little courage.
> It seemed easy when I thought of it. Weak women have done it. It takes humility, not pride.

Of course, the poetry and writing of these poets are often projective of their *unbearable* pain. Cesare Pavese, for example, once wrote that he fought the pain "every day, every hour, against inertia, dejection, and fear." He and the others in the end chose death, leaving their poetry as documents to their suicidal mind. And like diaries, biographies, and suicide notes, they are open to analysis by placing them in the context of broad theoretical formulations about suicide and personality functioning in general. The theory presented in Chapter 2 will allow us to understand the warrior's self-directed violence better. Of course, there are many other personal documents that can be used for case histories. *Therefore, documents may contain special revelations to why the soldier died by suicide.*

## THE PROBLEM IN STUDYING CASE HISTORIES

A problem in the use of case documents is the one that is the ubiquitous issue of psychology itself: the mind-body problem or the admissibility of introspective accounts as opposed to objective reports. This resonates to Windelband's (1904) division of two possible approaches to knowledge, between the nomothetic and the idiographic. The tabular, statistical, arithmetic, demographic, population, nomothetic approach deals with generalizations, whereas the idiographic approach, which largely uses qualitative analysis, but as my studies of suicide notes show, not only, involves the intense study of individuals via clinical methods, history, biography. In this latter approach, personal documents are frequently utilized, such as chats, letters, poetry, e-mails, memoirs, diaries, autobiographies, and suicide notes. Let us, before addressing the topic at hand, explore the views on the idiographic approach in more detail; the other—the nomothetic approach—is well ingrained in suicidology, psychology, and science in general.

Allport (1942, 1962) has provided us with a classical statement on the advantages of an idiographic approach. Allport (1962) notes that psychology is "committed to increasing man's understanding of man," humans in general, and humans in the particular. What concerns psychology deeply is individual human personality. This is true about *the* soldier who dies by suicide. John Stuart Mill proposed that we make distinctions about the general and the individual in science, both being critical for science's development, although some scientists—and that includes suicidologists—object strongly to the study of individual cases. Allport (1962) provides the following:

> Suppose we take John, a lad of 12 years, and suppose his family background is poor; his father was a criminal; his mother rejected him; his neighborhood is marginal. Suppose that 70 percent of the boys having a similar background become criminals. Does this mean that John himself has a 70 percent chance of delinquency?

Allport (1962) answers, "Not at all. John is a unique being." (p. 411)

Allport (1962) noted that the real concern about the idiographic and nomothetic is developing methods that are more rich, flexible, and precise, that "do justice to the fascinating individuality" of each individual (see Runyan, 1982). This is what we need in the study of military suicide today if we are ever going to control it.

## SOME CONCLUSIONS ABOUT PERSONAL DOCUMENTS

Allport (1942) has noted that personal documents have a significant place in psychiatric/psychological research. Although Allport cites some shortcomings in the use of personal documents in psychological science—unrepresentativeness of sample, self-deception, blindness to motives and errors of memory—he makes a clear case for the use of personal documents, citing the following: learning about the person, advancing both nomothetic and idiographic research, and aiding in the aims of science—understanding, prediction, and control. One is here reminded of Maslow's view (1966) that much of scientific psychology is "mechanistic and ahuman." Most psychologists only know the controlled quantitative experiment, and many actually believe that this is the *only* golden road to the truth. But, according to Maslow (1966)—and Shneidman (1980a)—if we want to know the person, then we have to be more open-minded. The experiment (such as used in statistical studies, third-party interviews, and studies of attempters) has an important place, but so do other methods (and questions). Maslow (1966) noted, for example, that we can use subjective reports in psychology as well as "covert communications, paintings, dreams, stories, gestures, etc.—which we can interpret." Personal documents provide this invaluable source of data. As an interesting footnote to his trailblazing work, Allport wrote about diaries, memoirs, logs, letters, autobiographies, but, it did not occur to his capacious mind to think of perhaps the most personal document of all: suicide notes.

We do not need to be defensive about the use of personal documents or pleading for their occasional admissibility in suicidology and psychology in general. On the contrary, documents emphasize their special virtues and their special power in doing the main business of psychology—the intensive study of the person—in this instance, *a soldier*. Furthermore, the use of case documents does not mean that the essential method of science, Mill's method of difference, has to be abandoned, as comparison between genuine and simulated suicide notes clearly illustrates (see Chapter 2). Personal documents provide a unique place in human science wherein maximum relevance can be mated with acceptable precision. There are not many marriages like that in psychology.

Before we proceed, I need to comment a little further on suicide notes. Suicide notes are ultrapersonal documents. They are the unsolicited productions of the suicidal person, usually written minutes before the suicidal death. They are an invaluable starting point for comprehending the suicidal mind and for understanding the special features of the people who actually die by suicide and what they share in common with the rest of us who have only been drawn to imagine it. Suicide notes are a way through the looking glass to suicide, although unlike Alice, we will not find "beautiful things" there, but unbearable pain—psychache (Shneidman, 1985, 1993).

There are the perpetual questions. How representative are suicide notes? To what extent can one generalize from the note writers to those who do not leave a note? Regarding the latter, it seems that approximately 12% to about 40% of people who die by suicide leave notes. (We do not know what percentage of soldiers who die by suicide leave notes). Erwin Stengel provided the best answers to these questions in *Suicide and Attempted Suicide* (1964) when he stated,

> Whether the writers of suicide notes differ in their attitudes from those who leave no notes behind it is impossible to say. Possibly, they differ from the majority only in being good correspondents. At any rate, the results of the analysis of suicide notes are in keeping with the observation . . . common to most suicidal acts. (p. 44–45)

Research in suicide notes has been useful and can be made more so, even though suicide notes by themselves do not give a complete account of the suicidal mind—and no data will. Like any other data in a case study, the notes must be put in the context of people's lives (Shneidman, 1980a). This is the very aim of this book. Even more important, the notes and other personal documents must be put in context of broad theoretical models—theory—about suicide and personality functioning in general (Leenaars, 1988).

As a source of data, suicide notes are not uncontroversial. Some see suicide notes as the key to unlocking the mysteries of suicidal phenomena—the looking glass into suicide. Others believe that suicide notes, unlike other personal documents, can never be illuminating. Suicide notes are no looking glass at all; indeed, the very person, who because of his/her constriction, "knows" least why he/she is doing it, writes suicide notes. Shneidman's point of view (1985) is that these notes may not be bountiful, but they are rarely banal. Like other documents, suicide notes can, indeed, have a great deal of meaning if, as suggested by Shneidman (1980a), they are put within the context of the details of that person's life. Furthermore, if put in the context of different formulations or theories of suicide and personality functioning in general, these documents can allow us to understand our suicidal warrior (e.g., a General) better. It is all about construct validation (Cronbach & Meehl, 1955; Meehl, 1986, 1990; Millon, 2010; Zachar & Kendler, 2010). I will next illustrate the PA in the armed forces.

## THE PSYCHOLOGICAL AUTOPSY TO MILITARY COURT

Shneidman's paper, "An Example of an Equivocal Death Clarified in a Court of Law" (see Leenaars, 1999) is an example of a military psychological autopsy. This one was done in the adversarial setting of a court of law; specifically, an Army court martial. Everything here, except a few names, is verbatim. The bare facts were that an Army officer was charged with the murder of his wife and faced a lifetime sentence in Leavenworth Federal Prison. The prosecution claimed that it was a homicide, citing that his wife had died in the nude and, after the testimony of some expert, who stated that since suicides do not occur in the nude, it was therefore a homicide. The defense, for which Shneidman was an expert witness, believed this death was a suicide and that the officer had been wrongly accused. It is instructive to read the various experts' different opinions, to take in Jerome Motto's model report, and to note Shneidman's testimony on the state as he holds ground on behalf of the accused. Next, I present some excerpts from Shneidman's testimony (a true window into Shneidman's mind.) This chapter is an illustration of the principle of *res ipsa loquitur*: "The facts speak for themselves."

## The Charge

I shall begin by reproducing a local newspaper account.

### CAPTAIN CHARGED IN DEATH

ARMYVILLE, Sept. 22—A five-man court-martial panel was seated Wednesday in the trial of an army officer who was accused of slaying his wife in their Armyville apartment last July.

Capt. Joseph P. Campbell, 33, is charged with premeditated murder in the shooting death of Peggy Scott-Campbell on July 12. The 30-year-old woman was killed by a single blast from a 12-gauge pump shotgun.

The case, being heard before Judge (Col.) George D. Maris, is expected to last two weeks and involve dozens of witnesses and a trip to Los Angeles to hear testimony from an expert on suicide.

In his opening statement, Dan R. Hyatt, a civilian attorney who is representing Campbell, said he would prove that Scot-Campbell had suicidal tendencies. Also representing Campbell is Maj. James Purdich.

Prosecuting attorney Maj. Stanley Bates told the panel that the Campbells, who had been married about five years, were having marital problems and had discussed separation and divorce only hours before the shooting.

Bates said he would represent evidence to show that following the argument, Scott-Campbell went to a nearby pizzeria where she was well known. She stayed late, he said, to help with the dishes. She left for home, but returned a few minutes later, telling the owner she had too much to drink. She asked that a cab be summoned to take her back to the couple's apartment.

Bates said he would introduce evidence that would indicate that Campbell beat her after she returned home and cut her five times with a knife. Bates told the court that Campbell later retrieved a shotgun from the bedroom, took her into the bathroom near the living room, and shot her in the chest.

Campbell woke up a neighbor and told him his wife had shot herself, Bates told the court.

In his opening statement, Hyatt said Scott-Campbell had made several prior suicide attempts On the night she died, Campbell had awakened to find her in the bathroom shower stall cutting her abdomen with a knife. His shotgun was sitting nearby, Hyatt said.

He said that when Campbell attempted to move the shotgun she grabbed the barrel with her right hand and pulled it toward her. The gun discharged.

Hyatt challenged the prosecution to prove that Campbell beat his wife. No witnesses, he said, ever saw Campbell strike his wife, although the bruises noted in the autopsy of the 130-pound woman could not be accounted for, he said.

Hyatt termed the cuts on Scott-Campbell's body as "hesitation marks" and said a doctor would testify that such marks are often inflicted by suicide-prone people.

"Why should he (Campbell) shoot his wife intentionally when she was trying to commit suicide?" Hyatt asked during his opening remarks.

Prosecutors told the court that the case would be based on circumstantial evidence from investigators and forensic experts.

In summary, the events were as follows: Peggy Scott-Campbell, age 30, is dead as a result of a shotgun wound to her chest. Her husband, Army Capt. Joseph P. Campbell, age 33, asserts that the gun was accidentally discharged as he sought to take it from his wife, who was in the process of attempting suicide by cutting herself in the abdomen (while seated in the shower of their apartment bathroom). The government charges that, following a domestic quarrel, he willfully murdered his wife by deliberately shooting her and then cutting her abdomen to make the event appear like a suicide attempt. He is being tried by a court-martial.

## The Issues

To either the ordinary observer or the trained investigator, the nature or mode of a death is usually self-evident. Each falls into one of the four modes: natural, accidental, suicide, or homicide (NASH). In some cases, however, the mode of death is not so clear; it is equivocal. Then what usually occurs is a psychological investigation that focuses on the decedent's intentions vis-à-vis his or her own death.

The main function of a psychological autopsy is to clarify the mode of death in cases wherein the mode appears unclear.

Putting the human drama aside, what we have in this case is a psychological enigma within a legal context.

## Pretrial Report of Military Special Investigator

What follows in this section is a paraphrased and excerpted presentation of the "Report of Consultation in the Matter of Peggy Scott-Campbell (deceased) dependent wife of Capt. Joseph Campbell." The report was written by Perry N. Olds, a special agent of the Army Office of Special Investigation. Below are a few publicly available sections.

4. **The Crime Scene**. The criminal investigation agents reported the crime scene as "a real mess." In the bedroom the agents observed what they believed to be bloodstains on the pillows, on the bedspread, on the fitted (bottom) sheet, and on the wall next to the bed. They also reported some damp washcloths with blood on the floor. In addition, they reported finding a trail of blood from the bedroom into the bathroom, and what they called "drag bloodstains" from the bedroom to the living room. They reported finding blood in places where Campbell denied having been subsequent to his contact with his wife's blood.

5. **The Autopsy**. An autopsy of the deceased was conducted by a forensic pathologist. He reported a number of findings, including a contact shotgun wound, which would indicate that the barrel was against her skin. The direction of fire was approximately level. On her lower abdomen there were four stab wounds. One of the cuts had numerous pricks around it. These the autopsy surgeon indicated might have been caused by "taunting" (tentative cuts on the body before a deep cut is made). In addition, there were multiple bruises on the extremities and the back of her head. He identified sixty bruised areas, with a majority being less than four hours old at the time of her death. There was a bruise on her chin. He also noted scars on both wrists and in each elbow. He reported her blood alcohol level was .10—actually it turned out to be .17—and no other drugs in her body. "The manner of death was ruled homicide."

10. **Deceased's Nudity at the Time of Her Death**. At the time of her death she was nude. What was the significance of this point? A review of 150-plus female active-duty and dependent suicides revealed not one in which the person was nude at the time of her death. "Her reported nudity at the time she was reported to be attempting suicide is thus highly unusual and correspondingly improbable."

13. **Summary**. "It is my professional opinion that Captain Joseph Campbell shot deceased with the shotgun and attempted to conceal the true facts of the event by staging the scene he subsequently reported."

/ signed /
Perry N. Olds
Special Agent
Office of Special Investigations
U.S. Army

## REPORT OF PSYCHIATRIC CONSULTANT FOR THE DEFENSE

Besides Shneidman, the Defense had contacted another great giant in the field, Jerry Motto. I present his report verbatim, which is in the public domain. Dr. Shneidman once told me that it was the best prototype of a forensic report in the field.

## Consultation Report

The problem at hand is the determination of the mode of death of Peggy Scott-Campbell. The question addressed is whether her death on July 23, 1990 was an accident in the course of thwarting her suicide attempt, or whether other issues played a role.

In order to form the basis for answering the question at hand, four preliminary questions are addressed: (1) Was Peggy Scott-Campbell the kind of person who would resort to suicide under adverse circumstances; (2) Was she being subjected to significant emotional stress at the time of her death; (3) Was there a precipitating event or circumstance that could reasonably have provoked a suicidal act at the time of her death; and (4) would Captain Campbell's character and behavior pattern be consistent with either premeditated or impulsive homicide?

1. **Was she vulnerable to suicidal impulses or behavior?** The record is very consistent as regards the emotional vulnerability of Mrs. Scott-Campbell. Her medical record indicates an ongoing affective disturbance since early adolescence. In 1984 she complained of severe dysphoria, anger with episodic aggressiveness, and suicidal thoughts. She was diagnosed at that time as suffering from a major depressive disorder with borderline personality traits. The recommendation that she have a psychiatric evaluation and follow-up treatment with anti-depressive medication was not implemented. Nor was the recommendation that she be seen for psychiatric and psychological treatment.

    Mrs. Scott-Campbell is said to have exhibited outspoken suicidal behavior on at least four occasions. . . . These events are given varying dates by different informants, but the episodes themselves appear to be discrete and separate experiences. Some were related by Mrs. Scott-Campbell to confidants, and others are noted in Army medical records.

    In addition to specific suicidal behaviors, there is virtual unanimity among those who knew her best that she was moody, emotionally volatile, prone to act impulsively and violently, and moved to intense feeling states by seemingly minor issues. The most provocative of these issues were apparently feelings of being criticized, ignored, or in some way devalued. Examples are numerous, such as Mrs. Scott-Campbell's own description (on tape) of the altercations with her friend Marlene, and the episode in which she is said to have kicked in the bathroom door after Capt. Campbell retreated there to escape her angry outburst.

2. **Was she under significant emotional stress?** Mrs. Scott-Campbell made no secret of her discontent with Army life, with "not fitting in" with the separations entailed or with other Army wives, or with the "homebody" nature of her husband, leading to repeated consideration of divorce. Her persistently heavy drinking underscored her inability to find a stable adjustment; at the same time she apparently felt guilty about her behavior. She was heard to repeat "I'm sorry, I'm sorry" when in an intoxicated state. She confided that Captain Campbell treated her well

and that she wanted to reward him with a party. Though emotionally very immature, she was intelligent enough to know that sooner or later her repeated drunkenness and violent outbursts would catch up with her—that it was only a matter of time, and that time might be getting short. For a person with intense rejection sensitivity, the threat of being abandoned would constitute an ever-increasing emotional pressure in addition to her underlying insecurity and dissatisfaction with her life.

3. **Was there a precipitating event?** A recurrent theme in the available data regarding Mrs. Scott-Campbell's behavior is the sequence of (1) argument, (2) withdrawal by Captain Campbell, (3) drinking by Peggy Scott-Campbell, (4) guilt, depression and contrition on the part of Mrs. Scott-Campbell. In one such previous sequence, Mrs. Scott-Campbell came home from a bar, told Captain Campbell she was going to cut her wrist, and when he tried to calm her down, she hit him with a chair. On the night of her death, this sequence seems to have been repeated.... Whatever the facts were, they went beyond the usual pattern of returning from a drinking episode feeling depressed, and calling someone to talk, or even making suicidal threats. Mrs. Scott-Campbell was unable to get Captain Campbell to respond to her that night, even to placate her, and combined with the progressive dissatisfaction of her life and the depression and guilt she was prone to express after drinking, this experience of "being ignored" could easily have triggered a characteristic violent act. In this instance the act was directed at herself, but her disorganization and ambivalence created only preparation for a lethal act (loaded gun, cutting abdomen).

4. **Would his character be consistent with homicide?** Any effort to determine the likelihood of a given behavior by a given individual must take into account that person's behavioral history, as well as the characterological elements that contribute to that history. In short, the best predictor of behavior is past behavior.

In Captain Campbell's case, those who knew him on a day-to-day basis could hardly be more consistent in their impressions; "loving and supportive," "good natured," "least prone to violence," "violence is not in him. He is not capable of violence," "Absolutely no way he could have done anything to hurt Peggy," and the like.

While it appears to be sophisticated to be skeptical of such unequivocal statements, it is difficult to document contradictory evidence. Mrs. Scott-Campbell's multiple bruises are only a mild example of what one sees in chronic heavy drinkers, as falls and bumping into objects are so common. Even cracked ribs severe facial bruises, and head trauma are not uncommon. She stated that some bruises on her arm were from his trying to stop her from swinging a broken glass at him. In short, Captain Campbell's apparent need to be protective of his wife under the most provocative conditions, and to reflect this demeanor toward the rest of the world as well, is a virtually unanimous perception, and the record gives us nothing substantial to question it.

## Comment on Report of Investigator Olds

Investigator Olds recapitulates the events surrounding Mrs. Scott-Campbell's background and death, and then asks whether she would have been "sufficiently depressed to become suicidal." The question is irrelevant. The nature of her pathology is not depressive but characterological. This is familiar to persons engaged in the field of suicide prevention, although lay persons and even some professionals continue to consider suicide a function of depression rather than psychological pain.

Investigator Olds's subsequent three questions are based on epidemiological comparisons that are interesting but likewise irrelevant. Though some of the numbers can be challenged, the most important issue in this analysis is that such issues as nudity, methods used, and the nature and extent of injury are unique to each individual. The degree of ambivalence, clarity of cognitive functioning, how organized and obsessive the individual is, his or her state of turmoil and agitation, and so forth, are the determination of behavior. What another or a thousand others have done plays no essential role in understanding a given individual's behavior. In the present instance, an intoxicated, frustrated, angry, disorganized individual is involved. Trying to create a scientific aura by judging how common a given pattern appears is simply sophistry.

In short, I believe Investigator Olds's data are interesting, as is his discussion, but they are of no significance whatever in the individual instance under consideration. Thus an opinion based on such considerations is little more than speculations and fertile imagination.

## Summary and Conclusions

In summary, Mrs. Scott-Campbell was a person with a lifelong emotional vulnerability, manifested by extreme of reactivity, a very low threshold for psychic pain, and limited behavioral control. She required a great deal of emotional support and in its absence had repeatedly demonstrated a pattern of self-destructive behavior. In suicide prevention language, she would be regarded as "chronically-suicidal," implying that even at her best level of functioning she was vulnerable to becoming suicidal in adverse circumstances, especially if a perceived rejection was involved.

Captain Campbell demonstrated an inexplicable ability, perhaps a need, to protect and nurture this emotionally vulnerable young woman. The record is replete with observations documenting his consistency in this regard. No evidence is available of traits that would contradict this.

The circumstances surrounding Mrs. Scott-Campbell's death are entirely consistent with her patterns of behavior and with Captain Campbell's efforts to protect her from her own lack of control. I find nothing in the record to indicate otherwise, and regard undocumented hypotheses to the contrary as speculative. Though the documented record can mislead us, I feel that it provides a better basis for sound judgment than hypotheses without documented support.

Jerome A Motto, M.D.
Professor of Psychiatry
University of California, San Francisco

## EXCERPTS FROM DR. EDWIN SHNEIDMAN'S TESTIMONY

The following is an excerpt of Shneidman's responses to questions from Mr. Dan Hyatt, counsel for the defense:

**Hyatt:** The Army investigator, Mr. Olds, has testified earlier in this case. He offered an opinion that it was unusual to see two instrumentalities when either attempting or committing suicide. Do you have an opinion with respect to that statement?

**ESS:** Yes, in a picayune way he is right. But in an overall way he is howlingly wrong. I'll tell you about each of those if I may.

**Hyatt:** Please.

**ESS:** Suicide itself, fortunately, is an event of infrequent occurrence. So that you can make tabulations of methods and all sorts of things. A lot of events are infrequent, but if the incontrovertible evidence is that the person has done it, then you can't say that the person has not done it simply because it is infrequent. Using two methods is much more infrequent than using one. That's true. But then to argue from that to this particular case is a tyro's error. It's a mistake that freshman undergraduates in my Death and Suicide course at UCLA make of going from statistics to an individual case. Statistics are an interesting background for a case, but they don't tell you about that case. Here we are talking about this case.

**Hyatt:** Mr. Olds also testified that he thought it was very rare based on his study of Army personnel and their dependents—his database—that it was very rare to find a dependent female to commit suicide or attempt to commit suicide in the nude. Do you have an opinion about that?

**ESS:** Yes. Well, I would say to him, "That's true. That's absolutely true. But you're really not seriously making an argument that that has a bearing on this case, are you?" And if he said, "Yes," my already low opinion of him would drop precipitously.

**Hyatt:** What value do you see of statistical information such as that offered by Mr. Olds in determining the cause of death?

**ESS:** In a particular case?

**Hyatt:** Yes.

**ESS:** None. It's background material.

**Hyatt:** Is the utilization of statistics in the manner testified by Mr. Olds a scientifically acceptable method, or is that data reasonably relied upon by other experts in your field as a means of drawing conclusion?

**ESS:** If your question is, is it a scientifically credited method the way he has done it, the answer is no.

**Hyatt:** And why would that be?

**ESS:** The technical response is that in these matters, in suicidology, the confusion of statistical-demographic-epidemiological-numerical data with the etiology or outcome of any particular individual case, to make a judgment about that individual case on the basis of statistics is a methodological error.

**Hyatt:** Why is it a methodological error?

**ESS:** Because it has things backwards. It isn't that the statistics generate the case; it is that the cases taken in long series or large numbers generate the statistics. To say that it is rare is not to say that it did not occur.

**Hyatt:** What do you say when you hear that Peggy Campbell was nude on the evening of her death?

**ESS:** I would say, "Gee whiz, isn't that unusual." But then to argue as he did that it couldn't be suicide on that account is a howler. It boggles the mind. Where did his logic go?

**Hyatt:** And would you have the same opinion as to the use of two instrumentalities?

**ESS:** Yes, sure. What is persuasive is the whole history of her lifetime. (Leenaars, 1999, pp. 433–434)

## Newspaper Account of the Verdict

### Captain Acquitted in Trial
ARMYVILLE, Oct, 1– An officer charged with the shotgun slaying of his wife last July was found innocent Friday by a five-man court-martial panel.

## A MILITARY PERSPECTIVE ON THE PSYCHOLOGICAL AUTOPSY

What is the military's perspective on the psychological autopsy? In the *Journal of Forensic Science*, Colonel Elspeth Cameron Ritchie and Michael G. Geller offered an answer. The 2002 paper, entitled, "Psychological Autopsies: The Current Department of Defense Effort to Standardize Training and Quality Assurance" states, beginning with the following abstract:

> Psychological autopsies have been gathered by the US military for a long time, both for lessons learned after a known suicide and to investigate an equivocal death. The term "psychological autopsies" is now being restricted to define an investigation by mental health to help determine, in an equivocal death, if the manner of death is a homicide, suicide, an accident, or from natural causes. The Department of Defense has developed policy, and is now implementing training and peer review. A sample model curriculum, report format and quality assurance standards are included. (p. 1370)

The Department of Defense (DoD), we learn, has developed clear policies and procedures on psychological autopsies and is in the process of implementing standards of training and quality assurance. The DoD covers the Army, Navy, Air Force, and Marines. Yet there are issues; most concerning is that "records were accessible to the media under the Freedom of Information Act (FOIA). In one situation, a reporter requested all the psychological autopsies performed at Ft. Bragg and then published that information on the front pages of a newspaper, to the consternation of relatives" (Ritchie & Geller, 2002). Colonel Ritchie goes on to state the current policy:

> This current policy governing psychological autopsies was issued in the summer of 2001, as a Department of Defense Health Affairs policy letter (Psychological autopsies, 2001). It mandates that the primary purpose of a psychological autopsy is as a forensic investigative tool to assist in ascertaining the manner of death. (Ritchie & Geller, 2002, p. 1371)

The American military's efforts, described by Ritchie, are works in progress. She also offers some important wisdom that the privacy of the information collected is essential. Ritchie and Geller (2002) give the order, "It is the authors' recommendation that the PAs be kept as discrete as possible. However, immediate family does have the right to know most details of the cases" (p. 1372). Yet there is a caveat: researchers, with military clearance should be allowed to dig into the archive for nomothetic study, both qualitative and quantitative. The archive is a vault of information that would save many soldiers. We need to learn, with the best method available, the psychological autopsy. Will the DoD allow it?

## CONCLUDING THOUGHT

In keeping with the mode of this chapter, I will add no more except to state that the PA, with special use of documents, is the best possible way (and probable, for that matter) to answer the question: Why did the General kill himself?

# PART TWO

# Historical Study

Wishing to be brief in my interstitial note, I want the reader to remember the words of George Santayana: "Those who cannot remember the past are condemned to repeat it." This part contains two chapters.

There is no question in suicidology that Emile Durkheim is an early pioneer in the nomothetic (population) study of suicide and military suicide. I reproduce some quotes from his few pages on what he actually wrote over a century ago on the topic of military suicide. One of Durkheim's most famous conclusions was that suicide in the military was often altruistic suicide. Altruistic suicides are a diverse group and include martyrs and terrorists. I explore the notion and present my recent study into the suicidal mind of martyrs today. They are soldiers too. We need to know the terrorist's mind better; Dr. Hasan, the Fort Hood killer, tells us at least that. We need to know the terrorist's mind.

Durkheim showed a century ago that there was a high incidence of military suicide in the United States after the Civil War. In the second chapter of this part, I bring home the enduring pain of suicide in the military; I present the peerless idiographic (individual) case of the great Civil War hero, General Emory Upton. An obvious question then arises, Why the military's secrecy? We also ask, Why did Admiral Boorda kill himself? There are no definitive answers. However, Upton's case, due to visibility, answers many questions.

# CHAPTER 4
# Military Suicide: A Classic Population Study

In the 19th century, science was rife. In that century, arithmetic had already set science apart from the earlier more philosophical approach to knowledge. In 1741, Johann Peter Süssmilch, a Prussian, showed the power of the science of statistics. Vital statistics was born, and it was thought that laws could now be found to explain all of nature, including people, mind, whatever. By the 1800s, what we now call quantitative science had set its mark (actually more like a seal) on understanding. The tabular, arithmetical, taxonomic, nomothetic approach ruled.

Carolus Linnaeus, in the 18th century, and Dimitri Mendeleyev and Emil Kraepelin in the 19th century, developed systems to explain living creatures, chemical elements, and diseases of the mind, respectively. They exemplified what science did in the last few millennia: bring order to the booming, buzzing mess. Durkheim did the same in sociology.

Linnaeus, the Swedish naturalist, developed taxonomy for all living creatures. He imposed an order on the living world, dividing it into plant and animal. Animal could then be further divided into one of 16 animal phyla and then classified further. (The schema was kingdom, phylum, class, order, family, genus, and species.) Linnaeus' template was impressive. Linnaeus had ordered the world, and everyone could learn to view living creatures through his system. It was scientific understanding at its most impressive. Mendeleyev, the Russian chemist, did the same with chemicals. He gave order to the chemical world, giving birth to periodic table. The table not only showed that order was possible, but it allowed scientists to predict the existence and properties of unknown elements. Understanding led to prediction and possibly control (the very aims of science, according to John Stuart Mill, Gordon Allport, and others).

Kraepelin, the Swiss psychiatrist, attempted to do the same, that is, bring order to mental syndromes. His 1883 book, *Psychiatry*, is still the basis for taxonomy in the *Diagnostic and Statistical Manual of Mental Disorders* (the DSM), although it can be questioned whether Kraepelin's structure is at the same level as Linneaus and Mendeleyev (Shneidman, 2001). Be that as it may, taxonomy was figural in science in the last few centuries and continues to be so. Durkheim is important because he did the same in sociology—suicide was the topic. The method was, in fact, the most important aspect of the book; he could have used almost any topic to exemplify the method, but we are fortunate that he used suicide.

## EMILE DURKHEIM AND SUICIDE

Emile Durkheim (1858-1917) is, indeed, the best-known sociologist in the field of suicidology (Shneidman, 2001). His book, *Suicide: A Study in Sociology*, first published in 1897 and first translated into English in 1951, is not only a book on suicide, but a book about the science of his day. This work is about a method combining empirical facts with sociological theory. Durkheim, like Linnaeus, attempted to show that there was order to suicide. He provided a template, or pattern, on suicide, which allows us to explicate (develop the meaning or implication of an idea, principle, etc.) In this enduring book, Durkheim presents his most famous typology of four suicides: egoistic, altruistic, anomic, and fatalistic (he added the latter in a footnote, to be found on page 276 in the 1951 English translation). Egoistic refers to the unhappy person who is not integrated in society. Altruistic is the person who is too integrated; he/she sees the death as a duty or honor. Anomie refers to the estranged person—one's relationship to society is changed, that is, no longer regulated by the social world. Fatalistic is when the person is too regulated, where the future is blocked—he/she is "choked by oppressive discipline" (the slave who kills him/herself is an example).

The key—being a good sociologist—in all four main types of suicide is the *intersection* (i.e., integration and regulation) between the individual and tha individual's society. Durkheim brought order to suicide; in fact, he literally reclassified suicide. Durkheim does not contemplate only the numbers, such as how many Catholics die and would be predicted to die in a specific region, say in France, but he imposes a system on them. Of course, all systems, whether Durkheim's or Mendeleyev's, are "heuristic." Put more simply, as people, we impose order on a mess, when it is a mess; the systems are methods of classification of events. It is meaning making. The order can be seen as classification seeking, although it arbitrarily imposes the structure (a good read of Kuhn's *The Structure of Scientific Revolutions* [1962] is a must). Durkheim's ability to explicate (over and over) and show us a useful system, makes his book a great volume.

All of Durkheim's typology is worthy of extensive study. We will here examine his altruistic type in more detail. As a footnote, there were sociologists before Durkheim who studied suicide. Yet it did not occur to most of their minds to view "altruistic suicide" as suicide, although there were exceptions. Durkheim's claim was maybe not a first, but the most influential. He said that the people who are classified under "altruistic" in his typology might be heroes or martyrs, but they are also suicides. This was a new frame; before that, these acts were rarely labeled suicides.

What does it mean that a suicide is altruistic? What is the meaning of altruistic? The *Oxford English Dictionary* defines altruistic as, "pertaining to, or practicing altruism; unselfishness." Altruism is to have regard for others in society as a principle of action. One acts unselfishly, based on the rules, dictates, expectations of others—of society. Can suicide ever be such an act? Can suicide be unselfish? Can there be altruistic suicide? Some would see the two words together as an oxymoron. Are altruistic suicides like other suicides? Are there typical types of altruistic suicide? Are these people martyrs? Can soldiers who died by suicide be martyrs? Can they be terrorists? Who are they?

## ALTRUISTIC SUICIDE:
## A FEW FURTHER BEGINNING REFLECTIONS

What was the altruistic suicide's intent? What was the altruistic soldier's intent or motivation? Who is she? Who are the altruistic suicides? We will meet some *altruistic* suicides in the volume, but let me first outline Durkheim's schema in some more detail in order to understand better a would-be altruistic suicide. It is essential. He ends, as we will read, with *the* most important question.

Altruistic suicide, argues Durkheim, is different from the more common, egoistic, as well as from anomie and fatalistic. In his book, *Suicide* (1951), Durkheim argues that in this type of death, the person has a right to do so. It may, in fact, be his/her duty. Indeed, if the person fails to follow the duty, he/she could be punished (Would the soldier?). Altruistic suicide is first and foremost a duty. (Is it the soldier's?)

Durkheim (1951) explicates that some societies compel some of their citizens to kill themselves. Thus, the individual has little value. The person must be absorbed, or integrated, if not enmeshed in his/her society. To associate to our words from Chapter 1, they are indoctrinated. The person only has value in the collective. There is no individualism. Altruism, Durkheim writes, is

> where the ego is not its own property, where it is blended with something not itself, where the goal of conduct is exterior to itself, that is, in one of the groups in which it participates. So we call the suicide caused by intense altruism *altruistic suicide*. (1951, p. 221)

He explicates further, identifying three subtypes: obligatory, optional, and acute. The obligatory is duty and differs from the optional wherein it is not a duty per se; society does not actually require the optional altruistic suicide to kill him/herself, but is favorable to the act. The acute are altruistic suicides that occur, "only with the concurrence of circumstances." Durkheim further writes,

> Either death had to be imposed by society as a duty, or some question of honor was involved, or at least some disagreeable occurrence had to lower the value of life in the victim's eyes. But it even happens that the individual kills himself purely for the joy of sacrifice, because, even with no particular reason, renunciation in itself is considered praiseworthy. (1951, p. 223)

However, what is critical to understand is that "the individual in all these cases seeks to strip himself of his personal being in order to be engulfed in something which he regards as his true essence" (Durkheim, 1951, p. 225). Altruism can be acute. The goal is only on the outside, whether it is for God, one's country, enlightenment, and so on. There is enthusiasm, but not in oneself, only the group. One may even be a hero.

Throughout the ages, people have refused to consider these acts as "suicides." Durkheim's genius lay in including this in his taxonomy. Yet there are problems; Durkheim recognizes some of this, raising *the* question,

> All these cases have for their root the same state of altruism which is equally the cause of what might be called heroic suicide. Shall they alone be placed among the ranks of suicides and only those excluded whose motive is particularly pure? But first, according to what standard will the division be made? When does a motive cease to be sufficiently praiseworthy for the act it determines to be called suicide? (1951, p. 240)

Is Durkheim's taxonomy useful? Edwin Shneidman (personal communication, July 1, 2002) suggested that Durkheim's order is not like Linneaus's or Mendeleyev's. It is more like Kraepelin's. Durkheim and Kraepelin are, according to Shneidman, reductionistic. They attempt to be like Linneaus, but do so poorly. He even believes that it is based on false epistemology (and, in Kraepelin's case—or the DSM taxonomy that it spurred—erroneous thinking). On Durkheim, Shneidman (2001) writes, "Durkheim did not just look at the data; he transmuted them into sociological magic by radically reclassifying their internal implication" (p. 54). The reclassifying, however, misses the most essential element. Suicide, according to Shneidman, is more complicated. Durkheim does not address intent. Whatever else suicide is, it is an intentional act. Of course, all templates or patterns have problems. They seek to discuss order, but actually they impose it—even Shneidman's or mine. Suicide is complex, and no one template can ever completely explicate its complexities. It should not be this or that system, but all of them that can be made to be useful, that is, understanding. I agree that intent is essential, but one does not *always* need to discuss intent to give some meaning to the event. (Do we always need it for the warrior?) We need to be open-minded; Durkheim was.

Some will agree with Shneidman's 2001 position. However, in his best book, *Definition of Suicide,* Shneidman (1985) stated that suicide, as discussed earlier, is a multidimensional event. There are biological, psychological, sociological, cultural, and so on, elements in the event. Therefore, sociology (or social psychiatry/psychology) is a legitimate avenue of study and knowledge. Durkheim gives us an order, a sociological order—maybe not *the* order—and he does it in a way that is most useful. This is why the book endures. Durkheim has passed the test of time; indeed, his book may well be the best known book in the field. Which student in suicidology does not know the name of Durkheim? Yet Durkheim's schema has limitations, but so does all of science. Durkheim's system may not meet everyone's test, but it is an order that has been shown for over 100 years to be useful and can be made more meaningful. Durkheim's theory is a social template; it is one possible, enriching frame, especially for the military, given its collective culture. It may well be a most useful construction on a green culture.

Does our study of altruistic suicide also allow us to understand a soldier's suicide better? Can suicidology bring some order to the events of September 11th? The Iraq and Afghanistan wars? War in general? These questions are interwoven. It is my hope that this chapter will allow us to understand these events better, to allow us to understand altruistic suicide better, and to understand the soldier's suicide better. We need to allow suicidology to understand *all* suicidal events, whether they are attributed to martyrdom or terrorism. This was Durkheim's genius; he allows us to assimilate diverse events into suicide (in that way he follows Linnaeus or Mendeleyev). Durkheim has done the wonderful thing of allowing us to organize and think about suicide in the military in new ways. He has advanced our science.

I had wanted to present next verbatim Durkheim's writing on military suicide in *Suicide: A Study in Sociology* (1951), Chapter 4, "Altruistic Suicide," Section II, pages 228–239. Copyright policy does not permit the reprinting of the complete text; it is an essential read, however, for every student of the subject. The policy does grant the privilege of extraction of generous quotations. Thus, I hereby offer a synopsis of the most essential read on the classic study of military suicide. I quote: "It is a general fact in all European countries that the suicidal aptitude of soldiers is much higher than that of the civilian population of the same age. The difference varies between 25 and 900 per cent."

Durkheim (1951) reports the incidence and rate of suicide for the United States; he does not report a rate for Canada. He does, for example, report a rate for England and France. The England rate (1876–1900) was 26.9 per 100,000, compared with a civilian rate of 7.9 per 100,000. The rate in France (1876–1900) was 33.3 per 100,000, compared with a civilian rate of 26.5 per 100,000. Thus (at least in 1897), Durkheim's position was that the rate of military suicide in the United States, England, and France was very high and often well above the civilian rate of suicide. Durkheim's surveillance, I believe, was credible. Of note, the military rate of suicide in the United States (1870–1884) was one of the highest among the countries mentioned. The rate was 68 per 100,000, compared with a civilian rate of 8 per 100,000. Yet the U.S. rate was not the highest; Durkheim reported a rate of 125.3 per 100,000 in Austria (1876–1900). Thus, suicide in the military was then seen as a very significant problem in and for the military. He asked, Why? Durkheim studied the causes, stating,

> The cause most often suggested is disgust with the service. This explanation agrees with the popular conception, which attributes suicide to the hardships of life; for disciplinary rigor, lack of liberty, and want of every comfort makes barracks life appear especially intolerable. Actually it seems that there are many other harsher occupations which yet do not increase the inclination to suicide. The soldier is at least sure of having enough food and shelter. But whatever these considerations may be worth, the following facts show the inadequacy of this over-simple explanation. . . .
> 
> The first quality of a soldier is a sort of impersonality not to be found anywhere in civilian life to the same degree. He must be trained to set little value upon himself, since he must be prepared to sacrifice himself upon being ordered to do so. Even aside from such exceptional circumstances, in peace time and in the regular exercise of his profession, discipline requires him to obey without question and sometimes even without understanding. For this, an intellectual abnegation hardly consistent with individualism, is required. He must have but a weak tie binding him to his individuality, to obey external impulsion so docilely. In short, a soldier's principle of action is external to himself; which is the quality of the state of altruism. Of all elements constituting our modern societies, the army, indeed, most recalls the structure of lower societies. It, too, consists of a massive, compact group providing a rigid setting for the individual and preventing any independent movement.
> 
> Therefore, since this moral constitution is the natural field for altruistic suicide, military suicide may certainly be supposed to have the same character and derive from the same source. . . .
> 
> Everything therefore proves that military suicide is only a form of altruistic suicide. We certainly do not mean that all individual cases occurring in the regiments are of this character and origin. When he puts on his uniform, the soldier does not become a completely new man; the effects of his education and of his previous life do not disappear as if by magic; and he is also not so separated from the rest of society as not to share in the common life. The suicide he commits may therefore sometimes be civilian in its character and causes. But with the exception of these scattered cases, showing no connections with one another, a compact, homogeneous group remains, including most suicides which occur in the army and which depend on this state of altruism without which military spirit is inconceivable. This is the suicide of lower societies, in survival among us because military morality itself is in certain aspects a survival of primitive morality. Influenced by this predisposition, the soldier kills himself at the least disappointment, for the most futile reasons, for a refusal of leave, a reprimand, an unjust punishment, a delay in promotion, a question of honor, a flush of momentary jealousy or even simply because other suicides have occurred before his

eyes or to his knowledge. Such is really the source of these phenomena of contagion often observed in armies, specimens of which we have mentioned earlier. They are inexplicable if suicide depends essentially on individual causes. It cannot be chance which caused the appearance in precisely this regiment or that locality of so many persons predisposed to self-homicide by their organic constitution. It is still more inadmissible that such a spread of imitative action could take place utterly without predisposition. But everything is readily explained when it is recognized that the profession of a soldier develops a moral constitution powerfully predisposing man to make away with himself. For this constitution naturally occurs, in varying degrees, among most of those who live or who have lived under the colors, and as this is an eminently favorable soil for suicides, little is needed to actualize the tendency to self-destruction which it contains; an example is enough. So it spreads like a trail of gunpowder among persons thus prepared to follow it.

Looking back, it seems important that Durkheim's discovery that war causes suicide among soldiers was forgotten, or more likely intentionally hidden (deception). Durkheim (1951) had asked, is it a pattern? When I did my last literature search on PsycINFO and PubMed/Medline, before sending the final manuscript to the publisher, I discovered an article by Gregory Lande (2011), entitled, "Felo De Se: Soldiers Suicides in America's Civil War," published in *Military Medicine*. Lande examined the factors associated with suicide and the American Civil War; he undertook a retrospective investigation of the Union Army's military medical records and state-by-state archives on military suicide. Data for only White troops exist. No data on Confederate troops exist. Despite notable reliability issues with the existing data, Lande concluded that there was an "epidemic" of suicide among soldiers. Durkheim had made the same conclusion. However, Lande found that the association of military service and suicide received little public attention. The Union Army kept it a secret; however, it was known that suicide occurred even among its top officers, for example, Brigadier General Francis Patterson, and General Emory Upton. And, although the obvious cause was war friction, the Army blamed the deaths only on "insanity." Durkheim's finding, over 100 years ago, raises a question for the mind: What if the Union Army had accepted the relation between service and suicide, would the Vietnam, Iraq, and Afghanistan soldiers and veterans have received the understanding and care that they need? There is a cause and a pattern. Wars cause suicide.

## A CURRENT STUDY OF ALTRUISTIC SUICIDE: AN INTRODUCTION

The altruistic suicides, sometimes called martyr suicides, were a unique inclusion in Durkheim's (1951) scheme; the classification includes soldiers, saints, martyrs, and terrorists—a diverse group of suicides. As you have learned, no single factor or event explains why so many people are violent. Violence is multidetermined. Suicide, homicide, and war-related deaths are the result of an interplay of individual, relationship, social, cultural, and environmental factors. Altruistic or martyr suicides are no different. Yet the question can be posed: Are suicide and altruistic suicide alike? How can we understand violent acts of martyrs? Altruistic suicide, argues Durkheim, is different from the more common, egoistic suicide, as well as from anomie and fatalistic. In his book, *Suicide*, Durkheim argues that in this type of death, the person has the right to kill him/herself. In a green culture, it may in fact be his/her duty. Altruism, as stated earlier, is "where the ego is not . . . *altruistic suicide* " (p. 221).

From the ecological model, there are factors beyond the individual and relationship; the more important reasons for altruistic deaths, in fact, may be in the community and society. This would be true in a green culture. Yet there are problems in examining such acts of martyrs from a societal view. Little, in fact, has been done since Durkheim's seminal work. Altruistic suicides-to-be are not available to attend clinical interviews, to fill out questionnaires, to be subject to laboratory tests; yet research is needed. Leenaars and Wenckstern (2004) presented some new concepts and issues on the complex topic and offered diverse perspectives on altruistic suicide in the classical Greco-Romans, the Christian Greek Orthodox Neomartyrs, the self-immolators in Vietnam and South Korea, Muslim suicide terrorists, and India's Jauhar and Sati. There was only one empirical study, the work of Ben Park (2004). Park presented a trove of personal documents of martyrs from South Korea, which can be called altruistic suicide notes or martyrs' last letters. On the basis of suicide notes, diaries, and letters left behind by 22 self-immolators, Park shed some light on the intentions and beliefs of those actors and social significance of the meaning of their acts. In addition to the unique geopolitical circumstances of the Cold War era, under which massive numbers of dramatic public acts of self-immolation took place, the symbolic message embedded in the acts of self-immolators was explored. Yet it did not occur to him to analyze the notes scientifically. That was the purpose of a study by Leenaars, Park et al. (2010) entitled "Martyrs' Last Letters: Are They the Same as Suicide Notes?" It was published in the *Journal of Forensic Sciences*. This was the first empirical study of the last letters of martyrs.

The question asked was whether these altruistic suicide notes are the same or different from other more common suicide notes. To be more empirical, as outlined in Chapter 2 of this volume, does the evidence-based theory have applicability to the last letters of "martyrs"? The aim of the study was not only to describe the last letters of martyrs (or suicide notes of altruistic suicides) but also to determine whether any psychological differences (and similarities) are evident in the suicide notes from a more general population of the United States. (The United States has been the basis for all comparisons to date in international studies of suicide notes.)

## ALTRUISTIC SUICIDE NOTES: METHOD AND RESULTS

The problem in the current area of study is obtaining the very data themselves. There are problems in sampling, generalizability, and the like. The suicide notes and other personal documents used in this study of martyr letters come from a variety of sources, including leaflets, newspapers, magazines, and secondary publications. Some 33 South Korean self-immolators left behind letters or notes. Many of the last letters left behind by Korean self-immolators came from an underground publication, *Everlasting Lives* (Sal 'Aseo Ma'nnra Rina'a), compiled by the Ad Hoc Committee for the Preparation of a Memorial Service for the Nation's Martyrs and Victims of Democratization Movement. The martyr notes were matched (age ± 3 yrs.) to an adult ($n = 33$) American sample of suicide notes (Leenaars, 1988). Gender and age were controlled. The notes were analyzed for the presence of the 35 protocol sentences, as outlined before. (Numbers below, say, No. 10, would refer to protocol sentence No. 10.)

Similar to previous studies on the more common suicide notes, there was substantial evidence for the presence of the protocol sentences and clusters in both samples of suicide notes. The verification was extremely large. Thus, one can conclude that the model is applicable to suicide notes of martyrs; yet there were also extreme significant differences with the more common U.S. suicide notes. Table 4.1 shows a comparison of the contents of suicide notes from both groups according to Leenaars' suicide model, with abbreviations. As shown in the table, all the clusters were significantly different, all more evident in the notes of martyrs. When one examines specific protocol sentences in each cluster, there were many significant differences (see Table 4.1). Most of these were in the direction of being more evident in altruistic notes, but not all. The, for lack of a better word, "common" suicide notes in the United States more frequently cited ambivalence (No. 10) and ambivalent feelings toward the government (No. 28). Due to the young age most of the "martyrs," we also redid the analyses with only notes from young adults (YA, ages 18–25). The results were no different.

## ALTRUISTIC SUICIDE: DISCUSSION

The findings provide further support for the multidimensional model proposed (see Chapter 2). There is considerable evidence of both intrapsychic and interpersonal correlates of suicide, whether altruistic or otherwise. This is true with martyrs. (Will it be true for soldiers?) Similar to previous studies, there seem to be commonalities among suicides. By virtue of our human quality, people who are about to kill themselves have a number of important psychological characteristics in common. Pain is pain. Mental constriction is mental constriction. Unhappiness is unhappiness. The suicidal mind is the suicidal mind; yet

Table 4.1 Frequency of Endorsement of Protocol Sentences (Abbreviated), Percentages, and Significance in Altruistic ($n = 33$) and United States ($n = 33$) Notes

| Cluster/protocol sentence | Altruistic | | United States | | |
|---|---|---|---|---|---|
| | n | % | n | % | p |
| **INTRAPSYCHIC** | | | | | |
| I. Unbearable Psychological Pain | 33 | 100.0 | 25 | 75.8 | 0.000*** |
| 1. Suicide as a relief | 33 | 100.0 | 18 | 54.5 | 0.000*** |
| 2. Suicide as a flight from trauma | 32 | 97.0 | 15 | 45.5 | 0.000*** |
| 3. Emotional states in suicidal trauma | 31 | 93.9 | 14 | 42.4 | 0.000*** |
| 4. Loss of interest to endure | 30 | 90.9 | 15 | 45.5 | 0.000*** |
| 5. Inability to meet life's challenges | 30 | 90.9 | 10 | 30.3 | 0.000*** |
| 6. State of heightened disturbance | 30 | 90.9 | 7 | 21.2 | 0.000*** |
| II. Cognitive Constriction | 33 | 100.0 | 13 | 39.4 | 0.000*** |
| 7. A history of trauma | 28 | 84.8 | 11 | 33.3 | 0.000*** |
| 8. Overpowering emotions | 31 | 93.9 | 7 | 21.2 | 0.000*** |
| 9. Locus only on grief topics+ | 32 | 97.0 | 3 | 9.1 | 0.000*** |
| III. Indirect Expressions | 31 | 93.9 | 23 | 69.7 | 0.000*** |
| 10. Ambivalence+ | 2 | 6.1 | 9 | 27.3 | 0.044* |
| 11. Aggression has turned inward | 30 | 90.9 | 13 | 39.4 | 0.000*** |
| 12. Unconscious dynamics | 31 | 93.9 | 23 | 69.7 | 0.011* |
| IV. Inability to Adjust | 32 | 97.0 | 15 | 45.4 | 0.000*** |
| 13. Feels weak to overcome difficulties | 23 | 69.7 | 14 | 42.4 | 0.026* |
| 14. Incompatible state of mind+ | 25 | 75.8 | 4 | 12.1 | 0.000*** |
| 15. Serious disorder in adjustment | 30 | 90.9 | 8 | 24.2 | 0.000*** |
| V. Ego | 26 | 78.8 | 11 | 33.3 | 0.003** |
| 16. Weakness in constructive tendencies | 7 | 21.2 | 7 | 21.2 | 1.000 |
| 17. A complex or weakened ego | 25 | 75.8 | 7 | 21.2 | 0.000*** |
| 18. Harsh conscience | 12 | 36.4 | 5 | 15.2 | 0.049* |
| **INTERPERSONAL** | | | | | |
| VI. Interpersonal Relations | 33 | 100.0 | 20 | 60.6 | 0.000*** |
| 19. Problems determined by situations | 31 | 93.9 | 16 | 48.5 | 0.000*** |
| 20. Weakened by unresolved problems | 31 | 93.9 | 11 | 33.3 | 0.000*** |
| 21. Frustrated needs | 31 | 93.9 | 13 | 39.4 | 0.000*** |

Table 4.1 (Cont'd.)

| Cluster/protocol sentence | Altruistic n | % | United States n | % | p |
|---|---|---|---|---|---|
| 22. Frustration to a traumatic degree | 30 | 90.0 | 14 | 42.4 | 0.000*** |
| 23. Positive development not forthcoming | 30 | 90.0 | 5 | 15.2 | 0.000*** |
| 24. Regressive, intimate, relationships | 30 | 90.9 | 8 | 24.2 | 0.000*** |
| VII. Rejection-Aggression | 33 | 100.0 | 16 | 48.5 | 0.000*** |
| 25. Report of a traumatic event | 32 | 97.0 | 12 | 36.4 | 0.000*** |
| 26. Narcissistic injury | 14 | 42.4 | 13 | 39.4 | 0.802 |
| 27. Preoccupation with person | 9 | 27.3 | 9 | 27.3 | 1.000 |
| 28. Ambivalent feelings toward a person | 0 | 0.0 | 6 | 18.2 | 0.024* |
| 29. Aggression as self-directed | 5 | 15.2 | 3 | 9.1 | 0.708 |
| 30. Murderous impulses | 33 | 100.0 | 0 | 0.0 | 0.000*** |
| 31. Calculation of negative effect | 33 | 100.0 | 8 | 24.2 | 0.000*** |
| 32. Revenge toward someone else | 33 | 100.0 | 8 | 24.2 | 0.000*** |
| VIII. Identification-Egression | 32 | 97.0 | 22 | 66.7 | 0.000*** |
| 33. Identification with person/ideal | 31 | 93.9 | 12 | 36.4 | 0.000*** |
| 34. Unwillingness to accept life | 32 | 97.0 | 14 | 42.4 | 0.000*** |
| 35. Suicide as escape | 32 | 97.0 | 18 | 54.5 | 0.000*** |

+Fisher's Exact Test statistic reported.
*$p < 0.05$, **$p < 0.01$, ***$p < 0.001$.

significant differences emerged, not whether present or absent, but the intensity of the state. The psyche of the martyr is different, and these differences were seen in all clusters.

Altruistic suicide is different from the more common suicides (egoistic, followed by anomie). The state of mind of the altruistic suicidal person is extreme in such characteristics as pain, mental constriction, depression, and rage, to name a few. This difference is our most important finding. Thus, despite the value of looking at common factors, it is useful to group suicides under Durkheim's rubric. In any analysis of suicide, it is useful to think of the suicide beyond the individual and to place it within the community and societal context. This is consistent not only with Durkheim's view but also the WHO's (2002) ecological model. We have to understand the community and society's very meaning of the act. Not all suicides are the same. We should not assume a suicide is a suicide; altruistic deaths differ significantly. Moreover, this may be most important in military suicides.

A key question is why martyrdom emerged in South Korea (and elsewhere). On a community and societal level, the principal element common in altruistic suicide in Korea, with respect to the act of self-immolation, is that these acts grew out of intense political turbulence and widespread violence, at least as one reads in the last letters of the Korean self-immolators. Altruistic suicide, in Durkheim's sense, became a *best* solution. Self-immolation is only one, albeit especially powerful, form of service. The self-immolators' last letters are written as the penultimate public spectacle of their martyrdom. Martyrs, of course, always need an audience.

In his classic study, *Suicide* (1951), Emile Durkheim does not deal specifically with the type of self-destruction (or self-sacrifice) committed in an effort to further the cause of a social or political movement. Still, these acts of self-immolation would seem to fit as instances of "altruistic suicide." Moreover, in these cases, "it is not because [a person] assumes the right to do so but, it is his duty. If he fails in this duty, he is dishonored and also punished" (p. 219). For Durkheim, altruistic suicides are selfless individuals who are mechanically compelled by society, or the community, to end their life, with little or no individual intentionality involved. This perception is very evident in the last letters. In other words, Durkheim largely, if not completely, ignores the will of the human agent in making a decision to kill oneself for, at least what is perceived to be, the potential benefit to others. Yet our results support the notion that the individual's mind is figural in the act, both intrapsychic and interpersonal. *There is intent*. The martyrs, in fact, state that they choose the death.

## THE RELEVANCE OF THE STUDY

The relevance of this study is both somewhat historical and immediately contemporary. Even though most of the altruistic suicides in Korea occurred in the 1960s through 1980s, they seem less a part of modern life than newer forms of altruistic suicide. We live in an era in which politically motivated suicides have taken on major importance as terrorist acts. At issue here, of course, are the Palestinian suicide bombers and the hijackers associated with al-Qaeda and related groups. A central question of our time is, Are they the same? First, it is obvious that suicide is not homicide-suicide. Again, due to the lack of study, we do not know empirically. We know that there is no *the* common personality type. Like the self-immolators in Korea, these individuals do not turn to homicide-suicide because of poverty, trauma, madness, psychopathy, education and ignorance (Sageman, 2004). There is, in fact, a history of the individual and his/her community/society to the death(s). There are social integrations, such as networks, to the acts (Durkheim, 1951; Sageman, 2004).

Social bonds are central, probably in all altruistic suicides (Leenaars & Wenckstern, 2004; Sageman, 2004; see Chapter 1, this volume), but also as the study suggests, deep intrapsychic and interpersonal factors are equally figural. There is a psychology to altruistic suicides and thus to martyrdom. What is central, as in all suicides, is the attachment, not necessarily to people per se, but, as Zilboorg (1936) had shown and as evident in almost all of the martyrs' notes, the attachment (identification) (Freud, 1974b) can be to any ideal—freedom being one, and community integration being an other example. (This is probably true for the soldier.) Suicide and homicide-suicide can be for service.

## MURDER AND ALTRUISTIC SUICIDE

Murder and suicide are interwoven (WHO, 2002). As Unnithan, Corzine, Huff-Corzine, and Whitt (1994) proposed, the choice between homicide and suicide depends on the attribution concerns. "Martyrs," faced with frustration, may choose suicide, but some choose both homicide and suicide. The Korean martyrs, in their last letters, attributed the cause of their problems to the governments, the United States, and the outside community/society. They were both angry and very unhappy about the oppression. The principal source of frustration was unequivocal, marked by angst and rage, and absolutely no ambivalence. They could not live without freedom (or for the global enemy, the United States, to be destroyed). In the altruistic suicides, there is an extremely constricted mind (basically one problem and *only* one solution, *martyrdom*), aroused by anger and depression at the same time. Based on the study, the Korean martyrs' mind contained one solution, but in the current martyrs (or terrorists), like al-Qaeda in the Middle East, the martyrs' violence is self-directed *by duty*; and in some, *by a duty*, both other-directed and self-directed. A martyr's homicide-suicide is a conscious act of other- and self-induced annihilation, best understood as a multidimensional event in a needful individual, who defined an issue, by duty, for which homicide, followed by suicide, is perceived as the best solution. *Only* the martyr's suicide, as in the Korean self-immolators, has a very different attribution style, only self-death. Other-directed death is against Buddhism, whereas in Islam, it is not always. A Muslim is also, however, not free to end his life. Since the actual "owner" of life is God, any suicidal or homicidal act, other than acts in the name of religion (such as *martyrdom*), determines the individual to be banished to hell. Allah says, "And do not kill yourself (nor kill another). Surely, Allah is Most Merciful to you" (*Surah An-Nisa:29*). Of course, as Sageman (2004) points out, there are very radical and lethal interpretations of these lines. Killing is not only unacceptable by Buddhism in Korea, but also by the vast community of Muslims, who believe that only God owns life and death.

An intense identification with a lost or rejecting person or with any lost ideal (e.g., honor, freedom) is crucial in understanding the egression (escape) of the suicidal person. This would be evermore likely in the military. The definition of identification is attachment (bond), based upon an important emotional tie with another person(s) or any ideal, such as one's institutions. If this emotional need is not met, the (vulnerable) suicidal person experiences a deep pain (discomfort). There is an intense desperation, and the person wants to escape and to be gone from a world with no ideal. The anguish must be stopped. The suicidal person wants to exit, escape, be elsewhere, and not be—anything but the abyss on earth. Suicide is then the only solution. The (altruistic) person plunges into the death, whether it is for freedom, martyrdom, or another ideal.

## ALTRUISTIC SUICIDES: ARE THEY THE SAME OR DIFFERENT FROM OTHER SUICIDES?

On a penultimate point, there are diverse perspectives on what is altruistic and what is not. There is a psychology to altruistic suicide. Although as Durkheim (1951) suggested, the social level allows us to understand martyr suicide and related phenomena, the differences in the mind on "I am in pain," "This is the only problem," "This is the best solution," and so on, allow us to know the martyrs differently. The study of rage, narcissistic injury, vulnerable ego, psychopathology, and especially unconscious processes may be as important. A question remains: On the continuum of martyrdom, how can one predict a suicidal martyr (or any person) also committing homicide-suicide (being a terrorist)?

Suicidology is not alone in wrestling with the question, "Who are the altruistic suicides?"—the suicide bomber, the terrorist, the martyr, the soldier who dies to save his friends, or the Hindu woman who climbs onto the funeral pyre to be with her husband forever. There are many more examples of "altruistic suicides" around the world. Indeed, some people are studying this topic more intently of late, but suicidology with the military should do so from a scientific point of view.

## FINAL THOUGHTS

This chapter offered several perspectives on Durkheim's (1951) altruistic suicide. Does it help to understand a soldier's or a veteran's suicide? Regardless of the answer, a question remains: Are these suicides different or the same? This was a point of the opening introduction on taxonomy (see, for example, John Stuart Mill, 1984). Somehow, we are back at the beginning and understand Durkheim's point: When is a motive particularly pure and "according to what standard will the decision be made? When does a motive cease to be sufficiently praiseworthy for the act it determines to be called suicide?" (p. 240). This remains the question—especially about military suicide—and it has been the question on this topic for millennia. Socrates and Seneca are two "altruistic suicides"; deaths ordered by the state. It was their duty. Plato (428–348 BC) asked the question, so did Emmanuel Kant (1724–1804). Kant held the view that life was sacred and should be preserved. He wrote little on the topic of suicide; yet it is of historical note that Kant, in these sparse writings, raised questions on what he called heroic suicide. He asked, Is heroic suicide, despite claims by the martyrs, to the contrary, suicide? Is deliberate martyrdom suicide? Are Socrates' and Seneca's deaths, ordered by the state, suicide? What is suicide and what is not? This is still a question in this millennium.

Many agree with the concept of altruistic suicide in their own culture/county, but find it not applicable to others. They are saints, warriors, and/or martyrs, the others are terrorists (Jung, 1964). The so-called "suicide bomber" in the Middle East today illustrates this "reality"; one views him/her as a martyr; the other, a terrorist. One's culture may be suicidogenic; it compels some people to kill themselves. It is service. Socrates had no escape route. Does Durkheim's sociological system help to explain *all* suicides, altruistic or otherwise, better? Does it help explain some or many suicides in the military, whether American or al-Qaeda?

On implications, September 11th is not new, nor are the wars in Iraq and Afghanistan; the world has not changed. We are reminded of Carl Jung's discussion on the projection of the Shadow (the terrorist), on the enemy during the Cold War years (Jung, 1964). It is always projected on the other; but, is this true? For example, during the Third Crusade, English king Richard I (the Lionheart) ordered the massacre of thousands of prisoners (POWs), whereas Saladin, the Muslim leader, acted more honorably. Who is the martyr? Who is the saint? Who are the warriors? Who is the terrorist? And whose suicide is altruistic?

# CHAPTER 5

# Military Suicide: A Historical Individual Case Study

Why did Admiral Jeremy Michael Boorda kill himself? He intended to be dead, that is unequivocal. The mode of death was suicide. We know how he died. Yet we do not know, "why he killed himself" and "why at that particular time." The most widely accepted evidence-based, peer-reviewed, published method to answer these and many more questions is the psychological autopsy (PA). (It is also not novel science.) It is a retrospective investigation. It is nothing less than a thorough study of the intention of the decedent, such as the Admiral's intent. This has not been allowed; the Navy has steadfastly refused to release a comprehensive public report, or the suicide notes (there were a reported two), and so on. Justified secrecy prevailed. What could we learn? In addition, how different this way of handling a suicide *is* from how the suicide of President Bill Clinton's deputy White House counsel, Vincent Foster Jr. was handled. There was no secrecy, but visibility. Well-respected experts undertook a psychological autopsy, and a public synopsis was released. Foster's suicide note was made publicly available. I here reprint the full text:

> I made mistakes from ignorance, inexperience and overwork
>
> I did not knowingly violate any law or standard of conduct
>
> No one in the White House, to my knowledge, violated any law or standard of conduct, including any action in the travel office. There was no intent to benefit any individual or specific group
>
> The FBI lied in their report to the AG [Attorney General]
>
> The press is covering up the illegal benefits they received from the travel staff
>
> The GOP has lied and misrepresented its knowledge and role and covered up a prior investigation
>
> The Ushers Office plotted to have excessive costs incurred, taking advantage of Khaki [White House Designer] and HRC [Hilary Rodham Clinton]

> The public will never believe the innocence of the Clintons and their loyal staff
>
> The WSJ [Wall Street Journal] editors lie without consequence
>
> I was not meant for the job or the spotlight of public life in Washington. Here ruining people is considered sport.

What can we learn from Vincent Foster Jr.'s note? Utilizing the model outlined in Chapter 2, the actual protocol sentences (see Chapter 2, TGSP) that were verified in Foster's note are as follows: Unbearable Pain: 1, 2, 3, 4, 5, 6; Cognitive Constriction: 7, 8, 9; Indirect Expressions: 11, 12—no ambivalence (10); Inability to Adjust: 13, 14, 15; Ego: 16, 17, 18; Interpersonal Relations: 19, 20, 21, 22; Rejection-Aggression: 25, 26 (narcissism), 27, 30, 31, 32; Identification-Egression: 33, 34, 35 (see Leenaars, 2004 for the full case discussion).

Like the suicide note analysis, the psychological autopsy verified the core psychological constructs of why people like Vincent kill themselves. The PA was made public. It helped with the aftershocks (the same was done after Cobain's death). Would the same not have helped with the consequences of the Admiral's death? He was a public figure. This should not mean that *any* soldier's private information from a PA be released to the public. There are rights of the deceased and kin. Admirals or Generals are in the public eye, especially ones who led NATO in war. Boorda was a hero! A psychological autopsy would help prevent military suicide. I believe that we need these heroes' stories told. The soldier's story, as discussed already in Chapter 3 on the PA, will allow us to do the main business of military psychology: the intensive study of the human person, the suicidal soldier.

Edwin Shneidman has served in the military; he was a Captain in the Army during World War II and was on staff at the Veterans Administration for years. He had spoken to soldiers suffering from posttraumatic stress disorder, depression, and unbearable pain. Some were suicidal, and Shneidman counseled them. He also was involved in the study of the suicide cases in the military; he would be asked to investigate the mode and cause of death, beginning the roots of the retrospective psychological investigation that he later called the psychological autopsy, still a gold standard in the method of investigating suicides, including in the military, he thought We asked, "Why does a soldier kill himself/herself?" This is the task ahead. The plan is to undertake a retrospective investigation of one soldier: General Emory Upton.

## GENERAL EMORY UPTON

General Emory Upton died by suicide on the evening of March 14, 1881. The entire Army was surprised and traumatized; no one had expected his death. It would be easy to conclude, from the documents available, that many soldiers were in shock!

Emory Upton was born on August 27, 1839. He was the 10th child of 13 of Daniel and Electra Upton (Michie, 1885; Morris, 2000). On the father's side, Emory was a direct descendant of John Upton, a Scotsman, who came to America around 1650 and settled in Danvers, then called Salem, Massachusetts. On the mother's side, Emory was a descendent of Stephen Randall, an Englishman from New Hampshire, born in 1782. Emory was born on a farm near the town of Batavia, New York. His reported early years were not unusual and little is recorded; he was described as a controlled person. The most enduring characteristic about Emory's family was their strong membership in the Methodist Church; they were zealous Christians. The most influential person was his mother; they had strong bonds, and she was seen as loving and preaching hopeful perseverance. Peter Michie, in his 1885 biography of General Upton wrote, "The name of *mother* was ever the tenderest and gentlest of words for him, for it awakened the memory of a pure and boundless love which had never failed him" (p. 3). His father was an industrious man. Emory is known to have studied under evangelist Charles G. Finny at Oberlin College for 2 years. His career aspiration from an early age was to enter the Military Academy at West Point. His hopes came true, and his history was marked beginning at age 17, when he entered the United States Military at West Point. On March 12, 1856, he received a letter from the House of Representatives that read, " Dear Sir: I have the pleasure of indorsing a notice signed by the Secretary of War, informing you that the President has conditionally appointed you a cadet in the military service of the United States." He was an excellent cadet (Barry, 1918) and graduated sixth in his class of 45 cadets on May 6, 1861 (Morris, 2000). This was a historical military time, the great American Civil War.

On the day of his graduation, Emory Upton was appointed Second Lieutenant, and within 8 days was appointed First Lieutenant. He was immediately sent to Washington, DC, and within a few days engaged in active military duty. Like in academics at West Point, he excelled in military practice, being one of the greatest American military men in history. He was known as the great military strategist. Not only in battle, but also in theory he was surprisingly gifted. His work, *The Military Policy of the United States*, which analyzed military policies and practice, was not only the first systematic writing of American's military history, but also had a major impact on the U.S. Army, the system, and the green culture. Emory Upton, in fact, is considered one of the greatest military officers/theorists ever.

On July 18, 1861, Lieutenant Upton engaged in his first battle, the first Battle of Bull Run. He was wounded on July 21st, yet he stayed in battle. He was considered by all accounts a brave soldier and true warrior. He was wounded three times, the last seriously. Yet he complained to no one (dissembling?). There are questions about a TBI (Hyson, Mosberg, Sanborn, & Whitehorne, 1990). He was known to have engaged in different harm's way, being exposed to endless trauma and death. He experienced, witnessed, and was confronted with events that involved actual and threatened deaths and serious injuries, and threat to physical integrity of self and others. (Was there any response involving intense fear, helplessness, or horror?). The list of battles that he fought included the Peninsula Campaign, the Battle of Antietam, the Battle of Fredericksburg, the Battle of Gettysburg (he commanded the 2nd Brigade, 1st Division of the VI Corps), Mal Share, the Valley Campaign of 1864, the Battle of Selma, and so on. His greatest contribution was at Spotsylvania Court House. Here, he developed and implemented new, very successful military tactics that helped win the battle, foreshadowing the very tactics used in WWI. He was wounded in that attack. Upton became a recognized brilliant warrior. He excelled and was promoted from Lieutenant to Major General at the age of 25, a most unusual distinction. His active military career was stellar.

On April 16, 1865, Upton led the 4th Division of the Calvary Corps at the assault on the Confederates at Columbia, Georgia, a week after the surrender of Robert E. Lee's army. Under Upton's command, the Division captured 1,500 soldiers and stockpiles of ammunition, and burned the ship CSS Jackson. That was the last large-scale battle of the Civil War. In May 1865, Upton led the arrest of Alexander Stephens, the vice-president of the Confederacy and later Jefferson Davis, the most wanted Confederate, was placed in his custody (North, 1899). He again was seen as serving with honor. Emory Upton served from the beginning to the end of the war with great distinction.

After the war, he served in several posts, and from 1870 to 1875, he, at the age of 30, led the cadets at the United States Military Academy, a prestigious military academic appointment. He was known as a superb teacher in infantry, artillery, and cavalry tactics. In 1875, he was transferred and ordered to a tour of Europe and Asia to investigate military systems in various countries. Upon his completion of the tour he wrote *The Armies of Europe and Asia*. Not only did this volume outline the advances of the world's militaries, but it also contained 54 pages of recommendations to change the U.S. Army. This effort resulted in changes in advanced military school, a general staff, a system of reporting evaluation reports, and promotion by examination. The work was widely read.

After his European tour, he was appointed instructor at the Artillery School of Practice at Fort Monroe, Virginia. He specialized in teaching arms tactics (Morris, 2000). In 1881, Upton then returned to the rank of colonel in 1880 was placed in command of the 4th U.S. Artillery at the Presidio of San Francisco This was to be his last command.

General Emory Upton's greatest impact was from the work called *The Military Policy of the United States* (Morris, 2000). The book called for radical change Needless to say, many of the old guard blacklisted Upton (not unusual in collective systems even today). The book was controversial, and Emory Upton never saw its publication. This greatly affected him. However, it was later circulated widely. It was controversial because it criticized the influence of the Secretary of War and the civilian control of the military. It also presented the U.S. military system as weak and full of deficits. He argued that it was based on fundamental flaws. This did not mean that General Upton opposed American democracy; Emory was a strong advocate. General Upton agreed that the president should be Commander-in-Chief; he argued for a strong army, controlled by the military system (see Chapter 1), supplemented by volunteers or conscripts in time of war. Upton's ideas were strongly influenced by what he learned from Asian and European armies. Thus, he was not adhering to the system of his day. Further, he argued for promotion based on merit and evaluation, advanced military education, compensatory retirement of officers, and so on. Many felt insecure and the work was kept from publication. It "was buried in oblivion" and the U.S. Congress refused to print it (Barry, 1918). It was not until 1904 that the Secretary of War Elihu Root read the manuscript and immediately ordered its publication. It became an instant bestseller and remains the most influential book or reference on the history of the U.S. Army. General Upton's work was and is matchless; yet in 1881, General Upton did not believe so. His core belief was that he was a failure and that "his efforts had failed" (Barry, 1918).

## THE LIFE AND LETTERS OF EMORY UPTON: INTRODUCTION

Much after the publication of his monumental work, General Upton became quite famous. Already in 1885, Peter Michie published a book, *The Life and Letters of Emory Upton, Colonel of the Fourth Regiment of Artillery, and Brevet Major-General, U.S. Army.* It is the best book on General Emory, full of personal documents, most notably a treasure trove of revealing letters. General James Harrison Wilson wrote the introduction, under whom Upton served. I will present verbatim a few excerpts:

> It was my good fortune to know EMORY UPTON . . . aside from patriotism on the one hand and religion on the other, he was a genuine military enthusiast, whose thoughts night and day turned to the art of war. No knight of old was ever more absorbed in dreams of military glory, nor more grimly determined to win it, as opportunity offered. He was tremendously in earnest, and whatever his hand found to do, that he did with all his might. . . .
>
> Upton had early become convinced that the first requisite to success in the profession of arms was unflinching and unhesitating courage, not only for its influence over his superiors, but over those whom he had to lead, and yet observation taught him that the most courageous were frequently the first to fail. . . .
>
> The fact is, that he saw much to condemn in the daily operations of the army, and the reader will not fail to note that his active mind poured itself out in criticism in his letters to his sister. It was to her that he . . . wrote during the overland campaign: "Our men have in many instances been foolishly and wantonly sacrificed. Assault after assault has been ordered upon the enemy's entrenchments when [the general ordering it] knew nothing about

the strength or position of the enemy. Thousands of lives might have been spared by the exercise of a little skill; but as it is, the courage of the poor men is expected to obviate all difficulties. I must confess that, so long as I see such incompetency, there is no grade in the army to which I do not aspire." It was also to her he wrote: "We are now at Cold Harbor, where we have been since June 1st. On that day we had a murderous engagement. I say *murderous*, because we were recklessly ordered to assault the enemy's entrenchments"; and again: "I am very sorry to say I have seen but little generalship during the campaign. Some of our corps commanders are not fit to be corporals. Lazy and incompetent, they will not even ride along their lines; yet without hesitancy they will order us to attack the enemy, no matter what their position or what their numbers. Twenty thousand of our killed and wounded should to-day be in our ranks."

Upton was as good an artillery-officer as could be found in any country, the equal of any cavalry-commander of his day, and all things considered, was the best commander of a division of infantry in either the Union or the rebel army. He was the equal of Custer or Kilpatrick in dash and enterprise, and vastly the superior of either in discipline and administration, whether on the march or in the camp.

JAMES HARRISON WILSON
WILMINGTON, DEL., May 2, 1885. (Michie, 1885, pp. ix–xxviii)

On June 16, 1918, the *New York Times* published an article, "Emory Upton Military Genius" (Barry, 1918). General Upton was then becoming recognized as an enduring Civil War hero. Upton's writings were described as "a living reservoir of military truth" (Barry, 1918, p. 2). What did General Upton, however, believe? How would the factors in 1918 have influenced his suicidal choice? Unfortunately, of course, we cannot interview General Upton or his comrades, something one would want in a psychological autopsy. However, Richard Barry (1918) interviewed General James Wilson, the same person who had written the introduction to Peter Michie's 1885 volume on Upton; he had an intimacy only known among fellow soldiers, and the interview sheds some light on who General Upton was, and I quote verbatim again some excerpts:

His physical characteristics? He was five feet eleven, blonde and blue-eyed, weight 170 pounds, always in perfect condition. He never wasted a word in conversation and never was interested in anything but his profession. . . .

Upton was the only General in our service that commanded (1) a battery of artillery, (2) a regiment of infantry, (3) a brigade of infantry, (4) a division of infantry, and (5) a division of cavalry. This represented all arms of the service and he was equally successful in all. . . .

He early recognized the danger of death in the Civil War, where the chances were greater than they have been at any later period, but he also knew that unflinching and unhesitating courage was the requisite of a soldier. . . .

Upton was not a machine soldier. He gave loyal and unquestioning support to his superior officers and especially to those who were chief in command, but he was too good a student to shut his eyes to the blunders committed around him. He studiously refrained from public criticism, but he did not hesitate, even when less than 25, to write to his intimates: "I am sorry to say I have seen little generalship during this campaign. Some of our corps commanders are not fit to be Corporals." (pp. 2, 12)

## THE SUICIDE OF GENERAL UPTON

General Upton died by suicide in San Francisco, March 15, 1881. He shot himself in the head. It was a shock to everyone in the military. He is buried in Fort Hill Cemetery, Auburn, New York.

Unfortunately, little was written on Upton's suicide. It remained unexplained. This was not by accident; it was intentional. His death was unequivocal; he died by suicide. The reason why, however, has never been answered. Fortunately, General Upton's letters were published in a book, written by Peter Michie in 1885, entitled *The Life and Letters of Emory Upton: Colonel of the Fourth Regiment of Artillery, and Brevet Major-General, U.S. Army*. They offer a unique window into his mind. They are the most personal of documents, and I will below present some letters, with brief comment. However, first I need to highlight what is known about General Upton's death.

In 1990, John Hyson, William Mosberg, George Sanborn, and Joseph Whitehorne published a paper, "The Suicide of General Emory Upton: A Case Report" in *Military Medicine*. It explains some of what is known. I present some highlights, verbatim or paraphrase, of the article:

> On the evening of March 14, 1881, Emory Upton inexplicably and unexpectedly committed suicide. . . . The reason for his death by suicide has remained for the most part unexplained. On March 15, 1881, Major-General Irvin McDowell, the Division of the Pacific and Department of California, informed the military of Upton's death. The post surgeon, Major Joseph C. Bailey sent a report to the Surgeon General stating: "Emory Upton. Colonel 4th Artillery, committed suicide this morning by shooting. The ball [from the Army pistol] entered the mouth and made its exit near the occipital protuberance. Nothing positive is known as to the cause. He was not on sick report. No necropsy made." (p. 445)

He avoided the issue of why Upton killed himself. "A formal inquest of the suicide was held the next day, March 16, with the San Francisco coroner presiding over a jury of civilians and Army officers. The details of Upton's death emerged from this inquest. Upton "lying on his back in his bed," his right hand resting on his chest, clutching an Army Colt .45 caliber revolver . . . had shot himself in the mouth. Henry Hasbrouck, a neighbor, noted "powder stains" on the left hand and a "slight abrasion" on the left forefinger" (Hyson et al., 1990, p. 446). There was a collection of letters at the death scene. There was an unfinished letter to the Adjutant General of the U.S. Army; a letter to Upton's sister, Sara Upton of Batavia, New York; and another letter addressed to the

Adjutant General of the Army. In it, he simply stated, "I hereby tender my resignation as colonel of the Fourth Artillery." A resignation letter from a commander was highly unusual. Some interpreted the death as a result of an "overtaxed mind." (Did he have a TBI?) Regrettably, no autopsy was performed (Hyson et al., 1990).

The Coroner's Verdict:

> On March 16, 1881, the coroner's jury arrived at the verdict that Emory Upton's death the previous day resulted from the "effects of a pistol shot in his head penetrating the brain, inflicted by himself with suicidal intent"; and that "the act was not premeditated but was caused by temporary unsoundness of mind brought on by a overtaxed brain and by his own disappointment at what he conceived to be a failure in the revision of a work upon which he was engaged."
>
> The next day, March 17, 1881, Gen. McDowell issued a general order announcing Upton's death as a result of his "suffering under great mental disturbance caused by an illness of long standing." Emory Upton's funeral took place at the Presidio on Saturday, March 19, 1881, at 2 p.m. On Sunday, his remains were sent east by train for burial in the Fort Hill Cemetery at Auburn, New York, next to his wife. (Hyson et al., 1990, p. 446)

There were informants. Lieutenant Alexander B. Dyer, the adjutant of the 4th Artillery, had last seen General Upton alive at about 10 a.m. on the morning of Monday, March 14, the day of Upton's suicide. Dyer stated that on the previous Saturday morning, March 12, Upton's behavior showed "every sign of the keenest distress." Upton, seemingly was "much excited" and his eyes were filled with tears. Upton stated, "I can't think: I've had a great deal of trouble in the last five or six days" (Hyson et al., 1990). Dyer concluded that Upton had "A great deal of trouble."

Captain Henry C. Hasbrouck, his neighbor and a West Point classmate, was perhaps Upton's closest friend. Emory Upton told Hasbrouck that he was "despondent about his tactics." Hyson et al. (1990) note,

> Hasbrouck had reported that Upton "had "great difficulty" in completing the revision of his infantry tactics manual. . . . He threw down his pen and clasped his hand to his forehead. His eyes were suffused with tears, and [he] told me he was ruined. In every way he showed great emotion." As Hasbrouck walked Upton back to his quarters, Upton "recovered his composure"; however, he continued to talk about "the failure of his system of tactics." At a later meeting, Upton continued to dwell on his failure to revise his tactics successfully. Upton feared that his "professional reputation" would be ruined . . . he, alone, would be blamed for the manual's failure in the event of a war. (p. 448)

Captain Hasbrouck also noted that Upton had "severe headaches" and they were getting "longer and more painful," sometimes impairing his ability to concentrate. Upton was fatigued and showed "some signs of loss of memory." Yet the suicide was not predicted. "William Conant Church, the editor of the *Army & Navy Journal*, and a good friend of Upton's, in his March 19, 1881, eulogy, referred to Upton as the 'least likely' officer in the Army to commit suicide." Church had reported that prior to Upton's death, he was "much depressed and complained of the trouble in his head." He concluded that there was maybe "temporary insanity" (Hyson et al., 1990).

## THEORIES ON GENERAL UPTON'S SUICIDE

Hyson et al. (1990) have offered a number of theories on Upton's death. One was *melancholia*. They state, "Upton was still saddened by the death of his wife, Emily, who had died of tuberculosis some 10 years earlier, after a brief marriage of 1 year" (p. 449). They further note that "Upton may have mused a reunion in the hereafter with his beloved Emily." This, of course, is not unusual in suicides. Hyson et al. go on to conclude, "The contemporary evidence seems to suggest that the immediate cause of his suicide was depression. The best medical evidence available suggests that the general's depression was related to, if not largely caused by, a medical problem, that of "chronic nasal catarrh" (p. 449). There has been a host of speculation, beyond the nasal catarrh, as to the cause of the depression, including intracranial aneurysm, brain tumor, nasal tumor, chronic sinusitis, frontal sinus mucocele, genetics, and so on. Yet they note, "One cannot categorically exclude a psychiatric cause for General Upton's mental symptoms, especially in light of the self-imposed pressures involving his tactics revision" (p. 450). Yet despite all the theories, the commonly accepted theory for his death was his fear of failure, disgrace, and dishonor. This psychological pain was the proximal cause of General Upton's death. Why did General Upton die by suicide? How can we decide? Some of the answers may lie in his personal documents, his last letters. Given the archive, we can offer a psychological theory. I next present an answer to the very question raised.

## THE LETTERS

As I noted earlier, Peter Michie had the great foresight to collect General Upton's letters; Sara Upton, Emory's beloved sister, to whom he wrote some of the most revealing letters of his suicidal career, supported and copyrighted the book. She is the "Dear Sister" in the letters. Sara was the third in the most important trio of women (read: people) in Emory's life: mother, Emily, and Sara. Michie's work is exceptional and foreshadows some core approaches in the psychological autopsy; yet Shneidman would quickly point out that it is not; he would call it a forensic investigation. Michie, however, offers us the most unique looking glass, Upton's letters. They are a most peerless window into Upton's mind—his suicidal mind. I will let the data speak with a few comments here and there. I have selected letters, besides a few, from the following topics: His wife Emily and marriage, the military policy of the United States, and death. (All page numbers refer to Michie's 1885 book).

I will begin with the most satisfying letter. From an early age, Emory was fascinated with the military, and he wanted to go to *the* military school in America, West Point. The most decisive moment in Emory Upton's life was when he received a letter on March 12, 1856, from a President's appointment to the Military Academy at West Point. It read, "Dear Sir: I have the pleasure of indorsing a notice signed by the Secretary of War, informing you that the President has conditionally appointed you a cadet in the military service of the United States." Upton was committed to war; he once wrote, *"I will fight before I will deliver my sword."*

Marriage

Emory Upton married Emily Norwood Martin, the fifth child of Enos T. Throop Martin, Esq., and his wife, Cornelia Williams. They lived in scenic Willowbrook on the shore of Lake Owasco, in central New York. Emily

> was a fair-haired, blue-eyed maiden, gentle in her ways, modest in her demeanor, and full of kindly affection. Her childhood was wholly spent under the watchful eye and tender care of a devoted mother, whose lovely Christian character and womanly sweetness not only cemented the family in the strongest bonds of love, but extended their influence beyond the home circle. (p. 217)

She became a communicant of the church.

Upton was an isolated soldier. Michie (1885) writes,

> Upton had had but few opportunities to cultivate the friendship of women. His active military career had filled the interval between his narrow life as a military student and the broader intellectual and social life he was then experiencing. His affections had been centered upon the members of his own family. (p. 221)

Yet possibilities of human affection were growing in him, and Emily had that "personal beauty." They first met on a summer's day in Willowbrook. She too was attracted and drawn to him by his soldierly manliness. They were instantly attached to each other; yet there were proper expectations in those years, and Emily was devoutly Christian. The intimate association that followed was of the earnest Christian. But during the relationship, Emory developed "doubts and questionings" of his faith. "In their intimate personal friendship the maiden soon learned that her hero seriously questioned the truth of those tenets of her faith which she valued" (p. 223). She soon recognized his danger, and "under God's providence," Emily decided to pray and bring him out of what she saw as a delusion. Emily shared her love of God and him. Emory was humbled and accepted Christ, and her. Emory "saw clearly the beautiful, child-like faith which animated this pure maiden, and he could not deny in his inmost soul the truths which she brought so vividly to his mind" (p. 225). Emory's acknowledgment of her love read, "She listened with a fitting blush; With downcast eyes and modest grace; For well she knew he could not choose But gaze upon her face." They married. Emily "had always had reasonably good health"; however, she soon developed tuberculosis, the scourge at the time. It was then a death sentence. The following passage, to which, after her early death, Upton frequently referred:

> Just at twilight I went into the library, and, sitting down before the lovely wood-fire, I gave myself up to my favorite diversion of building castles in the coals. I love to spend the twilight in this way, thinking of pleasant things in the past, of dear friends, and dreaming such bright, beautiful dreams of the future, full of high and noble resolves of doing for others, gaining (through efforts to make others happy) happiness to myself, pondering how I can make my life worth the living. Then as the ashes fall and cover for a time the bright coals into which I have been gazing, and obscure the light by which I have seemed to view my future, the thought comes to me that *thus* may some of the brightest of my anticipations be clouded and quenched by adversity and sorrow.

The ensuing death would become an unbearable sting! Emory never survived the loss! "She died at one o'clock on the morning of the 30th of March, in the full exercise of her Christian faith" (p. 235). Her attending physician, Dr. Kirkwood, who was also a valued personal friend, in writing to General Upton, said,

> MY DEAR GENERAL: I have had no heart to write to you before now, since the death of your dear, good, beautiful wife, as all commonplace condolence would, for such an irreparable loss, be out of place, and incomprehensible to you. I have no doubt that you have regretted extremely that you were not with her during her last days.

Upton was indeed bereaved. Nothing but his firm religious faith could have sustained him in the trying months that followed. The memory of his wife was kept fresh and pure, and her influence on his life never for a moment failed him. But his life after, although devoted to the conscientious discharge of his duties, lacked that rounded fullness that would have graced and perfected it had it been given him to live it with his chosen wife" (p. 237).

Upton

> kept her memory fresh and pure, dwelling on her virtues, her love, and her inheritance as a child of God. This led him to believe her still living. . . . His weekly letters to her mother down to the day of his own death are filled with the noblest sentiments and records of the purest conduct; all unconsciously told with the humility and sincerity of a man in whom "the peace of God" has found a resting-place. (p. 238)

Emory wrote letters to his parents and sister:

WILLOWBROOK, May 11, 1870.
MY DEAR PARENTS: ...

My dear parents, who, like you, have nearly run the race of life, ought to look forward with joy and thankfulness to the dawning of eternal life, and I pray that with you we, as a family, may all soon be partakers of the joys prepared for those that love God. With tender love, my dear father and mother, Your affectionate son, EMORY.

ATLANTA, May 22, 1870.
MY DEAR SISTER: ...

The feeling of desolation has again come over me, as, in entering my home, I realize that the loved one who made it so happy, my precious Emily, has gone from me forever. But God can help me to bear this sorrow, and, while now life offers no attractions, I know that when again in active duty, employed in instilling in the minds of the nation's future defenders ideas of devotion to duty and discipline, I shall experience consolation in the thought that I am again useful in the world. Here I am in the midst of a thousand evidences of Emily's love for me. It was at this desk my heart flowed out to her daily in the letters which used to comfort her poor heart. But all is changed. She is hidden from me, and already violets, blooming over her sacred form, offer their daily fragrance unto Heaven. I am not tempted to arraign the goodness of God. I can humbly thank him for lending me, even for so short a time, his angelic child, who, under his chastening hand, brought me back to a knowledge of the truth, and with her I can say, "Bless the Lord, O my soul, and forget not all his benefits."

Upton was bereaved!

## The Military Policy of the United States

The Mount Everest of Emory Upton's military career was his book, *The Military Policy of the United States*. Here are some letters, *res ipsa loquitor*:

September 30, 1877.—I am going to trace our military policy from the beginning of the Revolutionary War to the present time, and, if possible, expose its folly and criminality....

November 6, 1878.—General Sherman is very anxious to have me go on with the work, but he tells me that I will receive much abuse. He has read up to the War of 1812, and says I arraign the politicians as "extravagantly blind."...

January 13, 1879.—Until our Representatives appreciate this responsibility, we shall witness no improvement in our military policy . . . neither by the Constitution nor the laws is the Secretary of War entitled to exercise command.... The Constitution, laws, decisions of the Supreme Court, and of the Attorney-General, nowhere give him the authority of command. In administration he is independent of the President, and ought to be, as thereby the purse is separated from the sword.

June 19, 1880.—Sometimes, I am free to admit, I get discouraged.

"Upton accomplished his arduous labor within the short period of two and a half years, and this is the more striking when we remember that he could only devote himself to it in the intervals occurring between his official duties" (p. 425). While engaged in the preparation of his great work, General Upton asked General Garfield, then occupying a most prominent political station, to examine his work. The book was never published, and this narcissistic injury multiplied Emory's unendurable pain. His death followed.

Death

General Emory Upton died by suicide on the 14th of March, 1881. Upton's letters to his beloved mother are most revealing. "Upton's life, in all of its minutest public, professional, and private details, offered not the slightest clew, to the ordinary observer, that could throw any light upon the true cause of his death" (p. 475). "Upton for several years before his death appeared outwardly to be a splendidly developed man, in the full possession of robust physical health.... He had gained by merit alone a reputation and standing second to none in the army.... Not one of his intimates suspected that death by his own hand could ever be possible" (p. 476). There are letters to his beloved mother. "The following letters express his sincerity and actual hopes at this near approach of death" (p. 476).

NEW YORK, August 27, 1879.

MY DEAR MOTHER: I am reminded that this is my fortieth birthday, and can not let it pass without expressing the gratitude that is in my heart for all the loving acts and sacrifices which you have bestowed upon me, who at best am but an unworthy son. Father, too, I must include, with full forgiveness for the many times he took advantage of my weakness to chastise me for acts which to a juvenile mind appeared perfectly proper.

FORT MONROE, February 1, 1880.

My mental condition is much improved. Last year at this time I was much depressed. I had malaria, and did not mistrust it till Dr. Robinson told me of it at Newport. Now, whenever I feel pains in the back of my head, I take from thirty to forty grains of quinine in thirty-six hours, and at once feel relieved. To-morrow I shall begin the revision of the tactics—a work of three months, and shall then again begin the Policy.

Michie (1885) believes that "These letters show a total ignorance of any serious danger which threatened his life." Michie discerned "not the slightest evidence is found to warrant the suspicion that he was aware of his condition." Michie goes on to conclude, "Even in his home letters during the war, there is but the barest mention made of his severe wounds received in battle, and he dismissed them with a sentence or two, simply expressing his hope of speedy recovery" (Dissembling? PTSD?).

Colonel Henry C. Hasbrouck, a classmate of Upton's and at the time of the latter's death an officer of his regiment and under his command, writes,

FORT ADAMS, NEWPORT, R.I., February 10, 1882.

It was Sunday morning, March 13th, that I first realized how much he was suffering. I happened to be alone with him in his office, and, in answer to my inquiry about his headache, he broke down completely, laid his head on his desk, and sobbed. After he was composed I walked with him to his quarters, and was much with him all that day and evening, and also Monday evening. He was very despondent, talked of the loss of his will-power, and of the respect of the officers of the regiment, spoke much of the failure of his tactics, and particularly of the system of deployment as skirmishers, said if his system was adopted it would involve the country in disaster in the next war.

My impression, when he first spoke to me about his troubles, was, that he was in a state of nervous depression, partly owing to his catarrhal troubles, but principally to overwork and hard study.

It seems to be the accepted opinion . . . the mental disorder that made him falsely imagine that he had lost the respect of his officers, that his tactics were a failure, etc., and, finally, that night when he retired, resulted in suicidal mania.

On March 13th, Emory wrote his sister:

PRESIDIO, SAN FRANCISCO, March 13, 1881.

MY DEAR SARA: Since writing to you, last Sunday, I have been in no little distress over the revision. It has seemed to me that I must give up my system and lose my military reputation. God only knows how it will eventually end, but I trust he will lead me to sacrifice myself, rather than to perpetuate a method which might in the future cost a single man his life. Whichever way it may turn, I know I shall have your sympathy, and may our heavenly Father bless and keep you and our precious father and mother!

I need all your prayers, for I would keep my integrity.

Friday I went over to Oakland to a luncheon. The city is flat, and has beautiful lawns strewed with flowers. I don't feel like writing any more. Only let me feel that I have your love and sympathy.

With a fervent kiss for you all, ever your affectionate brother,
EMORY.

Emory penned at the last moment, before the citadel of his mind was surrendered, the following penultimate note:

PRESIDIO, SAN FRANCISCO, March 14, 1881.

To the Adjutant-General U.S.A.
SIR: I hereby tender my resignation as colonel of the Fourth Artillery.
Very respectfully, your obedient servant,
    E. UPTON,
    Colonel Fourth Artillery

Thus, we return to the question raised, Why did General Emory Upton die by suicide? I next offer some thoughts on this question.

## ANALYSIS OF THE LETTERS

We are most fortunate to have General Upton's letters; they provide his personal narrative—his life history about such topics as marriage, his military policies and tactics, and death. My thesis is that documents, such as letters and suicide notes, can be very informative. Upton's letters present a unique archive. They allow us to present a very personal psychological theory. They allow us to answer the question, Why did Upton kill himself?

In my analysis, I examined Upton's letters. The most important is his last letter to his sister, Sara, to be found on a previous page. In performing my protocol analysis, I read the letter over numerous occasions. I needed to know what Upton wrote. The letter was analyzed, based on my peer-reviewed system of theoretical analysis. I used the TGSP, found in Chapter 2, Table 2.2. Below is my analysis of this letter and some of his other letters.

## OPINIONS REACHED AND
## BASIS THEREFORE

Upton's last letter to Sara, dated March 13th, has the following characteristics of a suicide note:

Intrapsychic Drama

- I–*Unbearable Psychological Pain*: 1) Suicide as a relief, 2) Suicide as a flight from trauma, 3) Emotional states in suicidal trauma, 4) Loss of interest to endure, 5) Inability to meet life's challenges, 6) State of heightened disturbance.
- II–*Cognitive Constriction*: 7) A history of trauma, 8) Overpowering emotions, 9) Focus only on grief topics
- III–*Indirect Expressions*: 11) Aggression has turned inward, 12) Unconscious dynamics
  (Re: 10. There is no ambivalence, none!)
- IV–*Inability to Adjust*: 13) Feels weak to overcome difficulties, 14) Incompatible state of mind, 15) Serious disorder in adjustment (f)
- V–*Ego*: 16) Weakness in constructive tendencies, 17) A "complex" or weakened ego, 18) Harsh conscience

Military Stage

    VI–*Interpersonal Relations*: 19) Problems determined by situations, 20) Weakened by unresolved problems, 21) Frustrated needs (e.g., abasement, autonomy, harm avoidance, infavoidance), 22) Frustration to a traumatic degree, 23) Positive development not forthcoming

    VII–*Rejection-Aggression*: 25) Report of a traumatic event, 26) Narcissistic injury

(Re: 27–30. There are no aggressive themes)

    VIII–*Identification-Egression*: 33) Identification with person/ideal, 34) Unwillingness to accept life, 35) Suicide as escape.

Based on the foregoing, I have reached the following primary opinions:

1. Based on the analysis, General Upton's writings had almost all the characteristics of a suicide note. Not only are the content and themes highly predictive of a suicide note, the writing contains the essential differentiating protocol suicide themes that discriminate genuine notes from simulated ones. Upton's letter has all the "richness" of a suicide note. It lacks any ambivalence and expressions of outward aggression, only inward. Yet that is not rare in soldier's, martyr's, terrorist's suicide notes. (It did, however, add to the very lethality of his mind.)
2. Based on my training, experience, and research in the field and my review of approximately 2,000 suicide notes, there is no doubt in my mind that this *is* a suicide note, that is, a letter written by someone with suicidal intent in mind.
3. Because the note does have almost all common characteristics or factors of a suicide note, General Upton's mind at the time of writing was suicidal.
4. Because the note does have many of the commonalities of suicide, we can postdict a psychological theory of Upton's suicide. We can explain his death psychologically.
5. The second note, to the Adjutant-General, dated March 14th, offering his resignation from the military is more banal, but screams the reason for his self-directed violence. The note has the following characteristics 12) Unconscious dynamics, 33) Identification with person/ideal, 35) Suicide as escape. There is no question Upton wanted to escape. However, through Michie's biography and letters, we now know better. Most important, I believe, are the letters about Emily's death and the death of his monumental hard work, his policies and tactics. In his suicidal mind, *everything was lost.*

Most revealing is his letter, dated May 22, 1870, to his sister, Sara, on the death of his "precious Emily." He was bereaved. The protocol sentences are as follows: 3, 6, 7, 12, 15g, 19, 20, 21, 22, 25, 27, 30. The letter is full of howling pain—"The feeling of dissolution." There is loss and deep bereavement, but also no suicide intent. Emory was traumatized, but not lethal. He was understandably perturbed, in pain. This was a survivor's story. What changed? What was the proximal cause?

The letter on the military policy of the United States is largely banal in terms of suicide protocols, except they are central to his suicidal beginning. He was very attached to his book, *The Military Policy of the United States*. I scored the following items of the TGSP: 7, 17, 19, 21. This writing lacks the richness of a suicide note. It does, however, outline the history of a trauma, that is, the intentional suppression by superiors of his writing. This was, to use the proverbial metaphor, the straw that broke the camel's back—a General's back. It was a death sentence. (Was the bullying intentional?) It is of interest that a page is missing at the end in Michie's book; what did it say? Michie was outlining the known causes of Upton's death.

There was one more letter found at the death scene. Upton, at that point, according to Michie (1885), "felt he could no longer do his whole duty." In a green culture, this is loss of *everything*—one is *not* a warrior. One is a burden, and in Upton's case, a possible killer, through his tactics, of his own men (a catastrophic delusional belief). He was then lethal.

> To the Adjutant-General U.S.A.
> SIR: In my effort to revise the tactics so that they might apply to companies over two hundred strong, I discovered that the double column and the deployment of numbers, when compared with the French method, was a failure. The fours, too, I was forced to admit.

It was an unresolvable problem. There is no question about the narcissistic injury. His needs were frustrated: attachment, affiliation, achievement, dominance, and infavoidance (see Chapter 2, Table 2.1). There was such shame and disgrace.

Based on my analysis, there is no question, General Upton was depressed at the time of his death. Of course, my opinions dealing with Upton's death were based on documents available—interviews, letters, Michie's biography, and so on. They were reviewed and considered. Of course, I never examined Emory so I could not make a diagnosis. Yet based on the information I had, it is reasonable to conclude that General Upton was perturbed (distressed) and depressed. There is, in fact, a history of depression—he actually used the words. Earlier, I presented some letters to his dearest mother. On February 1, 1880, he wrote, "Last year at this time I was much depressed." Did he experience recurrent cycles? Was there PTSD too? Quite probably, I believe, are the needless (he believes) deaths of his soldiers, Emily, and his book. There was at the end hopelessness!

The scores of the letters to his mother were 3, 7, 12, 15f, 21, 25. The letters are not suicidal, but they are letters of a perturbed, depressed mind. In my opinion, I have no doubt that Emory had a mood disorder, most like depression, recurrent (major depressive disorder, recurrent; posttraumatic stress disorder [probable]).

## OPINIONS REACHED AND BASIS THEREFORE

Based on the forgoing, I have reached the following opinions:

1. Further to Upton's last letter, the contents and themes in the note and earlier letters illustrate the suicide note, and many details of his life are tragically illustrated by the contents of his letters.
2. There is a history of a mental disorder.
3. There is a history of depression and probably PTSD.
4. There are abundant precursors to his lethality; yet it is only in the last letter that there is suicide risk. We do not know his history of suicide risk.
5. In addition to the mental, physical, and spiritual imbalance, he had a most traumatic relationship—a deep attachment to his policies.
6. General Upton did experience a number of traumas in his life (e.g., experiencing and witnessing the injury and death of others; threat to self; war wounds [TBI?]; death of Emily; the suppression and military bullying about his book). The publisher "not willing to assume responsibility for publishing narcissistically injured him." This should not be a surprise. He questioned the authorities, the rights of the Secretary of War, the entitlements of the officers, and so on. General Upton, to use another common metaphor, yelled, "The Emperor had no clothes." (Of course, the part that they don't tell you of that story is that the Emperor afterwards had the boy put in jail for life.) He questioned the Representatives in the House He questioned the policies of the time. He called some Generals, Corporals and he paid dearly for the truth. (Does this occur often in green cultures?)
7. In the green culture, there is absolute obedience. There was military bullying. (Is this common in suicides of officers?) Only after his death one of the greatest military masterpieces ever was published. Upton never knew this—one more reason to wait when you are suicidal. Heightened lethality does not last.
8. General Upton was treated for a number of painful physical impairments such as severe headaches. The treatments were unsuccessful.
9. There were proximal signs of unbearable pain, depression, and suicide risk Were there distal ones?

Thus, there is no doubt in my mind that in all probability that, at the time of his death, General Upton did intend suicide. We knew that before. However, can we also answer why? However, before I offer my theory, I need to address one more point.

## THE CONTEXT OF UPTON'S LIFE

General Upton's letter on March 13, 1881, to Sara is a *suicide note*. The note to the U.S. military on March 14th is too. These suicide notes, which, after all, are the penultimate act of that person's life, can be made more informative when they are placed within the context of the details of that person's life. We can do this, thanks to Michie's retrospective investigation. Almost every word in the suicide notes is illustrated by Upton's life and documents, and many details of his life are tragically illustrated by the content of the notes. The documents, especially the letters of his family members, fellow officers, and doctors made every one of the key words of the last letters come alive and take on special meaning. The story is told. The analysis, I believe, reveals a special theory of why General Upton killed himself that would otherwise have remained hidden and lost. Was it an intentional secret? (Why was no autopsy, for example, performed?)

Of course, suicide notes, by themselves, cannot tell all of the story for the plain and simple reason that they are written by a person with a very constricted mind (see Chapter 2 on cognitive style). They are usually constricted, illogical, overfocused, tunneled and often to a point of delusion (Upton was). In *Voices of Death*, Shneidman (1980a) cites Alfred Alvarez on the topic of tunneling:

> The contemporary English poet, novelist and critic A. Alvarez, who wrote an excellent book on suicide, *The Savage God*, has described what he calls "the closed world of suicide" in the following way:
> Once a man decides to take his own life he enters a shut-off, impregnable but wholly convincing world . . . where every detail fits and every incident reinforces his decision. . . . Each of these deaths has its own inner logic and unrepeatable despair. . . . [Suicide is] a terrible but utterly natural reaction to the strained, narrow, unnatural necessities we sometimes create for ourselves. (Alvarez, 1972; Leenaars, 1999, pp. 288–289)

Suicide notes can, indeed, have a great deal of meaning (and give a great deal of military information). They can be more so when they are put in the context of the life history of the individual. In Upton's case, we are most fortunate, we have *both* suicide notes and a *detailed* life history. Thus, in the psychological theory of why Upton killed himself, the notes will illuminate aspects of the life history, and conversely, the life history makes the key themes or commonalities of the suicide note come alive and take on a special psychological meaning. The secret is out! This is close to the art of autobiography. I will next answer the question, Why did General Emory Upton kill himself?

## A PSYCHOLOGICAL THEORY OF GENERAL UPTON'S SUICIDE

Psychologically, General Emory Upton's suicide was complex. It was an intrapsychic drama on a military stage. Utilizing the evidence available, and an analysis as discussed earlier, I propose the following psychological answer.

## Intrapsychic Drama

### Unbearable Psychological Pain

Emory Upton's mind, based, for example, on his suicide note, was permeated with pain. Shame and disgrace were paramount. He had lost everything; he even feared that he would cause death and suffering. *All* his work was a loss. Although I believe that his suicide was much more, his suicide was a flight from these believed specters—the loss of his wife, his flawed military policy and tactics, and his disgraced military reputation (he resigned in the end). He was in great distress. Deprivation, distress, and grief are evident. He felt boxed-in, hopeless, and helpless. He arrived at what he perceived to be the end; suicide became *the* solution, a relief from intolerable psychological pain. Death was an inviting solution; it kept, he thought, his integrity intact.

### Cognitive Constriction

Emory Upton's mind was a constricted mind. There was constriction of perception, logic, reason, and conscience. He reported only permutations of a history of trauma—his military system was flawed. He feared that the error would cost a soldier his life. He was overwhelmed and overpowered. There was a paucity of thought, focusing only on the traumatizing and grief-provoking topic.

### Indirect Expressions

Emory felt he was encircled; he submitted to his pain and irrational core belief. Emory became subordinate. He concluded that he had to sacrifice himself. He turned his anger inward. He was, however, not ambivalent (I suspect that in a trained soldier, they are indoctrinated to be determined). Although we know a lot about his psyche, I believe that there were unconscious dynamics in his death—even the Civil War culture was a deadly one, probably the most deadly by suicide ever.

### Inability to Adjust/Psychopathology

General Emory Upton considered himself too weak to cope (or adjust); he gave up the fight. Without a doubt, he believed that his death was best for his reputation, U.S. soldiers, and himself. From a 21st century perspective, I believe that Emory had a mental disorder. He says, in fact, that he was depressed; he may also likely have suffered from PTSD. Furthermore, I cannot rule out a TBI and/or chronic physical injury(s)/pain. There was, without question, *a deep psychological pain* and an inability to adjust to his life's demands.

## Vulnerable Ego

Emory's ego was vulnerable; he lacked constructive tendencies. With the loss of Emily (he kept her memory), he felt desolation. This was the beginning loss toward the end. Then, the core belief of the loss of his flawed tactics and policy was all that he could bear. (What is projection here?) The military, even by his superiors and administrators, all the way to Washington, made him believe that they were not only flawed but also lethal. He may have killed a soldier. There were unresolved problems, something that he could not change—the military. Emory became discouraged, to a lethal level. He thus felt that he must be punished with death (attribution inward). He was so insulated, illogical, and harsh.

Civil War Stage

### Interpersonal Relations

Emory's problems were determined by his history and the current situation that presented itself, even if much was delusional. He was weakened and felt defeated by the loss of his wife and the abuse of his superiors and administrators. He had trusted his commanders; however, there was no support forthcoming for his tactics. His needs were unsatisfied—at least attachment, perfection, achievement, autonomy, and control. The frustration of these core needs became exceedingly stressful—he was not capable of healthy action. He had hoped for the acceptance of his tactics and policy, but such he saw that as not forthcoming. Therefore, he believed that his tactics were due to his defective thought. He was hopeless about being able to revise the flawed tactics (a core belief). Emory was, by all eyewitnesses, under constant strain and frustration. He responded with fear, horror, and helplessness.

### Rejection-Aggression

In his suicide note, he wrote about a single traumatic event—the flawed tactics. The Mount Everest of Upton's life was lost. His ideal was rejected; in his eyes, he was harassed. He feared the loss of his perfect military reputation. He was so discouraged in the end and suffered a deep narcissistic injury. Yet his aggression was attributed inward; there was no aggression turned outward. He was defeated and severely depressed, not outwardly angry. (He should have been; yet few such wounded soldiers ever do. It is taboo.)

### Identification-Egression

General Upton was deeply identified with his system, policy, and tactics. They were rejected; he felt rejected (harassed). The tactics, he now thought, were flawed and lethal. His reputation was lost. He was unwilling to accept the loss of this ideal—his spotless military career. (The same can be seen in Admiral Michael Boorda and deputy White House counsel Vince Foster Jr. Is it common?) In similar illogical thinking, he reasoned that his beloved sister, Sara, would understand and give him support. God would bless. His beliefs were traumatizing. General Upton lost his integrity, and the only solution was suicide. He first escaped with his resignation to the Adjutant General and then by his death. He escaped!

Shame and Disgrace

One major theme in General Upton's suicide notes was shame and disgrace. These threads are entirely consistent with Emile Durkheim's speculations on the central role of honor in military suicide, especially among officers. These seem to be the core unbearable pain. Thus, I need to explicate this construct in more detail. On the topic, Shneidman writes, "Some suicides seem to be related particularly to a sense of shame, 'loss of face,' disgrace or a sense of dereliction of duty. Prideful people especially seem vulnerable to these motions" (Leenaars, 1999, p. 282). An example is the suicide of General Upton. His military tactics, he believed (and later proven to be true), would revolutionize warfare and services. He had done years of careful work, study, and observation of military tactics in the United States, Europe, and Asia. Yet, in the end, *the U.S. military and Washington rejected The Military Policy of the United States*. He was indoctrinated to believe that he was a failure, a ruined officer. He jumped to conclusions. He would needlessly now kill American heroes. He was the enemy. Therefore, he shot himself. There are many interesting details in his life and letters. The sense of shame, repentance, and restitution are evident. His attitude toward himself was of a defeat—he lost the battle. It seems to make most sense to view Upton's suicidal act as an overwhelming concatenation, in a disturbed individual, of several emotional surges in addition to the overarching shame: anxiety, anger, depression, hopelessness, guilt, rejection; in other words, heightened general perturbation. His, like all suicidal deaths, are complicated events.

Upton's suicide was multifarious, but also through the documents, *no* mysterious. He probably believed that there was no justification for living on yet his mind was incompatible with an accurate perception of what was going on. He was mentally constricted. His beliefs of disgrace and shame were delusional. His core beliefs were irrational. (He must have reasoned something like, "All soldiers that must give up their tactics ought to be dead. My tactics were a failure; it might cause a man his life. Therefore, I must resign and kill myself." I believe that he was a victim of harassment (bullying). Did he really die by his own hand? Are other suicides of soldiers due to these central factors in a green culture? Have other lives been ruined due to shame and/or disgrace?

## GENERAL UPTON: THE FUNERAL

I want to make explicit that in no way do I or did I intend harm to Emory Upton. On the contrary, my intent would be his: to save any warrior in harm's way and survivor of that battle. First and foremost, Emory wanted to save his soldiers lives. This way, he does again. His tragic choice need not be *the* reference. Indeed, General Upton's fervor was befitting a warrior. Michie (1885) writes,

> The funeral ceremonies at San Francisco were such as were befitting the well-beloved comrade and eminent soldier. His remains were brought to Willowbrook, attended by two of his former comrades, and after the final services of the Church they were placed by the side of those of his beloved wife in the cemetery of Fort Hill at Auburn.
>
> On the fly-leaf of his Bible, under the date August 31, 1879, is inscribed in his own hand, "Whatsoever you do, do it heartily, as unto the Lord, and not unto man." Exemplifying this text throughout, strong in faith, ardent in piety, firm in adherence to the Church, zealous in his official duties, loved for his personal virtues, and honored for his official integrity, his earthly life, rounded and complete, presents a fitting prelude and preparation for that heavenly life whose reality he now unquestionably enjoys. (p. 504)

General Upton's story is a sad one. Shame and disgrace are powerful factors. Are they common in soldiers and/or military officers who die by suicide? To test this hypothesis, I will examine the kept-secret case of Admiral Boorda. What do we know about Admiral Mike Boorda's life and death?

## ADMIRAL JEREMY MICHAEL BOORDA

Admiral Jeremy Michael Boorda (he preferred to be called "Mike") was born to a Jewish family November 26, 1939, in South Bend, Indiana. He grew up in Chicago. Little is known publicly about his early life. Sia (1991), in the *Baltimore Sun*, has reported that when a teen, "his world at home and school began to crumble." (What crumbled?) Mike is known to have dropped out of high school, lied about his age, and used a phony birth certificate (Sia, 1991) to join the Navy at age 16 in 1956. Mike is reported to have believed, "I was cool, I was tough." Yet much later he reported about these years, "I was scared to death. Hell. I was 16 years old, and I had nobody to help me." The U.S. Navy became his home, and he became the "Seaman to Admiral." This was an extraordinary achievement in the Navy. Under a commissioning program in 1962, Mike Boorda went to Officer Cadet School in Newport, Rhode Island. He attended the U.S. Naval War College and also earned a BA in political science from the University of Rhode Island. He also studied at the University of Oklahoma.

Mike Boorda's Navy career was sterling. He fought in the Vietnam War, the war in Yugoslavia, and the Bosnian War. Boorda was, in fact, in command of all NATO forces engaged in operations during the Yugoslavian wars. His awards, among others, included the Navy Commendation Medal and the Navy Achievement Medal, which he earned during the Vietnam War. These are great honors.

Admiral Boorda was married to Bettie May Mora and had four children. Two of his sons were in the Navy, and his Navy lineage has continued with grandchildren. His family is a true Armed Forces family.

The Arlington National Cemetery website presents a most unique publicly available biography (http://www.arlingtoncemetery.net/boorda.htm). I quote verbatim:

> Admiral Boorda, born in South Bend, Ind., 26 November 1939, enlisted in the U.S. Navy in 1956. He attained the rank of petty officer first class, serving at a number of commands, primarily in aviation. His last two enlisted assignments were in Attack Squadron 144 and Carrier Airborne Early Warning Squadron 11. He was elected for commissioning under the Integration Program in 1962.
>
> Following Officer Candidate School in Newport, Rhode Island, and commissioning in August 1962, Admiral Boorda served aboard *USS Porterfield* (DD 682) as Combat Information Center Officer. He attended Naval Destroyer School in Newport and in 1964 was assigned Weapons Officer, *USS John R. Craig* (DD 885). His next tour was as Commanding Officer, *USS Parrot* (MSC 197).
>
> Admiral Boorda's first shore tour was as a weapons instructor at Naval Destroyer in Newport. In 1971, after attending the U.S. Naval War College and also earning a bachelor of arts degree from the University of Rhode Island, he assumed duties as Executive Officer, *USS Brooke* (DEG 1). That tour was followed by a short period at the University of Oklahoma and an assignment as Head, Surface Lieutenant Commander Assignments/Assistant for Captain Detailing in the Bureau of Naval Personnel, Washington, D.C.
>
> From 1975 to 1977, Admiral Boorda commanded *USS Farragut* (DDG 37). He was next assigned as Executive Assistant to the Principal Deputy Assistant Secretary of the Navy for Manpower and Reserve Affairs, Washington, D.C. He relieved the civilian presidential appointee in that position, remaining until 1981 when he took command of Destroyer Squadron Twenty-Two.
>
> In 1983 and 1984, he served as Executive Assistant to the Chief of Naval Personnel/Deputy Chief of Naval Operations for Manpower, Personnel, and Training. In December 1984, he assumed his first flag officer assignment as Executive Assistant to the Chief of Naval Operations, remaining until July 1986.
>
> His next assignment was Commander, Cruiser-Destroyer Group Eight in Norfolk, Virginia; he served as a Carrier Battle Group Commander embarked in *USS Saratoga* (CV 60), and also as Commander, Battle Force Sixth Fleet in 1987.
>
> In August 1988, Admiral Boorda became Chief of Naval Personnel/ Deputy Chief of Naval Operations for Manpower, Personnel, and Training. In November 1991, he received his fourth star and in December 1991, became Commander in Chief, Allied Forces Southern Europe (CINCSOUTH – Naples, Italy) and Commander in Chief, U.S. Naval Forces, Europe (CINCUSNAVEUR – London, England). As CINCSOUTH, Admiral Boorda was in command of all NATO forces engaged in operations enforcing UN sanctions against the warring factions in the former Republic of Yugoslavia.

On 1 February 1993, while serving as Commander in Chief, Allied Forces Europe, Admiral Boorda assumed the additional duty as Commander, Joint Task Force *Provide Promise*, responsible for the supply of humanitarian relief to Bosnia-Herzegovina via air-land and air-drop missions, and for troops contributing to the UN mission throughout the Balkans.

On April 23, 1994, Admiral Boorda became the 25th Chief of Naval Operations.

Admiral Boorda's military awards included the Defense Distinguished Service Medal, the Distinguished Service Medal (three awards), the Legion of Merit (three awards), the Meritorious Service Medal (two awards), and a number of other personal and campaign awards.

## BOORDA AND HARASSMENT

As we have learned, harassment is a major stress for many military men and women. It is a risk factor. During the time of Boorda's tenure as head of the Navy, there were a series of investigations and issues of harassment. The most noteworthy was the Tailhook scandal. Rebecca Hansen, a female naval student, who had wanted to be an aviator but failed flight school, had filed a sexual harassment suit. She enlisted the help of Senator David Durenberger and a large-scale investigation called "Witchhook," the Tailhook scandal, was undertaken. The scandal revolved around a number of allegations of Navy indignations toward women. The Navy was attacked. Boorda was seen as not defending Navy culture. One person, James Well, the Navy Secretary under Ronald Reagan, believed that the Navy's top leaders needed to take action, separating Tailhook from Navy culture. Wells was not alone in the charge. During Bill Clinton's administration, Mike Boorda supported the call to reform the Navy, and many officers believed that Boorda betrayed them. One target of the inquiries was the Vice Chief of Naval Operations, Four Star Admiral Stan Arthur. Among others, Durenberger questioned Arthur's handling of the sexual harassment scandal. Boorda did not defend Arthur publicly, and Arthur retired from the Navy on February 1, 1995. This resulted, according to Senator John McCain, in a severe penance in the Navy. Some people, in fact, measure Tailhook against Pearl Harbor in its devastating impact on the Navy (Wikipedia, 2011). There were other pressures, conflicts, and battles in the Navy; Admiral Boorda was a target. There was, without question, harassment.

## BOORDA'S DEATH

Admiral Boorda died May 16, 1996 of an apparent self-inflicted gunshot wound to the chest (Arlington National Cemetery Website, 2011). He was reported to have been highly perturbed that the media was investigating his Valor device that he wore on his Navy Commendation Medal and Navy Achievement Medal. The small bronze "V" device has significant value in combat. It was claimed that he was officially not awarded the "V" and not entitled to wear them. Boorda was distraught, and the media pursued.

According to the Arlington National Cemetery Website (2011), Admiral Boorda was to have met at the time of the death with the Washington bureau chief of *Newsweek*. After the death, Rear Admiral Kendell Pease, who was with Boorda an hour before his death, reported that "Admiral Boorda was to meet with the bureau chief of *Newsweek* at 2:30 p.m. to discuss the question of the medals." Pease further said that "When he told Boorda at about 12:30 p.m., what the subject of the interview was, the admiral abruptly announced that he was going home for lunch," adding, "Admiral Boorda was obviously concerned." (A lunch had already been sent to him at his office.)

No one predicted the suicide. Navy Secretary at that time, John Dalton, said he met with Boorda a day earlier. "He was in great spirits," Navy Secretary Dalton reported. (Was Boorda dissembling?) Mike Boorda was found at about 2:05 p.m. in a side yard next to his home at the Washington Navy Yard. He arrived at the DC General Hospital a few minutes later with a gunshot wound to the chest. According to sources, a .38 caliber pistol was used in the shooting. At 2:30 p.m. EDT, he was pronounced dead.

The suicide was a shock, especially for the Navy. It was a huge loss. At the White House, President Clinton praised Boorda and stated,

> His death is a great loss not just for the Navy and our armed forces but for our entire country. . . . Mike Boorda was the very first enlisted man in the history of our country to rise to become the chief of naval operations.

He described the Admiral as a man of "extraordinary energy, dedication, and good humor." There were aftershocks, but a cloud of secrecy occurred. The military culture's system was evident; however, was it best? (As a postscript, Senator John McCain, himself a Vietnam hero, stated, "Possibly the tragedy of Boorda's death brought some closure to Tailhook").

The controversy over whether Boorda was entitled to wear the "V" prevailed, and still does. According to the Arlington National Cemetery Website (2011), on June 25, 1998, Navy Secretary John Dalton placed a letter in Boorda's file from Elmo Zumwalt, the Chief of Naval Operations during the Vietnam War, that asserts it was "appropriate, justified and proper" to attach the bronze "V"s to the ribbons on his uniform. The "V" stands for valor and significant service in combat. Boorda himself questioned it and in 1995, after being advised by the Navy's Office of Awards and Special Projects, believed that he was not entitled. Boorda was known to be very concerned over the controversy of the medals. Like Upton, he was a man of honor. He feared it would cause scandal and further tarnish the image of the Navy. (Yet, in some strange logic, he concluded that his suicide was not?) However, his commander, Zumwalt, believed that he was entitled. In 1998, Boorda's family requested a review of the recommendation. The Board of Corrections of Naval Records, the ultimate arbiter of whether Boorda was entitled to wear the "V," determined that he was not (Finding of the Board for Corrections of Naval Records).

## THE SUICIDE NOTE

An autopsy was undertaken; however, the results were not released. In fact, little was made public. Secrecy prevailed. Admiral Boorda apparently left two suicide notes, neither of which was released. It was known that one was written to his family and the other to the Navy. According to reports, the note to the sailors (addressed "my sailors") stated that he felt disgraced.

I believe that Admiral Boorda wrote two suicide notes, one addressed to his family and the other "to my sailors." The Navy has steadfastly refused to release the notes, and for that matter, any documents. Yet it is reported that Mike Boorda had typed the notes, explaining that he "took his life because of the questions raised about his wearing of "V" for valor medal on his combat ribbon from Vietnam" (James McIntyre, 1996). The CNN report goes on to state,

> According to a source who has seen Boorda's note to the sailors, Boorda wrote that he wore the Vs because he thought he rated them. [I don't believe this to be true—AL].
> 
> Boorda told the sailors how much he thought of them, and said that some people will not think he did the right thing, the source said. He ended the letter with a reference to "critics in the media" who have been "hard on the Navy," saying "I have given you more to write about," the source said.

I believe that the media was a threat. Rear Admiral Kendell Pease, who saw Mike Boorda last, had asked Boorda about how to address the *Newsweek* reporter. The Admiral had told him, "We'll just tell him the truth." (Pease also reported that Mike was perturbed.) In the *New York Times* on May 18, 1996, Philip Shenon also reported the known facts of the note. We learn *little* more, only that Boorda uses the words that wearing the "V" was "an honest mistake." Shenon goes on to report,

> In the letter to sailors, he wrote about how he cared about the core values of the Navy—honor, courage and commitment—and about how he would be viewed," the official said. "He tells the sailors that they are doing a good job. Don't let people get you down. You're a good Navy. There are times when they have to take care of other sailors, take care of themselves. It was vintage Mike Boorda."
> 
> Navy officials said today that they remained perplexed by Admiral Boorda's suicide, given his cheerfulness in the days before his death, and even more so given the tone of his suicide letter, which suggested that sailors gird themselves for adversity and fight on. "The tragedy for him, for all of us in the Navy, is that he chose not to fight on himself," said a former colleague in the Pentagon.

Based on my search, this is all we really know, and I question even these facts. However, in the December 1996 issue of *Washingtonian* magazine, the complete contents of the note are allegedly reported. In the *Los Angeles Times* on November 25, 1996, James Risen reported that Admiral Boorda wrote,

> "To my sailors," he wrote, "I couldn't bear to bring dishonor to you. . . . For those who want to tear our Navy down, I guess I've given them plenty to write about for a while. But I will soon be forgotten. You, our great Navy people, will live on. I am proud of you. I am proud to have led you if only for a short time. I wish I had done it better."

The Navy, however, has never released the note. To the best of my knowledge, the family has never discussed the note. I have read what I can glean in the *Washingtonian*, and I have decided not to present the content here. I made this decision based on my expertise on suicide note analysis. Is it simulated? I see no reason to perpetuate the question of authenticity.

I would like to offer one sad observation; what we think that we know from the note to his sailors is full of contradictions and illogical thinking. Boorda says to his sailors, "You have to fight for yourself" or something like that. Yet he also implies by his death the very opposite. He does not fight. On the very face of it, it is a contradiction. Admiral Boorda cannot face the future, but tells his sailors to do so. What was his logic?; something like: All sailors who are disgraced should be dead. I'm disgraced. Therefore, I should be dead. (All sailors disgraced for dishonorable behavior should be dead. I wore dishonorably medals. Therefore, I should be dead.) Mr. K. could be Admiral K. What was Mike's scratch on the Ferrari? What was his major *lethal* premise? What allowed him to jump in oblivion, and traumatize his sailors more than wearing the "V"?

## OPINIONS REACHED AND BASIS THEREFORE

Based on the findings, I have come the following primary opinions:

1. Admiral Mike Boorda's suicide is unequivocal. The Admiral died by suicide.
2. We know little about the reasons why beyond the history reported—only what the Navy released.
3. There is no question that the proximal cause was the reported questioning by the reporter for *Newsweek*. This story repeats itself. Vince Foster experienced the same concerns; he wrote in his suicide note that ruining careers was a media sport. The media needs to take responsibility in its reports, I believe.
4. There was a history to Mike's suicidal decision; there is no question the stressors, such as the Tailhook scandal and wearing the "V," are central to his suicidal career. His "crumbled" youth was. Otherwise, little is known.
5. There is, in all probability, a history and current situation to his suicide. Yet we know little. Our knowledge is banal.

6. There are serious questions about the authenticity of the note reportedly released in 1996. I will not comment further.
7. I will state unequivocally that shame and disgrace were central and are recurrent themes in suicide among the armed forces—it was in Upton and Boorda's cases.
8. There is no question in my mind that the *secrecy* that prevailed after Boorda's death adds to the traumatization of our sailors and the people and the United States. I know this from my personal experience on May 16, 1996. (I discussed earlier my experience on the very day of Boorda's suicide with Naval officers.) It, in all likelihood, based on our general knowledge, adds to the pain, PTSD, and in all probability, to more suicides. The secrecy is deadly.

Thus, I can offer nothing further, except to hope that the military, including the Navy, will review its policies on reporting suicides, but only for top officers—our Generals, Admirals, and Secretaries of Defense. They are public figures. President Clinton got it right with Foster. However, I continue to argue that the facts of soldiers' suicides remain the right of the families, not the military or government.

## CONCLUDING THOUGHT

One final remark: If the Navy now had the foresight of Peter Michie and Sara Upton on General Upton, then what would we have learned about Admiral Mike Boorda? Would it have helped in saving the lives of our soldiers? Upton's did. Would we have prevented more suicides? My answer is "Yes." We need to be more visible, and once we are visible, we can do something about the problem. *Would the Armed Services have the courage to do so in the future?*

# PART THREE

# Current Study

Back in the 1800s, Dr. John Snow, the father of public health, already believed in surveillance. The first step is geared to answer the question, "What is the problem?" In this section, I examine and describe the current state of knowledge of the problem and measure its magnitude. I specifically examined the American and Canadian problem. Moving beyond the individual (idiographic), the first chapter presents the general (nomothetic) literature on the American Armed Forces. The U.S. studies show that suicide in the military is high and increasing. Today, there is one military personnel who dies by suicide every day (Thompson, 2012). The second chapter in this section presents the literature on the surveillance of suicide among the Canadian forces. There are, regrettably, no reliable data. Very briefly presented is what we know in a few more countries, France, the United Kingdom, and Yugoslavia. More questions arise on military data. The last chapter looks at the whole issue of surveillance and concludes that there are major validity and reliability problems in military suicide statistics. This was also true in the earlier literature on this topic. As Dr. John Snow, however, showed in the 1850s, you need accurate and reliable data to prevent a disease or a problem.

## CHAPTER 6
# Suicide Among the American Armed Forces

Durkheim (1951) provided the earliest figures available for suicide rates within the U.S. military; Durkheim reported a rate of 68 per 100,000 for the U.S. Army during the period from 1870 to 1884. The suicide of General Emory Upton was *not* an unusual event in the military in those years. There was a matchless epidemic, compared with any suicide rate in the U.S. military since. Why? The reported U.S. rate for civilians has never been reported to be so high. Why was it so high? Was it the American Civil War? Civil war, we know, causes more suicides than international wars. General Upton's case offers some insights, I believe. During the first half of the 20th century, the U.S. Army suicide rate remained between 20 and 50 per 100,000. However, during the years of the first and second World Wars the suicide rate was reported to have dropped below 15 per 100,000 or less. Throughout the 20th century, suicides as a cause of death within the U.S. Army were second only to accidents (unintentional injuries). Today, suicide is still the second leading cause of death in the military (Centers for Disease Control, 1999). That is significant.

Does war cause suicide? Although it is commonly believed that the military suicide rate decreases during war, this is not always true. The recent high rates during and after the Yugoslavian War, as we discussed earlier, is an example. The U.S. rate decreased significantly during both World Wars. Is that true? Were the data reliable? There are no official data available on the impact of the Vietnam War on the Army suicide rate. Overall, veterans believe that the rates were high. Secrecy in the military prevailed.

Cassimatis and Rothberg (1997) provide a historical overview of suicide rates within the U.S. military. They report that the U.S. military crude suicide rate (number of suicides divided by the population number) was stable during 1980–1990 at approximately 12 per 100,000. This rate, they report, is slightly lower than the crude suicide rate in the United States for the general population during the same time period, which was also stable in those years at 12.25 per

100,000. Cassimatis and Rothberg further reported on standardized suicide rates (adjusted for age and gender) and reported that for the segment of society that provides the predominant source of soldiers (young adult males), the suicide rate within the military between 1980 and 1990 was significantly lower than in the comparable civilian population. Yet were the data reliable? This question is most important.

## RECENT RATES OF SUICIDE

What are the reported more recent rates of U.S. military suicide? Hourani, Coben, and Warrack (1999) reported on the suicide rate in the U.S. Marine Corps. They compared the observed Marine rate with the expected number based on rates for the employed general U.S. population. Hourani et al. noted that overall, there were fewer suicides in the Marine Corps ($n = 213$) than expected and showed no increasing trend. This did not mean that the numbers were insignificant. Scoville, Gubata, Potter, White, and Pearse (2007) examined the epidemiology of suicides among U.S. Air Force, Army, Marine Corps, and Navy recruits from 1980 through 2004. Scoville et al. calculated crude, category-specific, and age-adjusted mortality rates as deaths per 100,000 recruit-years. The overall recruit suicide rate was 6.9 deaths per 100,000 recruit-years. They reported 46 onsite suicides by gunshot (39%), hanging (35%), fall/jump (22%), and drug overdose (4%). Suicide accounted for 20 additional recruits from 1980 through 2004 after leaving the military training site. Only three (5%) suicides occurred among females, resulting in a 3.5 times higher risk for males compared with females. They concluded that suicide rates among military recruits were lower than those of comparably aged U.S. civilians. Clarke and Sutton (1992) reported the same conclusion; yet they only sounded an alarm on the increasing use of firearms among soldiers during the Gulf War.

Opinion became divided about whether U.S. military soldiers and veterans, the vast majority of whom are middle-aged or older, are at increased risk of suicide. Miller, Barber, Azrael, Calle, Lawler, and Mukamal (2009) examined suicide among U.S. veterans. The authors assessed the risk of suicide associated with veteran status by conducting a prospective cohort study of 499,356 male participants in the Cancer Prevention Study II. Participants reported their veteran status and other characteristics in 1982 and were followed for mortality through 2004. During follow-up, 1,248 veterans and 614 nonveterans died by suicide. In age-adjusted analyses, the risk of suicide did not appear to differ by veteran status. The authors concluded that the risk of death from suicide among middle-aged and older U.S. males is independent of veteran status. Is it? Kaplan, Huguet, McFarland, and Newson (2007) have convincingly shown, in fact, that U.S military veterans are twice as likely to die by suicide as civilians. That would be an epidemic. Kaplan et al. (2007), in a sterling research report, "Suicide Among Male

Veterans: A Prospective Population-Based Study," found that veterans were twice as likely (adjusted hazard ratio 2.04, 95% CI 1.10 to 3.80) to die of suicide compared with nonveterans in the general population. *This is an epidemic!*

Studies to date, which means almost all, focused on samples derived from patient populations in the Department of Veterans Affairs system (e.g., Miller et al., 2009). They have enormous limitations. Kaplan et al. (2007) note that

> the reliance on VA clinical samples is particularly limiting from a population-based perspective because three-quarters of veterans do not receive healthcare through VA facilities. Consequently, little is known about the risk factors for suicide among veterans in the general U.S. population. . . . In light of the high incidence of physical and mental disabilities among veterans of the conflicts in Iraq and Afghanistan (Gawande, 2004; Hoge, 2006; Kang & Hyams, 2004), it is important to examine the risk of suicide among veterans in the general population. (p. 619)

Mark Kaplan and his team (2007) examined the risk factors for suicide among veterans in the general population. It had never been done before. In pursuing this goal, Kaplan et al. undertook a prospective follow-up study with data from the 1986–1994 National Health Interview Survey (NHIS) (U.S. Department of Health and Human Services, National Center for Health Statistics, 1996). The NHIS study was conducted by the National Center for Health Statistics. The NHIS 1986–1994 data were obtained and linked to the Multiple Cause of Death file (1986–1997) through the National Death Index. The sampling was 320, 890 men, aged = 18 years at baseline. The participants were followed up with respect to mortality for 12 years. Follow-up personal (face-to-face) household interviews were conducted, with response rates ranging from 94% to 98%. The data were thus extensive and had large enough sample to allow for complex statistical analysis.

What was the consequent (dependent) variable? Kaplan et al.'s (2007) main outcome variable was death by suicide. Suicide cases were identified using the International Classification of Diseases, Ninth Revision, Clinical Modification (ICD-9 E950-3959). One key predictive variable was veteran status. Respondents were identified as veterans if they answered in the affirmative to "Did you ever serve on active duty in the Armed Forces of the United States?" Kaplan and his group undertook some complicated statistical analyses. Their results were striking:

> Veterans represented 15.7% of the NHIS sample but accounted for 31.1% of the suicide decedents. Over time veterans were twice as likely (adjusted HR 2.13, 95% CI 1.14 to 3.99) to die compared with male non-veterans in the general population. Conversely, the risk of death from natural (diseases) and the risk of death from external (accidents and homicides) causes did not differ between the veterans and the non-veterans after we adjusted for confounding factors. (p. 620)

What were the predictors of suicide risk among veterans? "The results indicate that whites, those with > 12 years of education and those with activity limitations (after adjusting for medical and psychiatric morbidity) were at a greater risk for suicide completion" (Kaplan et al., 2007, p. 620). Kaplan et al.'s crucial results revealed, thus, that *male veterans are at increased risk of suicide* relative to nonveterans. Dr. Snow would call that an epidemic! Kaplan et al. stated,

> Veterans were at greater risk of dying from suicide compared with a non-veteran cohort. The results of this study are particularly noteworthy because they are derived from a sample representative of all veterans in the US general population, whether or not they sought care in the VA system. (p. 620)

They also noted that a critical factor was physical injuries. These impairments often result in limiting disabilities (WHO definitions). These soldiers were limited in daily functioning, social skills, intellectual/memory functioning, and/or ability to adjust. Kaplan et al.'s (2007) results showed that activity limitation, due to physical impairment, such as a TBI, is an important risk factor for suicide among veterans. The implication is obvious. Health providers are needed to identify at-risk veterans who have physical and/or mental disabilities.

Who are these soldiers? Stander, Hilton, Kennedy, and Robbins (2004) reported on surveillance of completed suicide in the Department of the Navy. They report that in 1989,

> the Department of the Navy (DoN) began a suicide surveillance program using the DoN Suicide Incident Report (DONSIR) to collect data on completed suicides in the Navy and Marine Corps. . . . A DONSIR has been completed on 98% of the 200 DoN suicides that occurred from 1999 to 2001. Most DoN suicides occurred outside the military work environment and involved the use of a firearm. Most decedents were men, had experienced a recent relationship problem, and did not use any military support services in the 30 days before suicide. (p. 103)

Little more is known. What else do we know? Kaplan is no stranger to the study of gun control. Kaplan et al. (2007) noted the frequent use of firearms as another important finding. There was the higher probability that U.S. veterans used firearms as a mode of suicide compared with nonveterans. They almost always use firearms. Guns are lethal. A common belief is, "All soldiers die with their gun in hand." The preponderance of evidence suggests that the availability of a gun in the house (or base) increases the risk for suicide (Leenaars, 2009b). Kaplan further notes that veterans are familiar with and have greater access to firearms. They know how to use a gun effectively. They have been trained to be deadly in the kill.

## Firearms and Suicide

Are firearms relevant? Helmak (1996) examined occupation and suicide among males in the U.S. Armed Forces. He reported that during the period 1980 to 1992, some 95% of the 3,178 military suicide victims were men, and 92% were enlisted; of the men, 71% were 20 to 34 years of age, 82% were White, and 61% used a firearm. Helmak extracted information from the Worldwide Casualty System maintained by the Department of Defense to describe the occupational risk among military men. Occupations related to the use of or access to firearms were associated with a significant risk of suicide when compared with other military occupations. Helmak reported that, collectively, military security and law enforcement specialists had a significant occupational rate ratio (1.25; 95% confidence interval: 1.02, 1 S3; $P < 0.05$). Availability is a factor (Leenaars, 2009b). Acceptability is too. This corresponds to findings from my own studies on firearms and suicide among police (Leenaars, 2010a). Guns add to the risk.

Kaplan, Bentson, McFarland, and Huguet (2009) examined firearm suicide among veterans in the general population. Kaplan and his team noted that scant attention has been devoted to the problem of firearm suicide among veterans, particularly women. They examined the rate, prevalence, and relative odds of firearm use among veteran suicide decedents in the general population. The analyses are based on data derived from 28,534 suicide decedents from the 2003 to 2006 National Violent Death Reporting System. They found that across age groups, male and female veterans had higher firearm suicide rates than nonveterans, and younger veterans (18–34 years) had the highest firearm and total suicide rates. The male and female veteran suicide decedents were, respectively, 1.3 and 1.6 times more likely to use firearms relative to nonveterans after adjusting for age, marital status, race, and region of residence. This is most important. Kaplan et al. (2009) reported that

> although violent death and the use of firearms are generally associated with men, the results reported here suggest that firearms among female veterans deserve particular attention among health professionals within and outside the veterans affairs system. In addition, the focus should not be exclusively on the Operation Enduring Freedom/Operation Iraqi Freedom military cohort but also on men and women who served in earlier combat theaters, including the Gulf War, Vietnam Era, Korean Conflict, and World War II. (p. 503)

Availability and acceptability of firearms may be a significant factor (Leenaars, 2009b; Leenaars et al., 2002). Yet capability, defined as the degree to which an individual is able to enact a lethal suicide attempt, is a growing concern in the military (Bryan, Cukrowitz, West, & Morrow, 2010). The green culture makes it so. Combat training and exposure may cause habituation (Bryan, Curkrowitz, et al., 2010). Does it make death normal?

## IRAQ AND AFGHANISTAN WARS AND THE RATES OF SUICIDE

Is Iraq and Afghanistan making a difference? This is the question today.

In 2007, Castro and McGurk (2007) noted that there had been 72 confirmed U.S. soldier suicides in the Iraq War to date. The majority of the suicides involved single, White, male, junior enlisted soldiers, with the cause of death being a self-inflicted gunshot wound. Guns are available. The soldier suicide rates in 2003 and 2005 were significantly higher than the U.S. Army 10-year average suicide rate; the rates were 18.8 and 19.9 per 100,000, respectively. This compares with 11.6 suicides per 100,000 in the general population. The rates are, therefore, increasing and are finally being reported. The system was changing. Of course, some suicides did receive attention early on, notably soldiers who died by suicide in incarceration. Reay and Hazelwood (1970) noted decades ago that suicide of soldiers in prison already resulted in alarm, and deep concern was "created that extends beyond the immediate interests of law-enforcement officials" (p. 135). The military was paying attention to this group toward the last part of the 20th century. Today, suicides in the military generally are getting that attention.

Simply being in harm's way is risky. As noted earlier, suicide is the second leading cause of death in the U.S. military (Ritchie, Keppler, & Rothberg, 2003). Military service has recently been accepted as a risk factor for suicide and suicidal behavior. Male veterans are twice as likely to commit suicide compared with male nonveterans in the general population (Kaplan et al., 2007). There is a recent rise in the rate of suicide among active-duty military service personnel (Kang & Bullman, 2008). Traditionally, military suicide rates had been reported to be lower than those of the general population; however, the military service may have not reported the rates accurately (Kang & Bullman 2008). Unfortunately, there is very little research on the process through which military service influences the visibility of suicide risk; however it is beginning in the United States.

## SUICIDE IDEATION AND ATTEMPTS

In Chapter 2, I discussed a range or spectrum of suicidal behavior (or as some prefer, suicidality). It does not refer only to suicide but can also refer to suicide attempts and suicide ideation. What is the most current knowledge about these types of suicidal behavior in the U.S. military? As is obvious, suicidal ideation and nonfatal suicide attempts are far more prevalent than completed suicide. There is a continuum, much associated with the concept of lethality. One can start with suicide ideation, to suicide attempts, to dying by suicide. They are overlapping populations, but they are also different. Suicide ideation is not rare in the general population; attempts are more so. The National Survey on Drug Use and Health (Substance Abuse and Mental Health Services Administration

[SAMHSA], 2009), which included questions assessing suicidal thoughts and behavior for the first time in 2008, estimates that 8.3 million adult Americans (3.7% of the adult population) had serious thoughts of suicide (suicide ideation), and 1.1 million (0.5%) attempted suicide, in the previous year. What about in the military? Little data exist. What are the potential risk factors of active-duty soldiers suffering with suicidal ideation and those who have made nonfatal attempts at the population level of a military service? Jeffrey Snarr and his colleagues provide us with some beginning answers.

Snarr, Heyman, and Smith Slep (2010) presented a paper on the topic, "Recent Suicidal Ideation and Suicide Attempts in a Large-Scale Survey of the U.S. Air Force: Prevalences and Demographic Risk Factors." In their method, Snarr et al. sampled U.S. Air Force (USAF) active-duty members ($N = 128,950$). The sample came from 82 USAF sites worldwide. They completed the 2006 Community Assessment. The survey was anonymously conducted online. The response rate was approximately 45%. (This return rate is not unusual in online samples.) The final sample ($N = 52,780$) was deemed to be representative on all sampling variables at the USAF level. As discussed in Chapter 2, one major question in suicidology is assessment; how do you determine suicide ideation and attempts validly? In this study, the Centers for Disease Control's 5-item measure was used. The questions were (1) During the past 12 months, how often did you have thoughts of ending your life? (never, rarely, sometimes, frequently), (2) During the past 12 months, did you ever seriously consider attempting suicide? (never, rarely, sometimes, frequently), (3) During the past 12 months, did you make a plan about how you would attempt suicide? (yes, no), (4) During the past 12 months, how many times did you actually attempt suicide? (never, 1 time, 2 or 3 times, 4 or 5 times, 6 or more times), and (5) [If applicable] Did any of your suicide attempts result in an injury, poisoning, or overdose that had to be treated by a doctor or nurse? (yes, no). Although not a perfect measure, this is better than the common one or two questions. One has to clarify the definition; what does it mean when we say, "The soldier made a suicide attempt?" Our answers will depend on what questions we ask.

The 1-year prevalence of suicidal ideation in the sample was 3.8% (Snarr et al., 2010). They found that, not surprisingly,

> women were significantly more likely to consider suicide compared with men. Women in lower enlisted ranks and non-Christians were in turn at higher odds than women in general, whereas Roman Catholics, Evangelical Christians, and senior enlisted women (i.e., noncommissioned officers) were at reduced odds. Among men, those of unknown race/ethnicity, unmarried individuals, non-Christians, junior enlisted members, and medical personnel were all at increased odds of suicidal thinking, whereas Protestants, lower ranking officers (i.e., company-grade officers), and pilots were all at decreased odds. (p. 546)

We are beginning to learn the details of who is at risk in the U.S. military.

On attempts, Snarr and his group reported,

> Of participants reporting suicidal ideation, 8.7% (representing .33% of the total sample) made at least one suicide attempt during the previous year. Among those reporting ideation, male junior enlisted members and female Hispanics were at significantly increased odds of attempts, whereas male support personnel and women with an occupation of "other" were at decreased odds. (2010, p. 547)

All suicidal soldiers are not at the same level of risk!

Therefore, at least in the professional literature, the reality of suicide risk (suicide ideation, suicide attempts, and suicide), American soldiers and veterans are finally being reported. Indeed, in the prestigious *Journal of the American Medical Association*, Kuehn (2010) calls an alarm, based on the military's own report of "a continuing epidemic of suicide among returning soldiers" (p. 1427). The wars in Iraq and Afghanistan have made a difference. The professional, academic literature, therefore, is sounding alarm and calling for action. *Suicide is visible.*

## CONCLUDING REMARKS

Therefore, in the United States, we know suicide is a significant problem in the military. Furthermore, we know the official U.S. Armed Forces position is the same. It has accepted the unacceptable. It was not always so.

Suicide in the military is not new. It was, according to Durkheim (1951), at an epidemic level during and after the American Civil War. We know that there were suicides during and after World War II. Edwin Shneidman, as discussed earlier, was a Captain in the Army during World War II and was for years on staff at the VA. He had counseled soldiers and veterans; some were suicidal. During those years, he also studied suicide cases in the military and began to collect suicide notes of military personnel. Shneidman would be asked to undertake retrospective investigations into the mode and cause of death. Suicide, he once told me, was a concern in the military then. Captain Shneidman knew that there was a price for service in WWII. Yet that fact was not published. It was no different after Vietnam; however, it was still a significant problem. After the Vietnam War, there was still secrecy, and it continued. In the late 1990s and early 2000s, I know of individuals called upon to assist in the U.S. Army, Navy, and Marines. Military suicide was raised discreetly. It was rumored that the administration had been covering up the problem. The statistics coming from Afghanistan and Iraq were kept a secret. There was denial. Shame and disgrace were reasons. However, a tipping point had occurred. History was telling tales. The fact that, over time, many Vietnam veterans died by suicide was becoming more and more problematic. One could not ignore the reality. The Vietnam veterans gave voice to the problem; they were the true pioneers in military suicide. After our recent wars, more veterans are crying for help. The veterans made the problem visible. It may well be a reason for the current acceptance. These veterans who died by suicide did not die in vain; their screams for help have been finally heard. How many of our heroes died for war? The veterans were also first to call for suicide prevention in the military. The survivors also gave voice. Finally, the voices were heard, and a more comprehensive investigation of the problem in the U.S. Armed Forces was called for. Surveillance was begun, and a more reliable truth was learned. Finally, not only was there acceptance, but of late, military efforts are being undertaken to predict and control the problem.

Armed forces personnel are *now* giving voice to the malaise in the United States. One such soldier is Colonel Elspeth Ritchie (2010). She is the Director of the Proponency of Behavioral Health at the Office of the U.S. Army Surgeon General. I have had the good fortune to hear her speak, once at the 2010 conference of the American Association of Suicidology (AAS). I here present my notes, one more type of personal document.

Colonel Ritchie has stated that suicide is a serious problem in the U.S. Armed Forces. There were, for example, 166 suicides in 2009. That is a lot of soldiers. The rate is higher than the general population, even if adjusted for age and gender. The current rate is 25 per 100,000. The rates are furthermore increasing.

Colonel Ritchie stated that before, people in the armed forces were *not* talking about what was happening. These were suicides; however, the deaths were secret. The problem was kept invisible. The main reason, she stated, was *stigma*. One would hear, "Oh, Private Joe is a basket case." It was not acceptable to be suicidal or to get help; somehow it was more acceptable to lie in a coffin. The U.S. military, she said, is intent on changing this iatrogenic attitude (belief).

There are some known risk factors, such as PTSD. Ritchie stated that it is not soldiers with major psychiatric disorders who are at greatest risk to die by suicide, but people with undetermined adjustment problems. TBIs were a very large risk factor, but not the only physical/medical one. Regrettably, psychological autopsy studies are not undertaken, except in exceptional cases. Thus, she reports that more research is now underway to better understand suicide in the military. Guns are a factor. Ritchie concluded that something has to be done about gun availability and acceptability. She asked, What could be done?

Beyond the individual drama, Colonel Ritchie pointed to relationship problems, both in the military and at home. She took an ecological perspective. Sexual abuse and harassment have been implicated. Legal problems (such as incarceration) can be relevant, she noted. All was not simple.

Colonel Ritchie stated the U.S. Army Surgeon General's Office view that soldiers are strong, resilient individuals, but that these soldiers had been traumatized and got lost. We need to help them, she stated. They can be helped. The military, she said, is doing suicide prevention training, and the United States plans to do more!

Colonel Ritchie is not the lone voice in the military. The Department of Defense, in 2010 issued its own statement. The DoD admits that

> more than 1.9 million warriors have deployed for Operation Iraqi Freedom (OIF) and Operation Enduring Freedom (OEF), two of our Nation's longest conflicts (IOM, 2010). The physical and psychological demands on both the deployed and non-deployed warriors are enormous. In the 5 years from 2005 to 2009, more than 1,100 members of the armed forces took their own lives, an average of 1 suicide every 36 hours. In that same period, the suicide rates among Marines and Soldiers sharply increased; the rate in the Army more than doubled. Numerous commissions, task forces, and research reports have documented the "hidden wounds of war"—the psychological and emotional injuries that have so affected our military members and their families. The years since 2002 have placed unprecedented demands on our armed forces and military families. Military operational requirements have risen significantly, and manning levels across the Services remain too low to meet the ever-increasing demand. This current imbalance places strain not only on those deploying, but equally on those who remain in garrison. In the judgment of the Task Force, the cumulative effects of all these factors are contributing significantly to the increase in the incidence of suicide and without effective action will persist well beyond the duration of the current operations and deployments. Heightened concern regarding this increase in suicides has led to development of scores of initiatives across the DoD to reduce risk.
>
> The Task Force acknowledges the significant efforts made by the military Services. The Services have substantially increased their focus and investments in suicide prevention over the years to meet current requirements. (pp. ES-1, ES-2)

Today, the military not only accepts the unwanted epidemic of military suicide, it has also started sharing some very personal soldiers' stories. This openness is new. Are the green walls coming down? I hereby quote verbatim some vignettes. There is honor in the stories:

> Age: 22
> Rank/Occupation: Specialist/Aviation Mechanic
> Service Branch: Army
> As one of six siblings, this Soldier proudly joined the Army as part of a long standing family legacy of service including his father's current duty as an ROTC instructor. He was an Army helicopter mechanic. He and his oldest brother were deployed to different parts of Iraq within days of each other. When he returned from his deployment, he was significantly troubled. He needed help but did not seek it. Others did not recognize that he was troubled and needed professional assistance. He spiraled downward and took his own life. "He felt that he could not get the psychological help he needed from the military for fear it would jeopardize his future career in the Army," his father said. "The [Army] wants its Soldiers to be mentally healthy but it's very hard for the Soldiers to get the help they need," he added.

Age: 22
Rank/Occupation: Private First Class/Infantry
Branch of Service: Marine Corps
This Marine was a very talented musician and had turned down a record deal with Atlantic Records in order to join the Marine Corps. He was incredibly proud to be a Marine, and wanted nothing more than to faithfully serve his country. In July, 2009, he was diagnosed with Bipolar Disorder and went to his Sergeant for help because he was concerned about how this diagnosis was going to affect his service. According to witnesses, his Sergeant berated him openly and called him "weak." The Marine went back to his barracks and, within two hours, hung himself. When he was found in his barracks, a well-worn suicide hotline card was found lying on his bed.

Age: 32
Rank/Occupation: Staff Sergeant/Security Forces
Branch of Service: Air Force
This Airman served honorably in the Air Force for 14 years and received numerous accolades throughout his career. His friends and fellow Airmen described him as "someone selfless" who was always looking out for his fellow Airmen and acted as "a friend, big brother, mentor and a leader." His family reported that he suffered from Post-Traumatic Stress Disorder and was overwhelmed by his impending deployment, his fourth to Iraq. He died by suicide in the basement of his own home with his family upstairs.

Age: 19
Rank/Occupation: Private First Class/Combat Support Battalion
Branch of Service: Army
This female Soldier and her sister joined the Army in hopes of building a rewarding career. Their father was an active duty Army Chaplain. During her initial advanced individual training, she was raped by a fellow soldier. She told her father that she was afraid to tell anyone about the sexual assault for fear that she would be "judged." She graduated from training and was transferred to another installation for her first duty assignment. A month after arriving, she was deployed to Iraq. Although the policy in Iraq was that female Soldiers would always have a buddy, she was alone most of the time because her "buddy" had been diagnosed with cancer. Seven days before she died, a female friend (also a sexual assault victim) died by suicide in Iraq. The Private did not leave a suicide note, but her journal was discovered lying open to an entry describing the torment, pain and impact of her rape. She died by gunshot wound in Baghdad, Iraq. (pp. D1-1, D1-3)

*These are sad stories!* It is, on a very different point, good to see that the Department of Defense supports both nomothetic and idiographic scientific studies. Science is both strands. The military needs both. And for many, the actual cases make the pain more real and normal. It will help the survivors. Their sons and daughters, wives, husbands, comrades, partners, brothers, sisters, and friends were soldiers, after all. Like our soldiers who died by war-related deaths, they are our heroes. They are casualties of war; we now have to be responsible to honor their deaths, no different from any other soldier's death. There is hope!

## CHAPTER 7

# Suicide Among the Canadian Forces

Unfortunately, there is very little published on suicide among the Canadian Forces. There is next to nothing in the professional literature, making the facts very questionable. Yet there are a few studies. However, I will first comment on a military study, namely, because the lead author is a well-respected Canadian researcher, Isaac Sakinofsky.

### THE CANADIAN FORCES' POSITION

In 1996, Sakinofsky, Lesage, Escobar, Wong, Loyer, and Vanier published a report for the Directorate of Health Protection and Promotion, National Defence Headquarters, entitled "Suicide in the Canadian Armed Forces with Special Reference to Peacekeeping." They noted that, "No independent, academic study of suicide in the Canadian Armed Forces (CAF) has been published to our knowledge to date." This is regrettable, but also it is not new. Durkheim published nothing on Canada. Little has changed. Sakinofsky and his team were specifically examining suicide after the Gulf War. They investigated the suicides of the 66 military personnel from the Canadian Armed Forces that occurred between January 1990 and June 1995. They noted,

> The 66 suicide cases, all consecutive cases from the regular (non-reserve) forces, were predominantly male (95%), and over half of them primarily English-speaking. Their ages ranged from 19–51 years with a mean of 30.4 ($SD$ 7.4) years. Other ranks and non-commissioned personnel comprised 91% of the sample, but a colonel, a major, and two captains were also included. Land Forces personnel constituted more than half, Air Forces a third and the remainder were Navy personnel. Twenty-four persons had experienced at least one peacekeeping tour of duty (usually six months each duration). (p. 8)

Sakinofsky and his team concluded that the military service of peacekeeping did not emerge as a risk factor for suicide overall. Yet, were the Canadian data reliable? No questions are asked. They accept the rarity of suicide in the military and conclude that "The relative rarity of suicide in the Armed Forces in many countries may be explained on grounds of selection and screening" (p. 13). Rothberg, Bartone, Holloway, and Marlowe (1990) noted the following on these two factors, selection and screening: "Firstly, there is the selection and entrance criteria by which a civilian becomes and remains a soldier; and secondly, the psychical and social environment to which the soldier is subsequently exposed" (p. 13). Yet the screening and selection processes are far from infallible. Apter, Bleich, King, Kron, Fluch, Kotler, and Cohen (1993) described 43 consecutive Israeli male suicides, aged 18–21, during compulsory military service. Apter and his team found that there was a failure of military screening methods to detect deliberately concealed information on antecedent personal and psychological difficulties. There was dissembling, deception, and lies. Apter et al.'s posthumous investigations through examination of school records and interviews with schoolteachers and multiple other informants discovered the truth. Deception is a problem, and that has a great deal to do with real rates. Indeed, individual cases make up the population data.

Thus, do we know anything about military suicide in Canada? In a 1994 unpublished internal report, Lt. Colonel Martin L. Tepper of the Directorate of Health Promotion and Protection, National Defence Headquarters (NDHQ) analyzed 101 deaths by suicide between 1977 and 1986. He reported that suicide was *not* a significant problem. Is that true? Was that true? Of course, we cannot rely on the armed forces to tell us the facts. We need professional, independent study. Sakinofsky and his team (1996) stated the same to the Canadian military. Will the Canadian Forces allow such ever?

Professional Studies

Sakinofsky and his team (1996) did publish some data in academic/professional journals, albeit they used the very same military data, a limitation. Wong, Escobar, Lesage, Loyer, Vanier, and Sakinofsky (2001) published a paper asking if Canadian peacekeepers are at risk for suicide. In a case-control design, they retrospectively compared 66 suicides in the Canadian military between 1990 and 1995 with two control groups: (a) 2,601 controls randomly selected from the electronic military database and (b) 66 matched controls with complete personnel and medical data. They found no increased risk of suicide in peacekeepers except among a subgroup of Air Force personnel. They concluded that confounding individual factors, such as isolation from supports and possibly inadequate preparation for deployment, elucidated their suicides. They further stated that the "theater of deployment (e.g., Bosnia) did not affect the suicide rate" (p. 103). They concluded that military service is not a risk factor. War is not a risk factor.

However, Wong and his group (2001) did note that alcohol abuse might play a role in some suicides. Although the military system, they note, officially discourages alcohol abuse, social life on a military base is often centered on drinking. Alcohol abuse understandably is a factor. Alcohol excess, in addition seemed to play a part in some interpersonal problems. Alcohol abuse is a face of violence; one of many faces of violence that we will discuss. Soldiers, of course, keep their alcohol abuse a secret ("It is just the Sergeant; he always drinks"). Wong et al. also noted that psychopathology is kept secret. PTSD is. It is well known that applicants for the military with a known history of psychiatric illness are excluded from recruitment and released from service earlier (Fragala & McCaughey, 1991). "Consequently, psychiatric and social difficulties may be concealed from helping agencies within the services" (Wong et al., 2001, p. 111). Thus, in the professional literature, at least Sakinofsky and his team (1996) call attention to alcohol abuse and other problems in the Canadian Forces. These problems require direct confrontation of the secrecy in the green culture.

Therefore, in its official documents, the military in Canada assumed, stated, and so on, that suicide was not a problem. Is that currently true? Again, there is little study. Sareen, Cox, Afifi, Stein, Belik, Meadow, and Asmundson (2007) examined the relationships between combat and peacekeeping operations and the prevalence of mental disorders, self-perceived need for mental health care, mental health service use, and suicidality (suicide behavior). They undertook a cross-sectional, population-based survey of the Canadian military. The participants were 8,441 currently active military personnel (aged 16–54 years). They concluded that deployment to peacekeeping operations, even if adjusted for traumatization of combat friction, was not associated with increased prevalence of mental disorders, such as PTSD or suicide risk. Jetley and Cooper (2010) have also concluded that military suicides are not linked to service. *All is well!*

However, there were major limitations in Sareen and colleagues' 2007 study on how suicide was assessed. They measured past-year suicidal ideation and suicide attempts using two separate questions: (1) "Did you seriously think about committing suicide or taking your own life?" and (2) "Did you attempt suicide or try to take your own life?" What does that mean? This is a huge problem (see Chapter 2). Despite these problems, they reported a prevalence of past-year suicide attempts at less than 1.0%. Suicidal ideation was reported by 3.8% of the sample. However the method calls these Canadian data in question.

What about studies from other countries; do they report low rates in military personnel assigned to peacekeeping duties? Thoresen, Mehlum, and Moller (2003) found that suicide in Norwegian peacekeepers was also very low. Thoresen and his team concluded that "studies of military populations generally report a reduced risk of suicide in comparison to the general population; this is primarily attributed to the 'healthy worker effect,' or the *'healthy soldier effect'* [italics mine] (Kang & Bullman, 1996). The healthy soldier effect refers to the reduced risk of morbidity and mortality in military populations due to such groups being selected on the basis of good health" (p. 606).

Are there academic studies on suicide in the Canadian military in Canada? Yes, there are a few. Belik, Stein, Asmundson, and Sareen (2009) studied suicide attempts in Canadian military personnel. The study was undertaken within the military system. Data came from the Canadian Community Health Survey: Mental Health and Well-Being Canadian Forces Supplement (CCHS-CFS), a cross-sectional survey of currently active Canadian military personnel ($n = 8441$; aged 16 to 54 years; response rate 81.1%). Regrettably, suicide attempts were measured using a question about whether the person ever "attempted suicide or tried to take [his or her] own life." (Assessment of suicide cannot be reduced to one or two questions.) Belik and his team found that the prevalence of lifetime suicide attempts for currently active Canadian military men and women was 2.2% and 5.6%, respectively.

## WAR AND SUICIDE IN THE CANADIAN FORCES

*Suicide is low among Canadian Forces.* Is that true in wartime? Is this *all* true about the Afghanistan War? The truth is that we know nothing (maybe a little) about suicide in the Canadian military. Sakinofsky's 1996 statement regarding no academic studies still holds largely true. We are left with the military's own statistics. I will cite at least two helpful documents on our current question.

Despite the secrecy and deception, a study released by Wong, Le Gras, and Mains (1988) in *Suicide Behaviour Trends in Canadian Forces Training System 1980–1986* (CFTS COSP Report 2/88) states that "little is known about suicide within the Canadian Forces" (p. 1). Wong and his team go on to note that the results and their interpretation are restrictive because of missing cases, limited numbers of participants, and difficulties in the recognition, definition, and documentation of suicides and attempted suicides. They note dissembling. There was deception. They state, "On many occasions the incidents were labeled simply as injuries or wounds" (pp. 17–18).

In Statistics Canada's (2005) *Canadian Persian Gulf Cohort Study: Summary Report*, it is reported that

> deployed veterans had a higher prevalence of self-reported illnesses (including diseases of bones and joints, digestive system, skin, and respiratory system) as well as higher rates of depression, Post-traumatic Stress Disorder (PTSD) and generalized anxiety disorder in comparison to non-deployed veterans The deployed group was also more likely to report a broad range of physical and psychological symptoms. All of these findings were consistent with the results of surveys conducted in other countries. (p. 4)

Do we know more? Tien, Acharya, Donald, and Redelmeier (2010), in their paper, "Preventing Deaths in the Canadian Military," in the *American Journal of Preventive Medicine*, present a very different perspective than the military's. They begin with an explosive statement. Although the Canadian military continues with its story, *the leading cause of death in the military is suicide!* It i second in the U.S. military. The causes of death were as follows: 369 intentional, 289 suicide, 10 homicide, and 70 combat. Yet they ask how to quantify this and how to identify modifiable behaviors that potentially contributed to death. (They hold to the same questions as Dr. Snow, Dr. Satcher, and me). They conclude, "The profession of arms is dangerous" (p. 331). There is ongoing media coverage. We see the mortal threats that military members face during war in Iraq and Afghanistan. During the Vietnam War, we got a snapshot of the harm; who among us, who remembers that war, can forget the execution by a Vietnam commander of the Viet Cong man with a gun to his head? Today, we get a full-length movie/video instantly over social media. It is a massage. Homer and his team examined the causes of death in an entire military force—the Canadian Forces. They undertook a retrospective chart review of all Canadian Forces members who died from January 1, 1983, to December 31, 2007. They included autopsy reports, death certificates, coroner reports, hospital records, military reports, and other miscellaneous sources; a retrospective study (not a PA).

What is the problem? What are the cause(s) of the problem? Tien et al. (2010) examined the underlying causes of death. Modifiable behaviors potentially contributing to death were determined. Individual behaviors contributing to death were analyzed. These included smoking, alcohol use, physical inactivity and poor diet, certain sexual behaviors, suicide, illicit drug use, not using a life jacket in water-transport deaths, and seat-belt usage in motor vehicle-related deaths. Firearms and blast were examined. They noted that

> firearms and blast were the underlying mechanism causing death in about one in 12 deaths ($n = 147$); however, about 50% of these were due to inadvertent training incidents or suicide rather than actual combat. Of the 289 military suicides, 21% ($n = 60$) involved firearm use. (p. 334)

They further noted, "Almost one quarter of all military deaths can be attributed to individual behaviors, the three major ones being suicide, alcohol consumption, and tobacco use" (p. 335). Is there a major cause? As you have now read, alcohol abuse is a problem. Of 289 suicides, 70 were attributed to alcohol. Alcohol abuse is a major factor.

*Therefore, suicide is a problem!* Will we ever learn about suicide in Canadian soldiers and veterans of the Afghanistan War?

## SUICIDE IN THE MILITARY: UNITED KINGDOM, FRANCE, AND YUGOSLAVIA

What do we know about suicides in other militaries around the world? Little. Canada is not unique; *the United States is the current exception*. What do we know, for example, about the UK, France, and Yugoslavia today? Fear, Ward, Harrison, Davison, Williamson, and Blatchley (2009) studied suicide among male regular UK Armed Forces personnel, 1984–2007. It was concluded that despite the UK Armed Forces being subjected to a number of unique occupational stressors, they experienced lower-than-expected numbers of suicides in comparison with the UK general population. It was concluded that the findings were reassuring and little further needed to be done! Desjeux, Labarère, Galoisy-Guibal, and Ecochard (2004) studied suicide in the French Armed Forces. They concluded that suicide in the French military was low; yet maybe specific surveillance may be needed. What about other countries? I here look at one further country, Serbia and Montenegro, parts of the former Yugoslavia. I examined suicide and war in Yugoslavia in Chapter 1. It was concluded that suicide was high and an illegacy of the war. As discussed earlier, there was a history of propaganda in the Communist (collective) culture. There was no suicide; yet our study (Selaković-Bursić et al., 2006) uncovered that suicide was and is high in the region. Is this still true in the military there? Gordana and Milivoje (2007) discussed suicide in the army of Serbia and Montenegro. They report that suicide has been a subject of investigation in the army of Serbia and Montenegro (the former Yugoslav Army). They reported that suicide was and always was low. That is *good* to know! Although they stated that more scientific study is needed, nothing more needs to be done in prevention. Is this propaganda? There is no question, Selaković-Bursić et al. (2006) believe that suicide is, in all probability, high in the Yugoslav Armed Forces.

I do not know how reliable the UK and France data are. There is no basis to assume they are any more credible than Canada's. Are there limitations? However, I do know that Serbia's data are *not* credible. Gordana and Milivoje (2007) stated that the Yugoslavian military have kept records for decades. Yet we know that the data were unreliable and invalid (Selaković-Bursić et al., 2006). Are other countries'? Are suicide statistics in the militaries around the world credible? Reliable? The first step in public health and epidemiology is, "What is the problem?" (Satcher, 1998). Is the surveillance evidence-based? Are there deceptions? Cover-ups? Lies? What is real? How real is "real"?

## CONCLUDING REMARKS ON MILITARY SUICIDE IN CANADA

What is the official military position today? In October 2010, Lieutenant Colonel Rahesh Jetly and Lieutenant Commander Kenneth Cooper presented "The Canadian Forces Expert Panel on Suicide Prevention." Cooper was speaking at the conference of the Canadian Association for Suicide Prevention in Halifax, Nova Scotia. Here are my notes: Cooper reported that suicide rates were low in the Canadian military. He stated it was a male problem. He reported an age adjusted rate of 20/100,000; of note, he agreed that this may be higher than males of the same age in the civilian population. He did admit that in 2008, for example, there were 14 suicides (1 female). However, he argued that Canadians are fortunate because there is no increase in suicide in the Canadian Forces.

unlike the United States. We have no urgent problem. Cooper further added that other psychological problems, such as PTSD, were not an increasing problem in the Canadian military too. Again, unlike American soldiers, Canadian soldiers were insulated by the Canadian military from such traumatization. He noted some possible reasons: excellence of mental health care; effective leadership; military personnel are aware and engaged. Be that as it may, he stated that suicide risk is a leadership issue in the Canadian military, and policies and programs are being developed. He concluded that more could be done and is recommended to the government. Canadians, however, can rest easy, unlike Americans.

When asked about Canadian veterans, in light of the high rates of suicide in U.S. veterans, Lieutenant Commander Cooper stated that he did not know. Veterans are not the responsibility of Canadian Forces. However, he noted that unlike U.S. veterans, Canadian veterans are provided with direct adequate mental health services. He thought that all is well. That is *good* to know!

At the end, Dr. Cooper did call for more research; specifically, he called for a psychological autopsy approach. He stated that recommendations were made for more study and was awaiting the government to take action. They had not done so to date.

The official Canadian military position is that suicide is not a serious problem among soldiers and veterans. The rates are low; indeed, they are lower than the general population. On what basis, however, are the *facts* made?

Jetly and Cooper (2010), despite limited study, stated that suicide among Canadian soldiers is low and not increasing. However, I have some questions about their facts. Why is it that in the United States, the rates are high and increasing? We know that suicide in Canada is higher than in the United States (Leenaars & Lester, 1994). We know that in young adults, the age of most soldiers, the rate of suicide is high. Further, in Canada, the rate of suicide for this age group is 50% to 70% higher than in the United States; this is a significant difference. Would this not impact on the young soldiers?

So far my questions are at a population level; however, I can ask questions at the individual level too. As discussed in Chapter 2, soldiers with traumatic brain injuries are at increased risk for suicide. This is well documented. It is known in the United States (Department of Defense, 2010). What about Canada? Little is published in the academic/professional literature on the topic. However, in the media, Julian Sher (2011) reported, "Canadian soldiers in Afghanistan were hospitalized from traumatic brain injury between 2006 and 2009 at almost three times the rate of American fighters there" (p. A6). The Canadian rate was at about 71/100,000, compared with an American rate of 25/100,000. These figures are based on Canada's own National Defence; the Canadian Forces, in fact, acknowledged this fact and attributed this to "the risky nature of our Kandahar operation." Okay, so TBI is a significant suicide risk factor (Brenner, Ivins, Schwab, Warden, Nelson, Jaffee, & Terrio, 2010); Canada's soldiers have 3 times the incidence of TBIs; and Canada's rate of suicide is lower. It just does not add up. The logic goes something like the following: All soldiers with TBIs are at increased risk for suicide. Canada's soldiers have 3 times the TBIs than American soldiers. Therefore, Canada's soldiers are at decreased risk for suicide. This, for me, is the same constricted logic of a suicidal person.

Furthermore, Jetly and Cooper (2010) stated that Canada's rates are lower in the military than the general population. Are the data credible? They were not in the United States for decades; this has been rectified only recently. Why not in Canada? Having taught statistics, something just does not add up for me. Why would Canadian soldiers be less traumatized? Less suicidal? Canadian soldiers were in significant harm's way in Afghanistan, with deaths and horrible casualties. Canada's soldiers had 3 times the incidence of TBIs. War is stress. It would be normal to be traumatized. So why the difference? It has to do with surveillance, I believe (see Chapter 8).

Further, do our veterans officially have low rates? In the United States, it is now concluded that veterans have a rate twice as high as the general population. But in Canada, they have low rates. I question the data, and I believe that it is, thus, premature to make any statements of facts in Canadian soldiers and veterans. Isaac Sakinofsky et al. (1996) call for professional study still must be met.

# CHAPTER 8
# Surveillance and the Reliability of Military Suicide Statistics

Public health is about surveillance and reliability of statistics. This is essential to predict and prevent a problem. This view is not new; indeed, advocates of this approach (Satcher, 1998) often point to a mid-19th century example as epitomizing the approach. In 1854, a major cholera outbreak occurred in England. Many people were dying; thousands were hospitalized throughout England. Physicians and other health workers responded with health care, but one doctor, Dr. John Snow, started to investigate the epidemic. He asked questions of the patients; he asked where they had been, what they ate, what they drank, and so on. After sampling hundreds of patients, he found a common link: They all drank water from London Broad Street pump. He then did something quite unusual: he went out into the field. He went to the Broad Street pump. He studied the pump; studied the water, and so on. Dr. Snow concluded that the water was contaminated. He discovered that a sewage line was contaminating the water, in fact. This was public health research at not only its first, but also its best. But Dr. Snow did more; he intervened, and, to the upset of many, he removed the handle from the pump. Soon after that, the cholera epidemic ended. Dr. Snow is a father of public health; his work is a classic in our field. How can we apply Dr. Snow's approach to surveillance of military suicide?

## A HISTORICAL LOOK BACK AT MILITARY EPIDEMIOLOGICAL DATA

In 1980, Edwin Shneidman published a paper, "The Reliability of Suicide Statistics: A Bomb-Burst" in *Suicide and Life-Threatening Behavior* (1980b). In this insightful article, the reliability of all demographic and epidemiological suicide data in the military is questioned. The results are judged to represent an unsatisfactory degree of reliability for scientific purposes. Because of Shneidman's conclusions, I present a few of his comments verbatim on William E.

174 / SUICIDE AMONG THE ARMED FORCES

Datel's (1979) article, "The Reliability of Mortality Count and Suicide Count in the United States Army," which appeared in *Military Medicine*:

Datel's article has implications of the first magnitude for the reliability of all demographic and epidemiological suicidal data—from the beginning of suicidal statistics (with Graunt and Süssmilch) and in every locale in the world. That four page article is a major bomb-burst.

In brief, Datel's succinct article tells us that, historically, there are two independent systems within the U.S. Army that report all U.S. soldier (officer and enlisted personnel) deaths, including of course suicide, to the Pentagon. The two systems are the medical arm of the Army, the Office of the Surgeon General (OTSG) and the forensic and legal arm of the Army, the Office of the Adjutant General (OTAG). Datel states: "The two data systems are separate and distinct from each other, arising from differing sets of requirements and performing different missions." One system was begun in 1971, the other in 1961.

Datel had the brilliant idea of simply comparing the two lists of Army deaths for two calendar years, specifically 1975 and 1976. . . .

Under "Findings," Datel indicates that the death count from the Army medical data base (OTSG) was 1,828; the death count according to the OTAG's accounting system was 2,178. "Common to both death lists was a total of 1,561 persons. Appearing on one list only was a total of 884 persons, 267 of whom were unique to the OTSG list and 617 of whom were present only on the OTAG list."

The first two paragraphs of Datel's "Discussion" section read as follows: "How many American soldiers died during the years 1975 and 1976? The Adjutant General reported 2,178 deaths, some 16 percent more than the Surgeon General count of 1,828. Together, both systems reported 2,445, a gain of 11 percent over the higher single-system count. Thirteen more deaths not recorded in either system were also discovered.

*"There was only 63.3 percent commonality in the mortality lists of the two independent electronic data processing systems studied. . . .* [italics added]

These results are judged to represent an unsatisfactory degree of reliability for scientific purposes. Presumably, they are also unsatisfactory for administrative, planning or legal use of the information as well." . . .

If mortality cannot be counted reliably, how then must go morbidity? As regards death, we have a quite hard criterion for case definition; case finding in morbidity is much more influenced by observer differences. How reliable are our counts of mental illness? Of hypertension? Of battle wounds?

Attribution of suicide is a cause of death question—not unlike matters of differential diagnosis. To ask how reliable was the obtained count on the suicide, then, is to inquire into the data reliability on a matter akin to that involved in the study of morbidity. The results give little cause for rejoicing.

That would seem to be the essence of Datel's interesting findings. If any reader has concluded that this extended reference to Datel's article here in this journal is meant as an indictment of the U.S. Army system—which admittedly needs to get its mortality-count act together—that reader would have missed the main point entirely. This article is a *caveat emptor* to all users of all mortality statistics, especially of statistics relating to suicide, however carefully those data seem to have been generated. The results of Datel's brilliant study would seem to have profound and systematic implications for all suicidologists as to the best possible tactics, in general, for us to pursue. (pp. 67–69)

There was little cause for rejoicing. Is this still true for military statistics today? Are the data in the United States and Canada reliable? That is the question. There are a few answers.

## RELIABILITY OF CURRENT MILITARY REPORTING

Carr, Hoge, Gardner, and Potter (2004) present a study, "Suicide Surveillance in the U.S. Military—Reporting and Classification: Biases in Rate Calculations," in *Suicide and Life-Threatening Behavior*. Carr et al. recognize that surveillance is everything, namely, because monitoring of suicide rates and prevention programs requires accurate reporting and classification of suicides. It is well known that there is an *a priori* problem in nonmilitary (civilian) data. Is this true in the U.S. military? Carr and the team evaluated whether suicides were underreported or misclassified under *accident* or *undetermined* manner of death (NASH) in the military system. They reviewed all 1998 and 1999 military deaths using official death reports and compared these data with additional sources, most importantly the Department of Defense Medical Mortality Registry. They assessed for evidence of expressed suicidal intent and past psychiatric history among deaths classified as *undetermined* and *accidents* due to gunshot, overdose, drowning, falls, or asphyxia. They used sources other than official records and found 17% more suicides than were reported and an additional 4% of deaths that were suspicious for suicide. They concluded, thus, that reporting and classification errors might account for 21% additional suicides in the military. This is too many. However, as Shneidman warned, the findings are comparable to rates seen in civilian studies. There are problems inherent in using administrative death classification data for medical surveillance purposes. There are problems with the NASH classification itself. Further, suicide rates in branches of the military fluctuate up to 30%–40% from year to year, making it difficult to monitor. There are potentials for classification bias, much like Shneidman noted 25 years earlier. Can something be done?

There are further issues. Carr et al. (2004) also did not investigate motor vehicle accident deaths, and some of these may be cases of self-harm. There are many faces of suicide. Overall, we can conclude that suicides were underreported in the U.S. military. We know, however, nothing about Canada. Yet it is a well-known problem. There is no basis to conclude that the inherent problems found in the U.S. military in the 1970s and 1990s would *not* occur in Canada. What can be done? Dr. Snow did something about such a problem 150 years ago.

## SURVEILLANCE, INVESTIGATIONS, AND RESEARCH: A MILITARY VIEW

Today, the U.S. military is following Dr. Snow. At least, the Department of Defense (2010) has called for better surveillance; they write,

> Suicide prevention requires a public health approach, and effective surveillance is essential to public health benefits. The Task Force strongly believes that much can be learned from standardized data collection and analysis that addresses risk factors and trends, and assists in identifying decision points for targeted interventions. Well-constructed surveillance that ideally leads to a predictive model can inform and shape future suicide prevention programs and efforts as well as improve public health efforts. Investigations must be standardized in order inform surveillance efforts. (p. 38)

Public health approaches offer answers. Eaton, Messer, Wilson, and Hoge (2006) suggested strengthening the validity of population-based suicide rate comparisons. They suggest that we generate precise estimates of suicide rates in the military while controlling for factors contributing to rate variability such as demographic differences and classification bias, and to develop a simple methodology for the determination of statistically derived thresholds for detecting significant rate changes. They suggest that direct adjustments can control for the demographics. They purposely used complex statistical techniques. For example, they demonstrated that by applying a Poisson-based method, more reliable data could be found. It is possible.

Next, I offer some personal criticisms in order to better understand the issues and problems of suicide among the military and veterans:

1. It would be an inaccurate comparison of the military with the general population (the general population would include high-risk groups, unemployed institutionalized, incarcerated, and mentally ill).
2. Armed services personnel are screened by psychological assessment medical evaluations and so on, at the time they are accepted into the military. If we assume that few of the military personnel and veterans committed suicide had a diagnosable preemployment mental disorder (or psychopathology), it may not be a fair comparison to look at their rate of suicide compared with the civilian population. The intent of the evaluation, of course, is to screen out many high-risk recruits, such as people with preexisting mental health disorders. If a similar screening were used with the civilian comparison group, the civilian suicide rate would be lower, of course.
3. Since death certificates are typically used, it is likely that the numbers of veterans and armed forces personnel who died by suicide are underreported

4. The review of the literature is inadequate; we need to do more. A comprehensive review would show the rate of suicide in the military is inconsistent and inconclusive, namely, because of methodological limitations; we need better surveillance.

Therefore, I believe additional reliable surveillance can help clarify the elements to suicide risk among armed service personnel and veterans. Can we do as well as Dr. John Snow? Or must we accept William Datel's mess?

# PART FOUR

# Beyond Suicide

In 1980, Dr. Norman Farberow, a pioneer in American suicidology, edited an insightful book, *The Many Faces of Suicide*. He speculated that there is a vast array of subintentioned deaths that occur beyond intentional suicides. The WHO (2002) stated the same about violence. There are many faces of violence; some are self-directed, and some are other-directed; and some are lethal, whereas most are not. Not only suicide, but also alcoholism, accidental deaths, self-harm, incarceration, and some homicides can be seen through the lenses of violence and suicide. In this section, I examine the many faces of violence.

In the chapter, I look beyond direct self-inflicted death, again from the WHO's ecological perspective. This chapter broadens the scope of the book. I examine the many faces of violence in the military, including some of the most concerning in soldiers and veterans; that is, homicide, accidental deaths (especially single-car motor vehicle accidents), self-harm, and incarceration. There is a stream of violence. There are, as you will read, a great many commonalities among the many faces of violence, and solutions. Suicide prevention, in fact, has to be included in a larger frame of violence prevention and well-being promotion in the military. Suicide has to be a specific target, but also, to only address suicide in health promotion is ineffective. The military has to look at suicide and beyond suicide. They need a larger well-being systems approach.

CHAPTER 9

# The Many Faces of Violence: Homicide, Accidental Deaths, Self-Harm, and Incarceration

War-related death is violence. Suicide is violence. Homicide is violence. Suicide is self-directed violence. Homicide and war-related death are other-directed violence. They are lethal violence. Suicide, homicide, war-related death, and other violence have probably always been part of the military experience. In this chapter, we will look into homicide, but also other forms of violence—especially self-directed ones. Suicide is not the only risk-taking behavior. Strom, Leskela, James, Thuras, Voller, Weigel, Yutsis, Khaylis et al. (2012) have reported that the two most frequently reported self-destructive risk-taking behaviors by veterans are suicidal ones and aggressive driving (e.g., accidents). Other elevated rates are alcohol/substance use, thrill seeking, aggression/violence, risky sexual practices, and firearm possession. We will look into accidental deaths, self-harm, and incarceration. I will begin, however, with other-directed violence: homicide.

## SUICIDE AND HOMICIDE

"What is homicide?" is an age old question (Allen, 1980; Henry & Short, 1954; Unnithan et al., 1994; WHO, 2002). Death is superordinate to homicide. Like suicide, homicide is one category of the four universally recognized modes of death—Edwin Shneidman's (1985) NASH categories: natural, accident, suicide, and homicide. There are different types of homicides, some of which can be unintentional, such as accidental homicide. Homicide, defined by intention, is murder. Murder is the intentional use of physical force or power, threatened or actual, against another person or community that results in death (WHO, 2002). Again, like in suicide, intentionality is central. This type of homicide is chosen on purpose. Although there is a problem in clear definition, in Chapter 2, we defined suicide. Following that definition, we can then define intentional homicide as a conscious act of other-induced annihilation, best understood as a multidimensional event in a needful individual who defines an issue for which the homicide is perceived as the best solution. Of course, the definition has complexities. Like in discussions on suicide, a question that can be asked is, When is homicide a conscious intentional act of other-directed cessation? These issues are beyond the scope here, and we advise the reader to look at the WHO's (2002) discussion on the topic.

As stated earlier, it is estimated that 1.6 million people die by violence each year; one-third are homicides (530,000). No single factor or event explains why so many people die by homicide. Like suicide and war-related deaths, homicide is the result of an interplay of individual, relationship, social, cultural, and environmental factors. Therefore, the WHO (2002) believes that if we also want to understand homicide, we must take an ecological view. Furthermore, from this conceptualization, a number of questions on our current topic of suicide and homicide include, Is suicide the same as homicide? Or is it different? How? Are people in the United States, from, say, the Marine service, so different from other Americans when they kill themselves? What about Canadian soldiers and veterans? Are there commonalities? What about homicide? Are there shared factors? And what about the other many faces of violence?

According to the WHO (2002), "While some risk factors may be unique to particular types of violence, the various types of violence more commonly share a number of risk factors" (pp. 13–14). There are associations between suicide and several other types of violence, especially homicide (Allen, 1980; WHO, 2002). They are associated in multifaceted ways, and by implication, because violence is a multifaceted problem with biological, psychological, social, and environmental factors, it needs to be confronted on several different levels at once. The ecological model guides us. And, on one more point, from the model's implications, violence is largely preventable (WHO, 2002, 2006). Yet the question can be posed, Are suicide and homicide related? How are they associated? How can we understand homicide and suicide among soldiers, even Majors?

The public typically distinguishes between homicide and suicide. They are seen as fundamentally different. Indeed, military commanders, professionals, coroners, politicians, the media, and such distinguish between them. Yet there is a different way, as we are already hinting at, to understand the behaviors. Homicide and suicide may be more similar than often believed. There may be commonalities (or common factors).

In the Western world, suicide and homicide were not always seen as different. Early Christian thought made no distinction. St. Augustine categorically saw suicide and homicide as the same; like homicide, suicide violated the commandment, "Thou shalt not kill." Only God had power over man's life and death. Homicide and suicide were greater sins than any other. This was not only true for Christians but also for many other religions, for example, Muslims, Hindus, and Buddhists. These views continue today. Suicide, in the Western world, in fact, is a relatively new term, probably dating from the 1600s. Before that, the words used were self-killing or self-murder.

In science, in the 1800s, Enrico Morselli and Enrico Ferri saw homicide-suicide as having the same underlying principle. Sociologist Emile Durkheim criticized this position (1951). Sigmund Freud (1974g), however, held the same view as Ferri and Morselli. Freud saw suicide as homicide turned inward on the self (see Chapter 2). Freud (1974g) noted,

> Just as the suicidal person may be attempting to kill an introjected ambivalently regarded other, the murderer may be killing an object of projection; the other, then is one in whom one sees one's own badness. The close linking of suicide and murder is seen in the mechanism of seeking to be killed, to be punished for one's own transgressions particularly for one's own murderous feelings. (as cited in Allen, 1980, p. 86)

Of course, many other scientists hold to Freud's (1974g) or Ferri's (1917) view. Yet it was probably Andrew Henry and James Short (1954) who constructed the best-known theoretical explanation of the relationship between homicide and suicide (Unnithan et al., 1994). Of course, there are differences between homicide and suicide, but there is much to be gained by following Henry and Short. Despite accepting unique differences, both general and specific, we need a unifying model, a best fit possible (not perfect) at this time. Unnithan et al. (1994) stated the basic argument as

> That although there are disagreements between homicide and suicide, there is much to be gained from revitalizing the theory developed by Henry and Short. Specifically, there are numerous issues related to lethal violence that can be better addressed—and, in some cases, understood—by working from an integrated model that emphasizes the similarities between self-directed and other-directed violence.... We are not, however, advocating a cessation of research that views homicide and suicide as distinct behaviors. Depending on the topic of investigation, this approach may be entirely appropriate and reasonable. Our contention is that for many research questions related to human violence, the goal of explanation will be better served by a theoretical model that explicitly takes into account the connections between homicide and suicide. (p. 5)

Suicide is self-induced annihilation. To understand suicide, as we have stated, it is useful to understand related topics of violence, especially homicide. In fact, suicide and homicide can be seen as interwoven expressions of the same stream, called the *stream analogy* of violence. Both can be conceptualized as a lethal response to experience of trauma and the frustration of needs (e.g., honor, victory), differing only in the direction in which the response is expressed.

The *stream analogy* of lethal violence is not new to this century. In the 1800s, the two Italian scholars, Ferri (1917) and Morselli (1882) held this belief. Many did not agree then, or today. Durkheim (1951) espoused that "suicide sometimes co-exists with homicide, sometimes they are mutually exclusive" (p. 355). In science, Copernicus was not the only great scientist silenced. For approximately a century, Durkheim's view dominated sociology and forensic study; Freud in psychology, however, held to a view consistent with the stream analogy. In 1954, Henry and Short resurrected the view.

Historically, Ferri (1917) and Morselli (1882) noted that different societies (or nations) had different rates of suicide and homicide and raised an important question. Why do persons in some social groups kill themselves more while other groups commit murders more? For example, why do Americans kill others more and themselves less compared with Canadians (Leenaars & Lester, 1994)? Given their close geographic and cultural proximity, that fact is worth remembering by all commanders in the Armed Forces in those two nations. (There are many other differences in suicide and homicide in those two nations.) Thus, do not assume, around the world, that a suicide is a suicide or that a homicide is a homicide (Leenaars, 2007).

Unnithan et al. (1994), following Henry and Short (1954), suggest that there is a stream of available destructiveness in a society (or culture). Ferri (1917) called this production. They propose that this can be measured by what they call lethal violence rate (LVR). LVR measures the size of the stream of violence (LVR = Suicide [S] + Homicide [H]). The direction of the stream, as Ferri called it, can be measured, on the other hand, by the suicide-murder ratio, SHR, which gauges the proportion of the total, expressed as suicide rather than homicide (SHR = S/[S + H]). Unnithan et al. suggest that these two measures allow one to understand the specific lethal violence in groups. From an integrated model, they write, "At the individual level, both forms of lethal violence result from a combination of negative life events (frustration, stress) with attributional styles that locate blame either in the self (suicide) or in others (homicide)" (p. 94).

Prediction is based, they argue, on attribution, which is a product of situational and cultural factors. The military culture would highly add to *attribution*. Ferri (1917) and Morselli (1882) had espoused biological factors (degeneration, impotence or decay of the organism), whereas Unnithan et al. (1994), like Henry and Short (1954), present a more social meaning view, and where Freud presented a more individual view (see ecological model).

The importance of this model is that the stream analogy of violence resurfaces for us to understand military suicide, which is something that allows us to better understand annihilation, whether homicide or suicide, or, for that matter homicide-suicide. On a footnote, it can be asked whether this is true for males and females. The answer appears to be yes (Stack, 1997). To conclude, a question raised by David Lester (1987) on homicide and suicide was, "Are they polar opposites?"

Lester (1987) concluded,

> All murderers are not alike . . . as distinguished between the overcontrolled murderer (who is calm, peaceable, and unaggressive much of the time, but who explodes occasionally into dramatically violent behavior) and the

undercontrolled murderer (who is continually assaultive to the least frustration or insult). Clearly, these two kinds of murderers handle their aggressive impulses very differently. Similarly, suicidal people differ. Some are chronically suicidal, making repeated suicide attempts before, in some cases, eventually killing themselves. Others who kill themselves have no history of prior suicidal behavior.... These two types of suicidal individuals may be appropriately characterized as undercontrolled and overcontrolled, respectively. Identification of types of murderers and suicides may eventually lead to a better understanding of which factors (intrapsychic, interpersonal, and situational) lead some individuals to direct their aggression outwards while others turn aggression against themselves. (p. 59)

Before I turn to the very sparse literature on hand in the military, I want to discuss homicide-suicide in more detail.

## HOMICIDE-SUICIDE

Suicide is a multidetermined event. Homicide is a multidetermined event. Thus, it follows that homicide followed by suicide is also not determined by one factor. I cannot here provide all of the literature, but allow me a few signposts. Here is what we have learned about homicide-suicide in the general population. (There is no study in the military.) The early theorists on homicide-suicide were West (1966), Wolfgang (1958), and Allen (1980). Allen, for example, studied 104 homicide-suicides in Los Angeles during 1970–1979. The rate of homicide-suicides was 2.0% of all homicides and suicides. However, West (1966) reported a rate of 33% of all homicides and Wolfgang (1958) reported a rate of 4% of all homicides (One has to read carefully; the classifications and comparative groups are not always the same). There are further findings of note in Allen's Los Angeles study: 93% of the offenders were male; 80% of the victims were female. (As an unusual tangent, in the West study, 40% of the offenders were female. That is high.) The majority (71%) of the killers were husbands/ boyfriends of the victim (2% of the women murdered their children, a subgroup of homicide-suicides.) About 20% of the victims (4 males and 7 females) and 21% of the offenders (16 males and 1 female) were intoxicated (having a blood alcohol level of 0.10% or higher). Traces of alcohol were found in 29% of the tested victims and 34% of the total offenders. Around 50% of the victims and murderers had been drinking. Therefore, once more, alcohol increases risk, not only for suicide, but also for homicide, and homicide-suicide (Allen, 1980). It can be explosive.

Steven Stack (1997), a prolific sociologist, has provided the best presentation on the known research on homicide-suicide, and I encourage a read of his study. Stack provides a long list of correlations: frustrated personal relationships, ambivalence, jealousy and morbid jealousy, separation, depression, helplessness, and guilt—an almost endless list of factors. The similarities between these observations and the ones that I made earlier on suicide is striking. There may well be a stream of violence. Homicide-suicide and suicide are not polar opposites. They are highly correlated, but also they may be less so to homicide. Homicide-suicide is more like a suicide than a homicide. Yet research on homicide-suicide in soldiers, like on military suicide, is lacking. I located no articles on homicide-suicide in soldiers. Thus, allow me to go into more detail regarding Stack's important study.

In 1997, Stack studied the records of all homicides in the murder files of the Chicago Police Department from 1965 to 1990. Stack attempted to correct previous methodological problems by having at least a homicide comparison group. Regrettably, he did not have a suicide comparison group. In the archives, Stack found 267 homicide-suicides. In his analysis, Stack controlled for socio-demographic variables (possible confounders). For the killers, these variables were male, age, and Caucasian race. For the victims, female, age, and Caucasian race. Statistical procedures simple bivariate analyses, multivariate and logistic regression analyses, were undertaken, representing a sound array of comprehensive statistical techniques.

The incidence of homicide-suicides to homicides in Chicago was 1.65%, low by worldwide standards. Stack (1997) concluded,

> The structural relationship that increases the odds of homicide-suicide the most is that of ex-spouse/lover. For ex-spouses/lovers, the risk of homicide-suicide is 12.68 times higher than in nonintimate homicides. In these cases, the bond that once held a couple together is officially broken. The loss is final; it is no longer just threatened or simply coming in the future. For persons very dependent on the old bond, this loss of their love object can be unbearable. For some ex-spouses/lovers jealousy intensifies after the breakup as they perceive that their former love object is involved with new partners, persons who have taken their place in love. [The bond (attachment) to the military may be no different—AL] As anticipated, the current girl-friends or boyfriends of killers are at less risk (6.11 time higher odds of homicide-suicide) than spouses and ex-lovers. Here the bonds are not quite as intense and so the loss of support is not as great. Further, since the couple was not living together, it is less likely that a killer would feel that "I can't live without you" relative to someone he or she is already living with or once did live with under the same roof. (pp. 447–448)

In relation to our previous discussion on Unnithan et al.'s (1994) work on homicide and suicide, Stack (1997) stated,

> This study has some implications for hypothesis formation in the general area of homicide-suicide. Unnithan et al. (1994) have proposed a "stream analogy," wherein the choice between homicide and suicide depends on attributional concerns. Persons and groups faced with frustration will choose suicide to the extent that they attribute the cause of their problem to themselves and to the extent that they are depressed and feel helpless. Other groups

and individuals will opt for homicide if they tend to attribute the cause of their problem to others and to the extent that they feel angry as opposed to depressed. Drawing on a dozen qualitative studies on homicide-suicide, this study contends that the principal source of frustration in homicide-suicide is a frustrated, chaotic, intimate relationship marked by jealousy and ambivalence. These relationships are marked by a feeling that one cannot live with the other person but cannot live without them either. A separation or threatened separation arouses anger and depression at the same time. The act of homicide overcomes a sense of helplessness. However, the associated depression and guilt over the loss of one's love object result in suicide. Perhaps homicide-suicide can be best thought of as containing both attribution styles discussed by Unnithan et al. (1994). (pp. 448–449)

Probably the best Canadian study on the topic is by Jacques Buteau, Alain LeSage and Margaret Kiely (1993), who studied homicide-suicide in Quebec from 1988 to 1990. Buteau and his team examined 39 consecutive cases of homicide-suicide. Sociodemographic data, circumstances surrounding the event, and clinical data were recorded. Yet they noted limitations to the data and called for psychological autopsies. One important homicide-suicide, the case of 14 victims of the mass murder-suicide at the Polytech in Montreal in 1987, had a large impact on the data. These events make a difference. The victims of the Columbine shootings in Colorado would be another extreme example. The thousands of victims of 9/11 make a huge difference. These mass homicides-suicide(s) are different, but also the same (see Chapter 4). However, Buteau and his team's study lacked utilization of proper control comparison groups. Buteau and his team further concluded that coroner's files, police reports, and so on, lack rich psychological data. The necessary, unequivocal, evidence is "scarce." They state, "Psychological Autopsies, well known in suicide studies, is the method of choice."

Suicide as aggression turned inward has been presented; it is a well-accepted fact. However, the intense role of aggression in homicide-suicide appears to be equally critical, but also not well known. As one homicide-suicide perpetrator said, "If I can't have you, no one can." This is common, aberrant thinking, a lethal major premise. It easily follows, "I can't have you. Therefore, no one will. I'll kill you."

I would be remiss if I did not highlight the importance of domestic violence in homicide-suicide. It may not be universal, but at least common. Of course, domestic murder-suicide is the ultimate example of domestic violence. The fact that it is almost always a male phenomenon needs to be better understood. Why is it that men, in particular, are so determined not to allow their partners to leave? Why would the male offender feel jealousy and hatred to such extent that they would rather kill the one they profess to love and die themselves than to accept that their partner no longer wishes to be part of their lives? (Why would this be in a veteran?) Like others, Barnes (2007) suggests that these are urgent women's issues. This is true in the military.

A final fact: suicide of a loved one has often been considered a suicide risk factor. This appears to be true for homicide-suicide as well. There may be a contagion. Did it occur after 9/11? The main conclusion from the research on homicide-suicide is that they are most like suicides. Homicide-suicide is an extension of suicide. I believe this would be true, of course, in the military.

## SUICIDE AND HOMICIDE IN THE MILITARY

We know almost nothing about homicide in the military. Although disappointing in detail on homicide, McDowell, Rothberg, and Lande (1994), in a beginning paper entitled, "Homicide and Suicide in the Military," noted that all members of the military learn to kill. The military is sanctioned to use lethal violence. Like suicides, homicides were, however, historically regarded as rare events in the military. This has changed since the end of the Vietnam War. McDowell et al. (1994) further noted that these incidents are not isolated from other faces of violence. They offer the following case:

> A 40-year-old O-4 was expected to appear in federal court to answer to charges of ordering and receiving child pornography. He was unable to keep this information from his superiors and feared public disgrace and the loss of his military status. Instead of appearing in court, he shot himself in the head with a 9mm pistol. (p. 107)

Are legal charges a risk factor? Admiral Boorda feared such. Is incarceration a major risk factor? McDowell et al. (1994) offer the following observation:

> A small number of U.S. Air Force victims (about 12%) were involved in difficulties with law enforcement agencies at the time of their death. About one half of those were under investigation for a suspected criminal offense, and about one half were involved in some fashion with local law enforcement agencies. Being under investigation for a suspected criminal offense, especially if the crime involves moral turpitude, is extremely stressful. This is because the legal outcomes are difficult to anticipate, and many suspects expect the worst. Legal problems almost always negatively influence one's career as conviction in court is also grounds for administrative action by the military. Thus, military members facing serious legal problems must also worry about public disgrace and a very real threat to their military careers. For many, this is simply too much to endure. (p. 107)

McDowell et al. (1994) offer further factors as well, such as interpersonal problems and financial ones, and they raise an important question for prevention of military suicide and homicide.

> Would it make sense to offer programs to help soldiers and survivors deal with failed relationships and financial, substance abuse, accidents, incarceration, and work related problems? Like almost all suicides, our heroes do

not die by suicide because they want to die; they kill themselves because they cannot cope with their problems, and suicide is a vehicle for making the problems go away. (McDowell et al., 1994, p. 112)

Is this true for homicide? McDowell et al. (1994) offer little insight.

Mahon, Tobin, Cusack, Kelleher, and Malone (2005), studied the suicide incidence and factors in a retrospective military cohort study comprising all deaths ($N = 732$) of regular-duty military personnel in the Irish Defence Forces between 1970 and 2002. A retrospective, case-control study using pair-matched military comparison subjects was conducted. Risk factors identified included the common psychopathologies, but also faces of violence, such as a past history of deliberate self-harm. This raises a question: Is suicide risk increased in the military because of possible self-selection of more aggressive individuals? Aggression is a common suicide risk factor, particularly in young male subjects. Are soldiers more aggressive than the general population? Mahon et al. hypothesized that suicide risk in the military, as in other occupations such as policing, involves constitutional, environmental, and occupation specific risk. It is multi-determined. Death due to accidental discharge of a weapon is another incident and so on, particularly alcohol abuse problems (alcoholism). These are faces of violence. Homicide is no different.

Allen, Cross, and Swanner (2005) agree; they note that there distal factors of violence that are longstanding and enhance vulnerability to suicide. They add the inability to adjust. Yet there are also proximal factors of violence that are more immediate and influence the particular timing for a suicide attempt. Accidents and incarcerations are two proximal risk factors. Sexual assaults and harassment are also proximal factors. There are many possible determinants, but some are more problematic, and I discuss this in this chapter and the one on PTSD specifically. Ritchie, Benedek, Malone, and Carr-Malone (2006) especially note that sexual assaults are a possible factor. Other examples of violence include drug and alcohol abuse, rape trauma, and battered spouse syndrome. There are many faces of violence associated with suicide and even homicide in the military. Ritchie et al. (2006) offer the following case.

> A soldier was found deceased in his quarters, when he did not show for work after a long weekend. An empty whisky bottle was found next to him. There were empty bottles of fluoxetine and zolpidem in his medicine cabinet. The soldier had no known major romantic, legal, or occupational issues, and there was no suicide note. The civilian medical examiner, who had jurisdiction because the death was off-post, ruled it a suicide. His family disagreed and requested a second opinion. The medical examiner requested a psychological autopsy to opine whether it was an accidental overdose resulting in death or a suicide. At the time of the request, alcohol toxicology levels had returned but other toxicology were pending. (p. 704)

Colonel Elspeth Ritchie and colleagues (2006) continue with the case:

> Because of the decomposition of the body, the alcohol level of 0.3 was hard to interpret. Toxicology showed no traces of fluoxetine or zolpidem. However, there were high levels of barbiturates. A repeat search of his room by CID found trace evidence of barbiturates in the pill bottles. There were also traces of mefloquine, an antimalarial he had been taking while in Afghanistan. An interview with an ex-girlfriend revealed that the soldier had been e-mailing her obsessively. This was corroborated by a search on his computer. His e-mails became increasingly bizarre, depressed, and desperate. Based on this additional informational, the death was ruled a suicide. The source of the barbiturates was never found. (p. 705)

War is trauma. There will be aftershocks. There will be many responses of violence. There is a stream of violence. Ritchie et al. (2006) note, "Some military members will develop chronic and disabling mental illness as a result of traumatic exposure and exacerbated by the demands of the austere and dangerous operational environment" (p. 705).

There will be violent and aggressive behavior in the aftermath of deployment. What about homicides? There has been little study among armed forces around the world. We need to understand the homicidal soldier better.

Hill, Johnson, and Barton (2006) offer an overview on military homicide and suicide in harm's way. They undertook a chart review of 425 deployed soldiers seen for mental health reasons. They found that 127 (nearly 30%) had been suicidal and 67 (nearly 16%) had been homicidal within the past month. *That is huge*! Who are these homicidal soldiers? I next present briefly the case of Major Nidal Malik Hasan.

## MAJOR NIDAL MALIK HASAN

The typical (if there is such) homicide among the armed forces may be like the following case. On February 12, 2010, Corporal Joshua Caleb Baker, a Canadian soldier, died at a training range in Afghanistan. It was reported to be an undetermined death. Suicide was ruled out. An investigation was undertaken. It was determined to be manslaughter, and this is a homicide under the NASH classification scheme. On Wednesday, June 29, 2010, Major Darryl Watts and Warrant Officer Paul Roversdale were charged in relation to the death. The charges were manslaughter, unlawfully causing bodily harm, and negligence of a military duty. It was alleged that the proper safety procedures were not followed. The case is now before the military judicial system. It is a typical homicide. Of course there are more atypical homicides in the military. Probably the most traumatic a perfect storm for PTSD, was the deadly massacre at Fort Hood, Texas, on November 5, 2009. These are the murders by Major Nidal Malik Hasan.

On November 5, 2009, a U.S. soldier shot and killed 13 people and wounded 33 at the Soldier Readiness Center at Fort Hood, Texas. The gunman shouted, as he opened fire, "Allahu Akbar" ("God is great"). Sergeant Kimberly Munley and Sergeant Mark Todd (American heroes) confronted Malik and exchanged shots. The terrorist was hit and felled by Todd and Munley's shots. They arrested the mass killer. It all took 10 minutes; it was one of the most deadly mass murders in the U.S. Armed Forces (Wikipedia, 2010).

The killer was Major Nidal Malik "Abdu Wali" Hasan, a U.S. soldier. Major Hasan was born on September 8, 1970, in Arlington, Virginia. His parents were Muslim Palestinians who emigrated to the United States from Al-Birch in the West Bank (McKinley, 2009). His father was Malik Awadallah Hasan (deceased April 16, 1998). His mother was Hanan Ismail Hasan (deceased May 30, 2001). (Were these distal factors relevant?) He has two brothers.

Major Nidal Malik Hasan was raised primarily in southwestern Virginia. He attended Arlington's Wakefield High School for his freshman years, and after his family moved to Roanoke in 1985, he attended and graduated from William Fleming High School. His parents operated a family restaurant in Roanoke; Hasan and his brothers worked at the restaurant. There is no known legal or unusual event in the history.

Nidal Hasan, after graduating from high school, joined the U.S. Army. He served 8 years as an enlisted soldier. He attended Barstow Community College in California, Western Community College in Roanoke, and Virginia Tech (an institution that is no stranger to mass murder). He graduated from Virginia Tech in 1995 with a bachelor's degree in biochemistry with honors and a minor in biology and chemistry. In 1997, he begun studies at the Uniformed Services University of the Health Sciences in Bethesda, Maryland; he graduated in 2003 as a military doctor. He did his residency in psychiatry at the Walter Reed Army Medical Center (Virginia Board of Medicine, Practitioner Information, 2010). He was licensed to practice medicine, specifically in psychiatry (License # 0101238630), issued July 12, 2005. In 2009, he studied Disaster and Prevention Psychiatry at the Center for Traumatic Studies (an irony for us in the field of trauma treatment); he is listed as a Fellow at the Center.

He was commissioned as a Captain and was promoted to Major. There were no records of physical or mental problems during his service (Washington Post, 2009a). Nidal Hasan was unmarried and was known to be a devout Muslim (Washington Post, 2009a). He was a godly Muslim, according to Faizal Khan, a former imam (Washington Post, 2009b). He attended prayer at least daily. He wore mainly his Army fatigues at the mosque. In 2001, Major Hasan attended the Dar al-Hijrah Mosque and Islamic Center in Falls Church; Anwar Nasser Aulaqi (aka al-Awlaki) was the imam. Anwar al-Awlaki was the spiritual advisor to two of the September 11 terrorists (Nawaf al-Hamzi and Hani Hanjour). Ahmed Omar Abu Ali, the conspirator in the planned assassination of President George W. Bush, also attended the same mosque. The imam was known to hold radical views. He preached Jihad, a Muslim holy war or spiritual struggle against infidels—and that was in the United States. Like these soldiers, Hasan was impacted greatly by al-Awlaki's teachings (Wikipedia, 2010). Hasan routinely visited al-Queda and Islamic websites, even just before the shootings. After the shootings, Anwar al-Awlak praised Major Hasan as "a soldier of Allah." Major Hasan was a hero. Hasan was a martyr. He was a soldier who killed for God.

Dr. Hasan was known to have extreme religious and ideological beliefs. From a very informative article entitled "Profile of Major Nidal Malik Hasan" by Douglas Hagmann (2009) of the *Northeast Intelligence Network*, excerpts are presented on Hasan's core beliefs on martyrdom, on non-Muslims, and on the meaning of Jihad. I present his view on Jihad:

> On the meaning of Jihad: Hasan reportedly provided his interpretation of "jihad" at the Muslim Community Center in Silver Spring, Maryland, a suburb of Washington, D.C., to Golam Akhter, 67, a Bangladeshi-American civil engineer, 67. Akhter stated that Hasan disagreed with his interpretation of the concept of jihad as "an inner struggle, fighting against corruption and injustice." Instead, Hasan stated: "That not a correct interpretation. Jihad means holy war. When your religion isn't safe, you have to fight for it. If someone attacks you, you must fight them. That is jihad. You can kill someone who is harming you." Source: http://www.thedailybeast.com/blogs-and-stories/2009-11-07/major-hasans-hidden-militancy/full/Summer of2009

Major Hasan was known to object to the wars in Iraq and Afghanistan. There was great discord, complications, concurrent contradictory beliefs and thrusts; one could even say constriction in beliefs and conscience. He strongly adhered to the Koranic World View as it relates to Muslims in the U.S. Military (Washington Post, 2009b). In the book, it states the objections to the war were "Identify what the Koran inculcates in the minds of Muslims and the potential implications this may have for the U S. Military"; "Describe the nature of the religious conflicts that Muslims may have with the current wars in Iraq and Afghanistan"; and "Identify Muslim soldiers that may be having religious conflicts with the current wars in Iraq and Afghanistan." The book espoused very strong holy beliefs for Muslims in the military. It states, "And whoever kills a believer intentionally, his punishment is hell; he shall abide in it, and Allah will send His wrath on him and curse him and prepare for him a painful chastisement. . . . And do not kill anyone whose killing Allah is forbidden, except for a just cause."

Major Hasan was known to have a great conflict with the wars in Iraq and Afghanistan. The wars were against Allah, he believed. Furthermore, he was awaiting deployment to Afghanistan, despite his objection; it was to be his first Army overseas service. After the shootings, Senator Kay Bailey Hutchinson stated, "Hasan was upset about his deployment" (Newman, 2009). And what seemed almost like suicidal-like behavior, he gave away his possessions from his home on the morning of the shootings. Was it to be a homicide(s)-suicide? This is the intent of the current martyr-to-be.

Major Nidal Malik Hasan had lethal intent. He believed that his fellow soldiers were infidels and should have their throats cut (Allen, 2009). In NASH, it was homicide. That is unequivocal. We know what he did; we don't know why he did it. A PA would answer that question and more. We need to know; every survivor of suicide, homicide, or war-related death has these questions. What I do know is that his personal documents read like the altruistic suicide notes that I discussed in Chapter 2. The documents are extremely suicidal-like and homicidal-like, I believe. They were his recipes for homicide-suicide.

On November 12, 2009, Major Hasan was officially charged with 13 counts of premeditated murder. It was intentional. On December 2, 2009, he was officially charged with an additional 32 counts of attempted murder. These charges were made in the military's legal system (McKinley, 2009, New York Times, 2009); he was found guilty of murder and attempted murder and sentenced to the death penalty.

On November 9, 2009, Senator Joseph Lieberman called for an inquiry by the Senate Committee on Homeland Security and Governmental Office, chaired by him. He stated, "If the reports that we're receiving of various statements he made, acts he took, are valid, he had turned to Islamic extremism. . . . If that is true, the murder of the 13 people was a terrorist act." Anwar al-Awlaki agreed; yet from his view, Hasan was a martyr and a hero, not a terrorist. It is the same old story retold.

Terrorists are often identified, by definition, as belonging to a network (Sageman, 2004). However, as I discussed in Chapter 5, terrorism is not simply about militants at the community or societal level, it is also an individual act at the intentional level. There is a psychology of martyrdom. The martyr's mind is most relevant. There is no question that Hasan had a martyr's mind. Major Hasan shows us the scourge of our time, global terrorism, and its intentional traumatization, beyond a scope ever imagined. We need a theory to better understand Hasan and the terrorist. A core belief: Once we know the terrorist mind and his/her intended traumatization, we can better control and prevent it—the aim of the armed forces and science. Hasan illuminates why people become extreme terrorists and how they intend to traumatize and kill us, leaving many with haunting wounds. This is the very aim of terrorists, martyrs, altruistic suicides, whatever you call them. They are all the same—heroes to some, terrorists to others.

Of course, the United States is not alone with its unusuals in the military. In 2010, Colonel Russell Williams, a Canadian Air Force pilot, who flew the Prime Minister and the Queen of England, killed two women. He sexually assaulted and broke into women's bedrooms to steal their underwear. He even broke into little girls' rooms, taking pictures of him with the girls' underwear on his head. At the same time, he flew Prime Minister Stephen Harper on the British equivalent of Air Force One. The irony is obvious. One more fact: Colonel Williams is known to have made one suicide attempt after his incarceration.

## ACCIDENTAL DEATHS

As discussed before, the NASH categorization, according to which deaths in the United States and Canada are classified, obscures many of the psychological dimensions of death. It treats death merely as a biological event and fails to address issues of intention and subintention (Shneidman, 1985). Moreover, it suggests that the categories in NASH are distinct, when in fact there may be considerable overlap in modes of death. For example, some deaths categorized as accidental may in fact be suicidally motivated. This is important to the topic at hand, because accidents are the leading cause of death among active duty members.

Freud (1974a), Menninger (1938), Shneidman (1963, 1985), Murray (1967), and others have speculated that there is a vast array of subintentional deaths that occur beyond intentional suicides. Alcoholism, drug addiction, mismanagement of physical disease, car accidents, self-harm, and many other behaviors can be seen in this light. Farberow (1980) edited an insightful volume on the topic, *The Many Faces of Suicide*, providing an array of studies on subintentional death. How can we study the faces of violence (and faces of trauma)? Can we show associations? Causes? What do we know?

Freud (1974h) speculated that such phenomena can be studied, not only at an individual level but also at a societal level. He argued, for example, that civilization and its discontents could be measured in the level of self-destruction present in a society. Subsequent views, such as those by Farberow (1980), support Freud's opinion. This would be expected because social mortality data are based on the population of individual cases, and if the individual cases are misleading, so are the aggregate data. Of course, we do not suggest that social analysis can do real justice to the psychological analysis of subintentional deaths. Farberow, Freud, Menninger, Shneidman, and others have, therefore concluded that self-destructive behaviors—suicide and its many forms—originate from psychological and social/cultural forces, therefore, our ecological model holds. In order to understand the act of suicide and its substitutes, it has been shown to be useful to relate the behavior to other kinds of violence (Henry & Short, 1954; Unnithan et al., 1994). Suicide, homicide, alcohol abuse deaths, and so on are seen as interwoven. Can we prove it?

Leenaars and Lester (1998) explored three correlates of suicide, namely, homicide, one form of accidental death (motor vehicle accidents), and one form of natural death (cirrhosis of the liver caused by alcohol abuse), which is sometimes assumed to be a subintentional death (Farberow, 1980). The question raised is whether the NASH classification obscures the complexities of suicide at a societal level.

Durkheim (1951) theorized that social integration—the degree to which people in a society are bound together in social networks—is associated with the self-destructive behavior of suicide, and research in Canada has indicated that measures of social integration (such as marriage, divorce, and birth rates) may be associated with suicide rates (Leenaars et al., 1998). Other research has suggested an association between unemployment and suicide. The primary question of Leenaars and Lester's (1998) study was, Are the social correlates of various kinds of death the same or different? Data were obtained for Canada for the period 1960 to 1985. The results showed that suicide, homicide, and death caused by alcohol are alike, but also that accidents are most different, although some are alike. These results, therefore, suggest that the impact of social integration on society may apply to self-destructive behaviors other than suicide. The results raised questions whether motor vehicle accidents, sometimes identified as a possible subintentional death (Farberow, 1980), can be grouped together with self-destructive behavior at a social (population) level. The answer is that it depends. Not all accidents are so. MacDonald (1964) reported a disproportionate representation of former psychiatric patients in accidents. He also noted that a striking feature was that half the patients made the attempt very impulsively following an argument with a lover, marital partner, or less commonly, with a neighbor or a superior at work. Each individual case must be evaluated separately, as it has been shown that occasional motor vehicle accidents may, in fact, be subintentional deaths (Farberow, 1980). The most likely are single-car accidents, especially if there was substance abuse.

On a different note: Can the NASH categories be distinguished from one another, as is frequently assumed? Do they blur the complexities involved in determining cause of death? Do they obscure our understanding of not only the rates of suicide but also complex social circumstances of death in the military? Death by automobile offers special opportunity for concealment of suicide and homicide. There may be dissembling. It would conceal disgrace and shame, for example. The extent of deliberate death on the highway is not known; this is true in the military. Pompili, Girardi, Tatarelli, and Tatarelli (2006) addressed the issue of a possible link between single-car-accident drivers and suicidal intent. They selected 30 single-car-accident drivers who had been admitted to emergency departments and then hospitalized for an average period of 10 days. They matched these patients with a control group of drivers who had never had a car accident. Pompili et al. concluded that, although suicide risk was low in the patients, they were engaged in looking for a solution to their problems in which the accident played a role in such a process. In fact, there are associations to constricted logic and cognitions. An example of core beliefs of suicidal individuals is "I need to escape. An accident is a solution." Thus, for example, are the single-road traffic deaths of a soldier suicides or accidents? Are they military suicides? There are many faces of suicide.

## Accidents and the Military

"I loved her so much. I'm so sorry but this is how its gotta be. I hate life. Life's done me wrong. Please God forgive me but I'm weak. My lifes finished." This is the suicide note of a 20-year-old soldier; he had physical problems resulting from an earlier automobile accident. He was distraught over the recent death of a close friend and told his girlfriend he was going to kill himself. Shortly thereafter, she broke up with him. He shot himself in the head with a .38 revolver (McDowell et al., 1994, p. 110). Was the automobile accident a trigger?

From 1978 to 1995, Thoresen and Mehlum (2006) investigated 43 suicides and 41 fatal accidents in Norwegian soldiers/peacekeepers. They undertook a PA. The groups differed. Mental health problems were the most important risk factor for suicide. Living alone and the break-up of a love relationship were especially more evident in the suicide group, even when controlling for psychopathology. PTSD was especially high in the suicides. There have been differences reported in the prevalence of PTSD. Most studies have pointed to a link between service-related stress exposure and subsequent health problems, including suicides. This was not evident in Thoresen and Mehlum's accidental deaths, keeping in mind these were not single-car accidents. The individual (case control) level is reflected in the social (aggregate) level (Leenaars & Lester, 1998). The aggregate, of course, should reflect the individual cases, although, as shown in Chapter 8, this is often not true in the military.

Thoresen and Mehlum (2004) examined risk factors for more specific types of fatal accidents and suicides in soldiers/peacekeepers, asking, Is there an overlap? It is commonly believed that there is an increased risk of fatal accidents in veterans of military operations and that such accidental deaths may be related to mental health problems. Thoresen and Mehlum conducted a study to investigate fatal accidents in Norwegian former peacekeepers. Unlike their previous studies, looking at all accidents, they looked at a subgroup of alcohol-related fatal accidents; a PA was undertaken (e.g., military records, police reports, interviews). Thoresen and Mehlum examined 17 cases of alcohol-related fatal accidents compared with 28 cases of other accidents and 43 cases of suicide. Thoresen and Mehlum appeared to confirm Freud's belief. They found that "the alcohol-related fatal accidents were found to share many common features with the suicide group, such as depression, alcohol and substance abuse, and various social problems, and were also found to differ significantly from the other fatal accidents" (p. 990). They further concluded that there was an increased risk of accidental death in military populations due to exposure to war and trauma.

Farberow (1980) coined the term *indirect self-destructive behavior* for the many faces of suicide. Freud (1974a) called it "unconscious" suicidal dynamics. Shneidman (1985) proposed the term *subintentioned death*. Whatever we call "it," Thoresen and Mehlum (2004), in their unique military study, showed the existence of the many faces of violence. They speculated that alcohol abuse, drug ingestion, excessive smoking, and self-harm are various forms of risk taking. Thus, the question for armed forces is how many accidental deaths are, in reality, hidden suicides? Secrecy is prevalent. (Why should it be different from the general population?) A small subgroup of motor vehicle accidents and drug-related deaths may be labeled a face of suicide (and thus, violence). A subgroup of fatal accident, namely, alcohol-related cases, is so. In the Thoresen and Mehlum study, alcohol-related accidents resembled suicides. They are suicides in many ways, and some could be homicides. This is true in the military. At the very least, Thoresen and Mehlum's findings suggest "the need for preventive measures directed at reducing the risk of premature death not only from suicide, but also from accidental death" (p. 990). The military needs to target all self-destruction and other-destruction.

Self-Harm

Norman Farberow (1980), among many other scientists, raised awareness of the problem of self-harm, or sometimes called self-mutilation. (Karl Menninger had already discussed the topic, of course; see Chapter 2) Farberow noted that there was an interest or curiosity about self-destructive behavior in our fellow man. He thought that some people's behaviors were motivated by what we would consider to be irrational. Albeit being taboo in study for centuries, Farberow thought it was essential that the secrecy be made visible (a common belief by now). He asked, Why do people inflict or harm themselves? Why defeat themselves? Why would a diabetic stop taking insulin? Why would a person endlessly cut him/herself? There are many behaviors that contain a high potential for serious physical injury and for serious damage. Yet there is no intention to be dead. Self-injurious behaviors or self-harm behaviors with no intentional cessation are not suicidal behaviors. These are behaviors that involve deliberate self-harm with no intent to die. However, it is also obvious that these people are more like suicidal people than more indirect self-destructive behavior in some ways; yet in others, they are not. They lack intent to be dead. (That is a critical criterion, as you have read, in suicide.) The self-harm has special meanings, often excitement and pleasure (Farberow, 1980).

> While self-mutilation takes many forms and may be inflicted on many parts of the body, the most familiar form is wrist-cutting, in which the act of cutting often serves the significant purpose of helping the person to regain contact with reality and to come out of a depersonalized state. Blood is often involved with a special significance, a kind of visible evidence of a sacrificial act which the person must carry out, perhaps to expedite some overwhelming guilt. (Farberow, 1980, pp. 9–10)

Shame and disgrace would be other motivators.

Farberow goes on to note that the pain experienced in the cutting becomes important too, as indicative of the *need* to transfer psychological pain to physical pain. Sometimes the physical pain numbs the deeper psychological ones—the unbearable pain of pain; the same motivation as for suicide. (There are commonalities.) Physical pain is more acceptable and maybe manageable; and in the military, cutters may be more tolerated. Every soldier knows about wounds; a hero has wounds. It is an old complex in a green culture, but still irrational. It may well be potentially suicidal, in fact. It is similar but not identical to a true suicidal condition (Shneidman, 1985). Both are inimical behaviors. Of course, there is great overlap between the two, suicide and self-harm. People who have a history of self-harm do kill themselves sometimes. However, not all people who are at risk of self-harm kill themselves. *Only some do.* It is a risk factor.

In Farberow's edited book (1980), Simpson (1980) presents a few insights; here are the words of a self-cutter:

> This time I slashed my wrist. I crushed a burnt-out light bulb and saved the piece of glass. But razor blades are better. I cut parallel lines across my arm, avoiding old scars. I watched myself cutting, not feeling a thing, as if it wasn't really my own arm. My head was full of images of blood and glass glittered and gleamed. The blood creeping stickily across my skin felt good and real. The blood is me. The cuts are neater than my handwriting. They're not too deep—just deep enough. I'm in control, in an odd way, just when I'd been getting out of control. I'm really alive again, when I'd been feeling so dead. I told them I flushed all the glass down the toilet. I lied. I've got the best bits hidden away. For next time. (Simpson, 1980, p. 257)

There are abundant studies on self-harm in the general population. There is little about self-harm in the military. There is, however, an exceptional study about the topic regarding armed forces personnel. Hawton et al. (2009) present a unique study wherein they investigated the characteristics of Armed Forces personnel in the UK presenting to a general hospital following self-harm. They compared these individuals with matched controls in the general population who had self-harmed.

The method called for investigation of armed forces personnel presenting to hospital between 1989 and 2003 following self-harm. They found that 166 armed forces personnel presented with self-harm (184 episodes) during the study period. Approximately 72.3% (120) were male. Almost two-thirds (62.7%) were young, under 25 years of age. Females in particular tended to be young, with nearly three-quarters (73.9%) being in the 16–24 age group. Fewer armed forces personnel than controls had evidence of current or past psychiatric disorders (psychopathology). Fewer soldiers had treatment or a prior history of self-harm. Their suicidal intent was judged to be lower (males only). Of 64 people in the armed forces who presented during the first 9 years of the study period, six had died by probable suicide (9.38%). Methods of self-harm varied: "Four out of five (80.1%) of the self-harm episodes by the armed forces personnel involved self-poisoning alone" (Hawton et al., 2009, p. 268). Self-cutting was also a frequent method and may be more prevalent in the U.S. than the UK samples.

There were diverse traumas reported. The most common type of problem was a relationship with a partner (62.0%). Next were employment problems (43.9%). The work-related problems were common military ones. "The most common type of employment problem concerned specific difficulties relating to the individual's job within the Forces. Including the job being stressful, disliking the job, the job being boring and repetitive, and failure to progress or be promoted" (Hawton et al., 2009, p. 269). Relationship issues with fellow soldiers within the forces were another source of difficulty. At the time of self-harm, more armed forces personnel were facing problems concerning a relationship with a partner and employment. Although self-harm is also more common in females in the general population, female soldiers were over-represented in the self-harm population.

Alcohol abuse was common (40.5%). Of course, alcoholism itself is a risk factor in suicide and in the many faces of violence among the armed forces. Disciplinary problems occurred in more than one in ten (11.5%, 16) individuals. These were mostly related to military disciplinary issues, although in other cases, consequences of civilian offenses (e.g., being convicted). Incarceration was a trauma. Shame and disgrace again seemed implicated.

Stigma was clearly evident. Hawton et al. (2009) note that, "minimisation of symptoms due to perceived stigma may have influenced these findings" (p. 271). There "may be a sizeable problem of self-harm in the armed forces that does not come to clinical attention" (p. 271). Once more, secrecy is a risk factor. Be that as it may, Hawton et al. "suggest that self-harm by armed forces personnel is often used as a means of communicating distress related to current personal circumstances" (p. 271). This is what Farberow (1980) had suggested is "a cry for help." Is anyone listening in the armed forces?

## Incarceration

Incarceration is obviously a very stressful event for anyone who experiences it. It is a trauma for many. A few inmates, in fact, die by suicide while incarcerated. Who then commits suicide in prisons, jails, and military detention areas?

Christine Tartaro and David Lester (2009), in an exceptional book, *Suicide and Self-harm in Prisons and Jails,* present the state of knowledge of this face of violence. It gives us some general answers. Males in the United States are 56% more likely than females in jails to commit suicide. However, this is not true in other countries; for example, suicide rates for males and females in UK correctional facilities are similar. Tartaro and Lester, however, suggest that these statistics should be viewed with caution. In the United States, the typical suicide is White. U.S. Black inmates have lower rates. Indigenous peoples from Canada and Australia have disproportionately high incarceration rates and suicide while incarcerated. Young people under the age of 18 have higher rates. Alcohol and drug abusers have higher rates. There are many factors.

Tartaro and Lester (2009) offer the following profile:

> A profile of the typical suicide in a jail or detention center.... The typical suicide was a twenty-two-year-old, white, single male, arrested for public intoxication, and this was the only offense leading to his arrest. The suicide was under the influence of drugs and/or alcohol at the time of his incarceration, and there was no significant history of prior arrests. The suicide was placed in isolation for his own protection or surveillance and was dead within three hours of incarceration. The cause of death was suicide by hanging. The profile for the sentenced prisoners is less consistent and varies. (p. 20)

There is no typical profile for a soldier. Indeed, it is suggested that in the military, one should move away from the identification of the "typical" suicidal inmate and instead focus on identifying incarcerated soldiers who are experiencing the onset of a "suicidal crisis." Mentally ill inmates are overrepresented among the suicides in correctional facilities. The adjustment to incarceration and the risk factors for suicide are especially important. In the green culture, there is shame and disgrace. Prisons, jails, and military lockups are dangerous places for our heroes. They are not heroes there. A strong predictor of suicides in custodial settings is previous acts of self-injury or suicide attempt, in fact.

Wortzel, Binswanger, Anderson, and Adler (2009) examined some important questions in their paper, "Suicide Among Incarcerated Veterans." They note the now obvious that both veterans and jail/prison inmates face an increased risk of suicide. There is a clear need for a better understanding of the incarcerated soldier and veteran populations and the suicide rate faced by these groups. They searched Pubmed/Medline/PsycINFO anchored to incarcerated veteran suicide, veteran suicide, suicide in jails/prisons, and veterans incarcerated from 2000 to the present. What we know is that suicide rates in correctional facilities are significantly higher than in the general population. The mix of the trauma, incarceration, and suicidality is a lethal combination. This is especially so at specific times of incarceration. Pretrial inmates are reported to face the highest risk. The first 2 weeks of the soldiers' transition into the jail environment is also especially a high-risk time. Transitions always are for suicidal people.

An important military fact is that "a significant proportion of the veteran population has faced incarceration in jails and prisons" (Wortzel et al., 2009, p. 85). Wortzel et al. (2009) go on to offer the following facts: "Substantial drug and alcohol use was noted in veterans" (p. 86). Furthermore, "Substantial proportions of VA mental health users had been incarcerated, particularly those who were young and those with substance use and mental health disorders" (p. 86). Exposure to trauma, and symptoms of are almost always evident. "Eighty-seven percent of the veterans reported at least one lifetime traumatic event, and 39 percent screened positive for PTSD. . . . Among those screening positive for PTSD, 70 percent reported witnessing death or injury" (p. 86). Is it combat experience that explains the frequency of incarceration among veterans? Wortzel et al. suggest an answer: "In 2004, only 54 percent of incarcerated veterans in state prisons had served during wartime, and only 20 percent reported experiencing combat duty" (p. 86). Thus, it is not simply war. However, it is a critical factor in most. What is relevant, we know, are diagnoses such as PTSD, traumatic brain injury, and substance abuse. Thus, we are back again at the importance of trauma and suicide, and this is no different in incarceration. It is not simply incarceration, but it may be related to the never-ending traumatization of a vulnerable veteran or soldier. Wortzel et al.'s main conclusion (something noted throughout this book) is the lack of research available on suicide rates among incarcerated veterans. Is it the secrecy? We need to know. Again, what is most relevant among our incarcerated soldiers and veterans: trauma and suicide.

## CONCLUDING QUESTIONS

In the highly recommended book, *The Many Faces of Suicide* (1980), Norman Farberow asked some important questions. I have rewritten some with a military focus in mind. I hope that they will resonate some commonalities: Why did the Marine veteran stop taking his insulin today when he knew he had to take it regularly to stay well, or even stay alive? Why did the veteran go on an eating binge of all his forbidden foods when the doctor at the VA told him over and over how important it was that he observe a strict diet if he wanted the hemadialysis to work? Why does the alcoholic soldier deliberately stop attending AA, knowing that it is a trigger for another binge, which she knows from bitter experience of the pain and depression that follow? Why did the Air Force pilot fly so fast and low through the treacherous mountain terrain of Afghanistan? Why did the soldier on the field of battle step out into the enemy's fire and deliberately not shoot back, bringing the margin of safety down to zero? And Why do our heroes have so many faces of violence?

# PART FIVE

# Military Efforts

Dr. David Satcher, the former U.S. Surgeon General, has stated that once we know a problem, the second step is to try to answer the question, "What is the cause or causes of the problem?" Echoing the great Prussian general and war theorist of the 1800s, General Carl Gottfried van Clausewitz, we find that war is friction. War is trauma. PTSD is a common cost. War stress is unforgiving. In this section, we analyze data, patterns, and associations in order to determine the probable causes.

The first chapter in this section looks at what is written about the psychology of suicide in the military. I present some of the best writings. Dating back to studies from the Vietnam War, PTSD is highlighted. Trauma and suicide are intimately interwoven. It concludes that suicide has been and can be prevented. The next chapter examines the central role of trauma and PTSD in military suicide. One must understand PTSD to understand military suicide. I begin with presenting the classic pioneers in the field, such as Figley, Freud, Janoff-Bulman, and Nash. Then I present some of the current writings. I believe that you have to understand trauma and its wake to understand suicide and to prevent it. The last and important chapter explicates the most current approaches to suicide *prevention* (prevention, intervention, postvention) among the armed forces. I will examine what really works, even in the combat zone. We need to ask, What can we do to prevent a soldier's death?

# CHAPTER 10
# The Psychology of Military Suicide

Suicidal ideation, suicide attempts, and suicide in the military have often been described as being manipulative acts. "It is just to get attention" is a common belief. The suicidal soldier is seen as just weak and is often described as a soldier who is actually just dissatisfied with military life. He or she is a coward, not a warrior. Gaines and Richmond (1980) note that suicidal soldiers are often described as "immature, inadequate, and sociopathic." There is stigma and very high green walls. (In such a culture, would you reveal your "cowardly weakness"?)

Like me, Gaines and Richmond (1980) argue that we have to see the suicidal soldier differently. It is not merely for attention. There are many ways to get attention in the military, just think of the massacre at Fort Hood. The question is why this lethal way? Gaines and Richmond, for example, show that suicidal soldiers experience significant feelings of depression, worry, and pessimism. They note that this is one key way that suicidal and nonsuicidal groups in the military differ—the *pain*. It is the nature and experience of *perturbation* that differ. The *pain,* such as anxiety, is *unendurable.*

Interpersonally, the suicidal soldier sees his or her ability to cope with life, such as basic training, much less favorably. They do not adjust well to the demands of military life. Gaines and Richmond (1980) note in suicidal soldiers such beliefs as "I never could tolerate much stress" or "Others can do it because they are stronger than me" (p. 264).

As discussed in Chapter 2, to understand the suicidal soldier or veteran beyond the mask or surface, you have to look deeper into the psyche. We have to know the suicidal soldier's mind. It is important to distinguish between the surface, obvious aspects of a suicidal act and the underlying reasons for it (Stengel, 1964). The most common situational drama for the young adult soldier is the loss and/or rejection of the girl/boyfriend or spouse. However, suicide is more than the situational aspects. A soldier does not simply kill himself because his girlfriend rejected him. This does not mean that the situation is not relevant; indeed, the soldier may be singly mentally constricted on that one event; however, the situational aspects in every suicidal act are only one facet of the complexity. The loss/rejection was a loss of a deep attachment (identification); the soldier is singly fixated on the loss; the loss is a narcissistic injury; the wounded soldier wants to escape (egress); and so on. The soldier, however, does not simply kill herself *only* because the loss of her spouse. That is too simplistic an explanation. The suicidal soldier is a traumatized soldier. Gaines and Richmond (1980) show this *most critical* fact in the psychology of military suicide in an evidence-based way. In their study, they found strong indicators of anxiety and depression in the suicidal soldier group versus nonsuicidal soldiers. There are facts on the cause(s) of suicide among armed forces. PTSD is a most significant cause. These psychological characteristics are more evident than characterological traits. They go even further and ask, Is the decision to see the suicidal soldier as weak and manipulative a way to be rid of the weakened soldiers? Of course, suicidal soldiers and trainees can be dissatisfied due to the walls, even by health personnel (white walls) and express dissatisfaction in "belligerent, disturbing and sullen ways" (Gaines & Richmond, 1980). Some military health providers are suicidogenic. Do the green walls promote suicide? Is prevention even possible in such a system? Gaines and Richmond wrote in 1980; is their perception still true in the U.S. military today?

## STIGMA, THE MILITARY, AND TACTICS

Steadfastly, the military declined to comment on stigma. However, veterans have been speaking out for decades about abuses. One of the most unbelievable tactics of the military carried out by health providers is labeling the soldier or veteran as having characteristics or traits of a personality disorder (no wonder many soldiers and veterans do not trust doctors and therapists), such as borderline personality disorder or antisocial personality (psychopathic, sociopathic) disorder (both are at risk for suicide; see Chapter 2). In the media, there are reports of the secret orders to use the misdiagnosis as a military tactic. Anne Flaherty (2010), from the Associated Press, carried the following story, "Hundreds of PTSD Soldiers Likely Misdiagnosed." I present some of the story verbatim.

> WASHINGTON – At the height of the Iraq war, the Army routinely fired hundreds of soldiers for having a personality disorder when they were more likely suffering from the traumatic stresses of war, discharge data suggests.
>
> Under pressure from Congress and the public, the Army later acknowledged the problem and drastically cut the number of soldiers given the designation. But advocates for veterans say an unknown number of troops still unfairly bear the stigma of a personality disorder, making them ineligible for military health care and other benefits. . . .
>
> The Army denies that any soldier was misdiagnosed before 2008, when it drastically cut the number of discharges due to personality disorders and diagnoses of post-traumatic stress disorders skyrocketed. . . .

According to figures provided by the Army, the service discharged about a 1,000 soldiers a year between 2005 and 2007 for having a personality disorder.

But after an article in The Nation magazine exposed the practice, the Defense Department changed its policy and began requiring a top-level review of each case to ensure post-traumatic stress or a brain injury wasn't the underlying cause.

After that, the annual number of personality disorder cases dropped by 75 percent. Only 260 soldiers were discharged on those grounds in 2009.

At the same time, the number of post-traumatic stress disorder cases has soared. By 2008, more than 14,000 soldiers had been diagnosed with PTSD—twice as many as two years before.

The Army attributes the sudden and sharp reduction in personality disorders to its policy change. Yet Army officials deny that soldiers were discharged unfairly, saying they reviewed the paperwork of all deployed soldiers dismissed with a personality disorder between 2001 and 2006.

"We did not find evidence that soldiers with PTSD had been inappropriately discharged with personality disorder," wrote Maria Tolleson, a spokeswoman at the U.S. Army Medical Command, . . .

Advocates for veterans are skeptical of the Army's claim that it didn't make any mistakes. They say symptoms of PTSD—anger, irritability, anxiety and depression—can easily be confused for the Army's description of a personality disorder.

They also point out that during its review of past cases, the Army never interviewed soldiers or their families, who can often provide evidence of a shift in behavior that occurred after someone was sent into a war zone.

"There's no reason to believe personality discharges would go down so quickly" unless the Army had misdiagnosed hundreds of soldiers each year in the first place, said Bart Stichman, co-director of the National Veterans Legal Services Program. . . .

A congressional inquiry is under way to determine whether the Army is relying on a different designation—referred to as an "adjustment disorder"—to dismiss soldiers. (pp. 1–2)

I (the author) ask, Can the military stoop any lower?

## WHY DO SOLDIERS DIE BY SUICIDE?

Why do soldiers die by suicide? That is the military's question. There are many answers. Sareen, Belik, Afifi, Asmundson, Cox, and Stein (2008) offer one answer, maybe a key one. They begin with the now well-established fact that "The current military occupations in Iraq and Afghanistan have created a substantial resurgence of international interest in the mental health consequences of combat" (p. 291). It is a military topic of the day. To examine the issues, Sareen et al. studied data from a large population-based survey of active military personnel, the Canadian Community Health Survey Cycle 1.2 Canadian Forces Supplement.

They examined such variables as PTSD. They measured PTSD using the World Mental Health version of the Composite International Diagnostic Interview (CIDI). They evaluated lifetime trauma. They measured exposure to combat and peacekeeping operations, and the following two questions were asked: "Have you ever participated in combat, either as a member of a military or as a member of an organized nonmilitary group?" and "Have you ever served as a peacekeeper or relief worker in a war zone or in a place where there was ongoing terror of people because of political, ethnic, religious, or other conflicts?" A total of 26 other traumas were evaluated and included sexual trauma, interpersonal trauma, disasters, and accidents.

Following the Method of Difference, Sareen et al. (1980) created two groups, either combat peacekeeping exposure or noncombat/peacekeeping. The question posed was, Do the two groups differ? The answer: "Soldiers returning from deployment are increasingly likely to have emotional problems, to have somatic complaints, and to use mental health services" (p. 2194). The highest association with combat or peacekeeping was PTSD. The number of soldiers with PTSD was well above any expectations. It is central in military suicide and thus its prevention. Many mental health problems are attributable to combat. This is not to suggest that other risk factors (e.g., genetics, childhood trauma, stressful life events, loss of social supports) are not important. They are important in military personnel. Yet the main proximal trauma is the war—and so much more.

What is the "more"? In their study, "A Chronological Perspective on Suicide—The Last Days of life," published in *Death Studies*, Orbach, Gilboa-Schechtman, Ofek, Lubin, Mark, Bodner, Cohen, and King (2007) begin to ask a core question. What do suicidal soldiers teach us? To answer the question, Orbach et al. undertook a psychological autopsy using a narrative/ personal document approach to study the completed suicides of 67 Israeli soldiers (a large sample). They looked at proximal clues and context. They examined the last 3 weeks of the soldiers' lives, examining triggers for suicide, emotional state of mind, army duty functioning, communication of suicidal intent, military responses to communication of intent, help provision, and help acceptance. The study was most revealing. Orbach et al. noted that if one looked, one could see many proximal clues. In 69% of the cases, triggers could be classified; they included environmental stressors, interpersonal difficulties, and achievement failures. The two most common environmental stressors were conflicts with army authorities (e.g., disobedience toward authority and the law, incarceration) and change of environment (e.g., being moved to another unit, a transition/change). This is no surprise! The interpersonal issues are diverse, but the two common interpersonal triggers were romantic rejection and separation (e.g., a break-up with a girlfriend) and conflict with a close associate (e.g., fight with a family member, insult by a friend). There are narcissistic injuries, whether on the military stage or the interpersonal stage (or both).

The emotional states were diverse; common pains were depression, agitation, mood swings, anxiety, and elation (positive manic mood). In many of the soldiers who died by suicide, the predominant pain was negative (depression, agitation, or anxiety). The pain was unbearable. Yet it was more than mood. Orbach and his team (2007) noted that there was "social isolation, decreased functioning, lethargy, expression of worthlessness, loss of appetite, disordered sleeping, and a tendency to cry" (p. 919). There were immense traumatizations; yet on the outside, often nothing was seen or expressed (or was there?). Most importantly, Orbach and his team found "no changes in army duty functioning were reported in the majority of the individuals who committed suicide" (p. 919). Why? Why the masks? (Or were they simply glossed over?) A wide majority of individuals (79%) expressed suicidal intent either directly (e.g., "I will shoot myself in the head"; "This bullet is intended for my head") or indirectly (e.g., "There is no reason to fear death"; "Somebody may kill himself"). That is a large percentage. Yet, in the majority of cases (64%), there was no psychological help provided within the military system. There was no help! Why? Further, for the few cases ($n = 14$; 21% of the sample) who referred themselves to a mental health professional, most had had an ambivalent attitude toward professional help. They believed that the help was useless, and after getting help, they became even more hopeless/helpless. They concluded, "No one can help, not even the Doc." Were there white walls? Often, mental health providers do not understand the emotionally wounded soldier; are not soldier-centered; have their own barriers (countertransference); or worse, are ordered to label the suicidal soldier or veteran with a personality disorder. Orbach et al. concluded,

> Perhaps the most important aspect of our findings pertains to the frequent asynchrony between the emotional state of an individual and his or her military duty functioning during the last three weeks of life. The analysis of emotional state, when juxtaposed with the analysis of military functioning, suggests that although most of the suicide completers exhibited clear signs of emotional distress their duty functioning was hardly affected: 83% continued to function with no change and some even improved. (p. 927)

The green culture demands it. They are soldiers, after all. Yet it is this expectation of the "warrior" that makes the suicidal soldier so lethal. She dissembles. The military created, supports, and reinforces the green wall. The suicidal soldier is not seen or heard or spoken about. Secrecy prevails.

Our topic—trauma and suicide in the military—demands that we dig deeper. The military needs answers. Bodner, Ben-Artzi, and Kaplan (2006) looked into soldiers who kill themselves, examining the contribution of dispositional and situational factors. They compared characteristics of combatant and noncombatant Israeli soldiers ages 18–21 who committed suicide ($n = 429$) with others who did not commit suicide ($n = 499$). Bodner et al.'s findings suggest that dispositional and situational factors had contributed to the death, but also that there are different contributions to suicide among combatant and noncombatant soldiers. Combatant soldiers differ. There was a higher rate of suicide among combatants as compared with noncombatants. They note,

> Whereas combatants are exposed to intensive physical and mental pressures, especially at the beginning of their military service, noncombatants serve in a non-stressful environment under conditions similar to civilian workers (i.e., stable working hours, sleeping at home), and do not experience these pressures particularly during their basic training. Our findings corroborate this explanation, but suggest that this difference can also be accounted for either by a priori differences in the quality of soldiers assigned to combat and non-combat units or by higher accessibility to weapons. (p. 39)

Like me, they argue for environmental control. They conclude, "Therefore, greater accessibility to weapons may account for the higher suicide rates among combatants as compared to non-combatants, especially as almost all suicides were carried out by weapon" (p. 39). Soldiers know how to be lethal.

## THE SUICIDAL SOLDIER IS A GOOD SOLDIER

The suicidal soldier, who served in harm's way, is not weak, limited, immature, and premorbidly aggressive. He or she is a "good soldier." However, they avoid seeking help in stressful situations. They do not get help today. The veterans do not differ; few go to the VA, for example. Few seek help. Psychotherapy or counselling is rarely sought. We need to know why, what the factors are, and what we can do about it.

## TRANSITIONS, REPATRIATION, INCARCERATION, AND SUICIDE

Suicide is not a random act. War is stress. It can be trauma. One war that I have discussed here and there in this book is the one in the former Yugoslavia (Serbia, Bosnia, Croatia). Many militaries, including soldiers from the United States, Canada, and Norway, provided peacekeeping duties in Yugoslavia during these bleak years. Thoresen, Mehlum, Røysamb, and Tønnessen (2006) examined risk factors for completed suicide in veterans of peacekeeping, specifically, a representative sample of Norwegian peacekeepers. They note the high "prevalence of PTSD (posttraumatic stress disorder) in the veterans." In addition to posttraumatic stress symptoms, they noted an increased risk of death by suicide and accidents. Not only suicide, but there is also an increased risk of accidental death, especially due to motor vehicle accidents. There are many faces of suicide.

Thoresen et al. (2006) found that repatriation, especially involuntary, may also result in problems related to mental health or behavior. Repatriation in the military is to return a soldier to his or her native land. This may be at the end of the tour of duty; yet it may be involuntary, that is, done by the military against the expectation or will of the soldier. It is a difficult transition. Transitions are always a heightened period of risk, including in the general population; a good example is during the transfer/change of a patient from a forward-deployed medical facility to a U.S. tertiary military treatment facility. Another potential traumatic example for many soldiers (up to 9%), such as in the Navy and Marine Corps, is transitioning from active duty to civilian life (Mansfield, Bender, Hourani, & Larson, 2011). Combat exposure, PTSD, alcohol/drug abuse, and resilience are some factors associated to suicide risk during transitions. In Thoresen et al.'s study, involuntary repatriation emerged as a major risk factor for completed suicide in veterans of peacekeeping. "Involuntary repatriation from peacekeeping service may itself represent a potentially traumatic, stigmatizing and socially disintegrative experience that includes: loss of comradeship, loss of honor, possibly sudden unemployment, and may awake strong feelings of shame, defeat or anger" (Thoresen et al., 2006, p. 356). It is a loss, it is traumatizing, and it is a narcissistic injury to a warrior.

Why is involuntary repatriation so lethal? Well, involuntary repatriation may include incarceration. In the military, it means shame or disgrace. Being charged with a crime, being arrested, charged with drunken behavior (alcohol abuse), and so on, are lethal traumas. Thoresen et al. (2006) state,

> Repatriation is not a random event. It is a standard and contemporary way military organizations deployed abroad deal with personnel who cannot comply with conduct codes or disciplinary rules or cope with the physical, social and psychological demands the service places on them. It would be natural, then, to assume that mental health problems or behavioral problems or other latent problems that existed prior to peacekeeping service could become explicit in a number of individuals when they are deployed in a highly stressful theater of operations, resulting in both an increased risk of being repatriated and of later risk of suicide. However, . . . repatriation in itself is often perceived as a major traumatic event, potentially causing or intensifying strong negative emotions such as anger, guilt, hopelessness and shame. (p. 359)

*Shame and disgrace are, thus, central.* It becomes unbearable, and suicide becomes an escape from the felt lethal wounds. Of course, involuntary repatriation (and in some, even voluntary repatriation) is only one possible detrimental trauma in the military. There are many losses, rejections, and injuries in the military. At minimum, Thoresen and his group (2006) show not only the event(s) but also why it is so traumatizing. The "therefore" suicide solution is easier to see.

## THE MANY FACES OF MILITARY SUICIDE

Colonel Elspeth Ritchie, as you have read, is no stranger to the field of military suicide. In one study, Ritchie et al. (2003) looked at suicidal admissions in the U.S. military. They examined 100 consecutive charts of suicidal patients at a tertiary military treatment facility. They reported,

> The findings included the following: 94% were admitted with a depressed mood; 67% had a history of previous attempts or gestures; 49% had been treated with psychiatric medication prior to admission and 88% were treated with psychiatric medications while on the ward; 47% returned to a full duty status; 29% were recommended for administrative separation; and 18% were recommended for a medical board. (p. 177)

Yet they call for better surveillance, something I have discussed before (maybe too frequently). Dr. Snow would do the same. Colonel Ritchie and colleagues (2003) noted in their study that only half of the suicides were reported. There were no reliable current figures. There is considerable difficulty collecting data, they noted. There is underreporting. There are privacy concerns. How do we get past these barriers in military suicide? Like me, Ritchie and colleagues call for credible data.

Incarceration is only one stressor, albeit a major one. Medical-physical evaluation boards present a variety of risks. Fragala and McCaughey (1991) examined the impact of Medical/Physical Evaluation Boards on soldiers' health. It is the law in the armed forces. They note,

> Psychiatrically hospitalized military members are at risk for suicide during Medical/Physical Evaluation Board processing. Following initial hospitalization and treatment, usually for major depression, members are entered into medical administrative channels. During this period they are often returned to duty, placed on "casual" status in a holding company, or moved to a "transition unit" within the hospital, often under the care of other than mental health personnel. It is in this situation that *they must face the reality of the loss of their military identity and career* [italics mine]. This stressor may occasion a reoccurrence of the presenting illness which may be missed by medical personnel because the member is viewed by them as treated, or no longer actually a patient, but rather in the process of leaving. Medical personnel are reminded that ongoing or de novo psychiatric illness, including the risk of suicide, can and does occur on the occasion of separation from the military as stressor-independent from the reason for admission. (p. 206)

Of course, soldiers are not sent to Medical/Physical Evaluation Boards only for depression and suicide risk. There may be a host of problems and behaviors for example, alcohol abuse. Unfortunately, there are no major articles in the military medical literature on the topic. There are only a few case discussions that are unique to a military environment.

Like me, Fragala and McCaughey (1991) discussed the central issues of separation and narcissistic injury. It is central (as my theory suggests). They are deep stressors in the intrapsychic mind and interpersonal stage. It can also be the proverbial last straw—*the unbearable* last straw that breaks the soldier's back. The stress can result in aftershocks of a depressive episode or posttraumatic reactions. As discussed, a narcissistic injury may not only cause suicide, it can cause other-directed violence (homicide). The wish for revenge (attribution) can be outward or inward. The loss of a soldier's military career becomes unbearable. What does it mean, for example, if the soldier comes from a military family? Suicide becomes the only solution: "No way but this."

We know that exposure to psychologic trauma in harm's way is associated with mortality from external causes, including homicide, suicide, drug overdoses, and unintended injury. However, the etiology of this association is unclear. Boscarino (2006), in an insightful paper, "External-Cause Mortality After Psychologic Trauma: The Effects of Stress Exposure and Predisposition," published in *Comprehensive Psychiatry*, examined the cause of death and survival time among a national sample of 15,288 U.S. Army veterans. The veterans had PTSD status 30 years after military service.

Like many others (see Chapter 1), Boscarino (2006) suggests that having a history of PTSD or exposure to traumatic stressors is associated with higher rates of morbidity. PTSD and traumatic stress exposures are also associated with mental health disorders and diminished quality of life (Centers for Disease Control, 1989). Boscarino further reports that military personnel returning from Afghanistan and Iraq have significant rates of psychologic problems among these veterans. These personnel may also be at higher risk for suicide after return. After deployment to Iraq or Afghanistan, soldiers are found to have significant increases in mental health disorders, including PTSD (Hoge et al., 2004). In addition, it is not simply mental health disorders; physical injuries, especially TBIs, may be figural. This, in part, is because of high survival rates from injuries, which speaks well for medical care in the military. Those recently wounded in harm's way were surviving injuries that were previously fatal. How does a warrior survive being an amputee? Having a brain injury (TBI)? Having PTSD? What did Boscarino (2006) find? In analyses adjusted for Army entry age, race, age at interview, volunteer status, discharge status, history of drug abuse, and intelligence, Boscarino's results "indicated that postwar mortality from external causes was associated with PTSD among Vietnam theater veterans ($N = 7924$), with a hazards ratio (HR) of 2.3 ($p = .001$)" (p. 504). Further, noteworthy deaths from other causes were also elevated, such as cancer and cardiovascular diseases. Like others before, Boscarino found that alcohol abuse is an associated factor of many deaths. Yet it is known to be underreported.

The deaths do not occur in isolation. Often there was a history of trauma (including early life history, being a member of a military family). There is a link between long-term (distal) psychologic traumatization and death from external causes. Suicide is no exception. The suicidal soldier's problems appear to be determined by the soldier's history and the present military condition.

## THE MANY COSTS OF MILITARY SUICIDE

Still, too few wounded heroes seek help. Yet there are effective treatments. For example, psychotherapy may often be helpful, especially for PTSD (Lambert, 2004). The burden of mortality can be reduced; I have no question about that. We just need the will to do so. And there are many more reasons than the deaths to do so in the armed forces. Herrell, Henter, Mojtabai, Bartko, Venable, Susser, Merikangas, and Wyatt (2006), for example, underscore the human and financial burden that psychiatric disorders place on the U.S. Armed Forces. As an illustration, the shared number of lost years to still serve in the military, and to live life, is high, simply because most of the soldiers were young adults. It is costly.

War is friction. Concerns about distressed U.S. military personnel predate World Wars I and II (Hitschman & Yarrell, 1943). It is known that the effect of the trauma(s) is cumulative. Recent studies suggest that psychiatric problems are a shock or aftershock. A most common reason that individuals leave the military is, in fact, psychopathology (Hoge, Castro, et al., 2004; Hoge, Messer, Engel, Krauss, Amoroso, Ryan, & Orman, 2003). Therefore, Herrell et al. (2006) concluded, "These disorders place a heavy burden—both human and financial—on the affiliated individuals and the military healthcare system" (p. 1410). This may motivate the leaders. They always ask, "Is it cost-effective?" Make no mistake about it, prevention and intervention in the military will save lives and money. What disorders place a soldier at risk? As an example, soldiers diagnosed with bipolar (manic depressive) disorder are especially at increased risk and can be effectively helped (Bronisch, Wolfersdorf, & Leenaars, 2005).

We know that PTSD, bipolar disorder, depression, alcohol abuse, schizophrenia and more (as described in Chapter 2) may moderate the association between negative life events and suicide. PTSD is, however, by far the main psychological problem. We need to know more about PTSD among soldiers and veterans. Therefore, researchers, officers, clinicians, and so on must be aware of both psychopathology (such as PTSD) and suicide. It is time that these two topics are examined in depth together for perhaps the first time—an aim of this very book. We must not keep it simple. Military suicide is complex.

We know more. Florkowski, Gruszcynski, and Wawrzynlak (2001) note that "the inability to solve and successfully cope with problems may lead to adaptive difficulties and provoke reckless reactions, including suicide attempts" (p. 44). They undertook a retrospective analysis of the origins and factors leading to suicide attempts undertaken by soldiers from 1989 to 1998 in the Polish Armed Forces: 163 professional soldiers (military service was their job) and 274 privates (obligatory military service group). The most common reason for suicides in the analyzed group was trauma. At this time, the military understand the psychology of suicide better. What can we do? *How do we help fight with the soldiers to live?* What is evidence-based?

## CONCLUDING REMARKS

I will keep the remark simple, but now obvious: There is a unique psychology to military suicide. First and foremost, we have to examine PTSD.

# CHAPTER 11
# Posttraumatic Stress Disorder

War is trauma. War-related death is trauma. Suicide is trauma too. General Carl Gottfried von Clausewitz noted the physical, mental, and emotional stress of combat and called it "friction." The nature of war can result in a traumatic reaction in any normal person. It is *trauma*. Charles Figley and William Nash (2007), in their introduction to the book *Combat Stress Injuries*, write, "The nature of war is destruction," and ask some questions:

> (1) What are the positive and negative short- and long-term consequences of war fighting for the warfighter? (2) What are the pre-combat factors that affect these consequences? (3) What are the factors during and following combat that affect these consequences? (4) What are the psychosocial and medical programs, treatments, and interventions that mitigate the negative consequences of combat and enhance the positive consequences? (5) What can be done to utilize the answers to these questions in order to more effectively educate, train, lead, and care for our future military combatants? (p. 3)

How does a soldier experience war-related events? Here is a soldier's story:

> Boom! . . . silence. Thank God we all survived, but my battle is not over. Since that battle I suffer from nightmares: I see the blood, I hear the screams. I am frustrated, I feel like a failure; nothing seems to interest me any longer. I have lost my ability to be happy. (Koren et al., 2007, p. 119)

Koren et al. (2007) offer an exegesis of the word trauma:

> The meaning of the Greek word *trauma* means "injury" or "wound." Accordingly, a psychological trauma is an emotional wound; an injury caused by a stressful extrinsic event that is perceived by an individual to be an actual threat to his or her life, or to his or her physical and emotional integrity. (p. 120)

Koren et al. (2007) also go on to ask some questions too:

> Given the frequency with which warfighters suffer both physical and emotional injuries following combative events, a crucial question arises: How do the two traumas—the physical and the psychological—interact? How does physical trauma affect an individual's chances of developing a posttraumatic reaction? Is the risk of developing psychological trauma higher among those injured during a traumatic event, or might the injury act as a buffer? Does the injury enhance an individual's ability to cope with the psychological trauma? And in situations in which a physical injury results in permanent disability, is it possible to even determine when the physical trauma ends? (p. 120)

We need to know the answers. What we do know is the relationship between combat and PTSD. War is not one stressful situation one time. We know "it is always the sum of all stressors over time that weighs down the proverbial camel to the point of damage" (Nash, 2007a, p. 18). How might the conscious, intentional other-directed violence, often lethal, affect military personnel? How do they perceive the trauma? How do they adjust to the war? Soldiers know about "war stress" or "combat stress." It is not a by-product or side effect (Nash, 2007a). The enemy knows this too. They inflict it intentionally—and so do we. The "friction" is real.

Although every soldier knows that war is stress, that stress is normal, and that friction is unavoidable, they perceive combat stress as a test of personal competence. They are warriors. Nash (2007a) writes,

> To the extent participation in war is perceived by warriors as a test of their personal strength, courage, and competence, admitting to combat stress "symptoms" may be tantamount to admitting failure. Even if some stress symptoms are understood to be due to unavoidable stress injuries, and not merely personal weakness or cowardice, developing stress symptoms can bring with it considerable shame. Warriors volunteer for, train for, and expect to conquer all the stressors of war, even the worst terrors and horrors of ground combat. Therefore, it is hard for warriors to not perceive stress symptoms of any kind as evidence of personal weakness and failure. (p. 18)

There is enormous stigma and fear of negative career consequences in the military culture. Sadly, a few soldiers choose death over dishonor, disgrace, and shame. Yet everyone is vulnerable, even an Admiral or General.

## Posttraumatic Stress Reactions

It has previously been argued that the aftermath of a war, suicide, or other trauma for the survivors is best viewed from a posttraumatic stress framework (Leenaars & Wenckstern, 1993, 1998). According to the Diagnostic and Statistical Manual of Mental Disorders (APA, 1994), the essential feature of PTSD is the "development of characteristic symptoms following exposure to an extreme traumatic stressor" (p. 463). War-related deaths and suicide, it has been argued (see Leenaars, 1988; Shneidman, 1985), are outside the normal range of human experience and evoke "significant symptoms of distress in most people." The criteria for PTSD in the DSM-IV (APA, 1994) include

1. The person has been exposed to a traumatic event in which the person was directly exposed, witnessed or was confronted with an event that involved death or threatened death or serious injury, or a threat to the integrity of self or other. The person's response to this event involved intense fear, helplessness, or horror.
2. Persistent re-experiencing of the trauma (e.g., recurrent recollection, recurrent dreams, associations that the event is recurring).
3. Persistent avoidance of stimuli associated with the trauma and numbing of general responsiveness (e.g., diminished interest, detachment, constricted affect, sense of foreshortened future).
4. Persistent symptoms of increased arousal (e.g., irritability, anger, survivor guilt, hypervigilance, difficulty concentrating).

Although survivors of trauma often fit such a description, the majority does not meet the diagnostic criteria of the disorder (Figley, 1985) but suffer from a posttraumatic reaction nonetheless. There is friction after a trauma. It should be anticipated that survivors develop posttraumatic symptoms that, although they may not meet full criteria of PTSD, are associated with significant distress and/or limitations in adaptive functioning. It is not known how many soldiers may develop PTSD, but it is a significant number. Sudden death is distressing. War-related death is too. This is even more so the more traumatic the deaths are. Thus, this is true after a suicide. Freud (1974b) thought the greater the loss, the greater the stress. Suicide among our heroes is a great loss. Yet it is important to remember that researchers studying the effect of traumatic events on individuals note that although most people tend to respond initially with anxiety and/or depressive symptoms, and such would be expected in bereavement, they generally resolve, and the majority of affected people are able to go on with their lives without experiencing persistent PTSD. However, some individuals do, and this is especially so after a suicide (Leenaars & Wenckstern, 1991). PTSD can be seen as a heuristic model (framework, schema), as it is presently the best approximation of the reactions of survivors to trauma, whether war-related, suicide, or otherwise, in the military. It helps to understand.

## ADJUSTMENT TO TRAUMA

The notion of posttraumatic stress reactions is not new. Freud (1974b) discussed what he termed psychical trauma instigated by a threatening, acute event. Adjustment to trauma is complex. Much has to do with a shattering of basic beliefs about the world. Janoff-Bulman (1992) noted that much of the psychological trauma produced by victimizing events derives from the shattering of very basic assumptions that victims held about the operation of the world. Freud (1974j) differentiated between the positive (e.g., remembering, repeating, and reexperiencing) and the negative (e.g., forgetting, avoidance, phobia, and inhibition) effects of a trauma. Both positive and negative responses are common in many victims after a trauma, even in military commanders. We do not believe that our warriors will die by suicide. What happens when she does? Further, Wilson, Smith, and Johnson (1985) highlighted that the victims of a traumatic event, such as war-related or a suicide, may be stuck in a no-win cycle of events:

> To talk about the powerful and overwhelming trauma means risking further stigmatization; the failure to discuss the traumatic episode increases the need for defensive avoidance and thus increases the probability of depression alternating with cycles of intensive imagery and other symptoms of PTSD. (p. 169)

Nash (2007b) notes a major change is in shattering core beliefs: Everyone interprets life events and makes life decisions based on a set of core assumptions about the world and one's place in it. Nash cites the wise Dr. Janoff-Bulman:

> Janoff-Bulman (1992) proposed three fundamental assumptions common to all people at all times: (1) the world is benevolent, (2) the world is meaningful, and (3) the self is worthy. All people also need to believe that they are safe—that their lives will not be snuffed out in the next few seconds—and that a moral order exists in the universe that discriminates right from wrong. The importance of these core beliefs is easy to take for granted because they all operate beneath our radar screens, until something violates one of these beliefs. (2007b, p. 53)

War violates many core beliefs. Regrettably, after traumatic experiences, a common response in the military is, "Snap out of it" or "Don't talk about it" or "Just get over it" or "Just go on with your duty." However, avoidance only exacerbates the problem. It has been concluded from previous research on PTSD (Figley, 1985) that the type of response provided largely affects an individual's adjustment to a trauma. In other words, it is the environment that plays a significant role in soldiers' reactions to trauma—this is true, whether it is war-related or suicide.

## INDIVIDUAL/CULTURAL DIFFERENCES AND STIGMATIZATION

As implied in the definition of PTSD, there are commonalities in individuals' responses to trauma, such as war-related deaths or suicides; however there are also individual and cultural differences that need to be considered. "It is unreasonable to believe that the psychological distress produced by a suicide will produce the same effect in everyone" (Leenaars & Wenckstern, 1991, p. 177).

In general, those most at risk are those individuals (e.g., family, fellow soldiers, close friends) who were closest to the person who died by suicide. However, those considered more distant (although not necessarily psychologically), such as commanders, have also exhibited symptoms related to PTSD. One recurring risk factor, regardless of closeness to the suicide, is whether the individuals have contemplated or attempted suicide themselves (Leenaars & Wenckstern, 1991). The green culture produces, in fact, some very unique differences.

When investigating cultural differences, a distinction, as we have learned, is often made between Western or individualistic cultures and non-Western or collectivistic cultures. The *Oxford English Dictionary* defines *individualism* as "a social theory favouring freedom of action for individuals over collective or state control," whereas *collectivism* is the "practice or principle of giving a group priority over each individual in it." For example, one of the major obstacles in getting help for suicidal behavior is stigmatization. Stigma surrounding suicide may occur for different cultural reasons, including religious sanctions and judicial laws, and may be expressed differently through discrimination from religious, social, or military communities. Religion plays a large role in how suicide is viewed. In the United States, which is predominantly Christian, it is still condemned. Traditionally, and in most cases currently, suicide, for example, is seen as a serious sin in Judaism, Christianity, Abrahamic religions (e.g., Islam), and Hinduism (with some exceptions). The stigma surrounding suicide has a profound effect and can bring with it shame, guilt, disgrace, and alienation.

In relation to trauma and PTSD in general, Johnson and O'Kearney (2009) argued that the psychological impact of trauma might be culturally specific. They found that trauma survivors with PTSD from independent (individualistic) cultures experienced more cognitive and affective difficulties than those without PTSD, whereas survivors with and without PTSD from interdependent (collectivist) cultures did not differ. On the other hand, survivors with PTSD from both cultural groups felt more alienation than those without PTSD. If suicidal, the most damaging common view is, however, that feeling suicidal is not acceptable and therefore often remains suppressed, denied, avoided, and can turn inward. This is especially so in collectivist countries such as India and Turkey (Leenaars, Girdhar, et al., 2010; Leenaars, Sayin, et al., 2010). Are people in these cultures denying PTSD too? Although there are common cross-cultural factors, suicidal people in collectivist countries express more indirectness (Leenaars, 2007). This is true in the military. The suicidal state may be more veiled, turned inward, ambivalent, and unconscious. This may be true for PTSD and mental disorders in general, in fact. This, as discussed before, is called *dissembling* or masking and is a lethal risk factor, especially in interdependent cultures like the armed forces (see Chapter 2 for details). We must be aware of cultural and religious differences, among other individual differences. We need to normalize the traumatizing process of service for soldiers including making emotional reactions, such as anger, an acceptable reaction. For many soldiers, it is not. All we hear is,

"Snap out of it. We have a war to fight." Yet it is so important not to normalize the suicide itself. It is pathological, a sign of a probable mental disorder or maladjustment.

## COMBAT STRESS INJURY AND RESEARCH

It is now widely accepted that soldiers exposed to combat event(s) are known to be at greater risk for developing PTSD. One aftershock is suicide. This was not always known. It was especially after the Vietnam War that combat experience and postservice psychosocial status became known to be predictive of suicide. One of the pioneers in this research was no stranger to suicidology, Norman Farberow. However, before we examine the association between trauma and suicide, I would like to discuss the more general combat stress injury or PTSD. What are the stressors in the military? Do we know what injures our warriors?

### Types of Military Stress

In a revealing article entitled "Military Stress: Effects of Acute, Chronic, and Traumatic Stress on Mental and Physical Health," published in Freeman, Moore, and Freeman's book, *Living and Surviving in Harm's Way*, Megan Kelly and Dawne Vogt (2009) present the best list of stresses in the military that I have ever read. They state, "Military personnel may be exposed to a number of stressors in their occupation. These stressors can be broadly grouped into three categories: job demands, work-family conflict, and sexual harassment and assault. In addition, military personnel may experience a number of stressors during deployment" (p. 90). I present here a very brief synopsis (see Kelly and Vogt for full text):

Job Demands
  Just as for other occupations, job demands are a significant concern for military members. Frequently cited stressors associated with military service include long work hours, heavy work load, low autonomy, and unpredictable schedules. . . .

Work-Family Conflict
  Sources of work-family conflict may include frequent separations from family, transfers, the threat of deployment and war, significant family stressors (e.g., illness, divorce), and long and unpredictable schedules. . . .

Sexual Harassment and Assault
  *Sexual Harassment.* Another concern that may be especially relevant for women in the military is sexual harassment . . . between 51 and 93% of female veterans have reported experiencing sexual harassment at some point during their military service. . . . Sexual harassment is also a significant issue for male service members . . . 42% of male military personnel report having experienced sexual harassment at some point during their military service. . . .

*Sexual Assault.* Sexual assault can also be a source of significant stress for some military personnel, especially for women. . . . The literature indicates that rates of sexual assault are significantly higher for women veterans compared to female civilians . . . with most studies reporting prevalence rates ranging from 30 to 45 %. . . . For male military personnel, rates of sexual assault range from 1 to 4%. . . .

Deployment Stress

Most military personnel experience one or more deployments during their military career, and in recent years military personnel have experienced more frequent deployments, longer deployments, and less time between deployments. Deployments, and war-zone deployments in particular, may place considerable stress on military personnel. . . .

*Combat-Related Experiences.* Combat-related experiences may include being physically attacked or ambushed, being fired on by enemies, being wounded, and witnessing injury and death of other military personnel, enemies, and civilians. . . . Rates of exposure for combat veterans deployed in support of Operation Iraqi Freedom (OIF) in Iraq and Operation Enduring Freedom (OEF) in Afghanistan are estimated at approximately 67 to 70%. . . .

*Perceived Threat.* In addition to these more objective deployment stressors is the subjective experience of perceived threat that often, but not always, accompanies deployment to a war zone.

*Difficult Living and Working Conditions.* Although combat exposure has received the majority of attention in the deployment stress literature, findings indicate that difficult living and working conditions during deployment (i.e., "malevolent environment"), though low in magnitude when compared to combat stressors, can cause significant stress and discomfort. . . .

*Interpersonal Stressors.* In addition to the stressors specific to the war zone are interpersonal stressors introduced by deployment, especially those associated with being separated from family members and other loved ones. (Kelly & Vogt, 2009, pp. 90–97)

The stress list is enormous! Is it any wonder that so many soldiers and veterans experience "friction"? Of course, there are protective factors toward the huge possible stressors. One protective factor discussed earlier is hardiness or resilience. There are many more (Leenaars, 2004).

## POSTTRAUMATIC STRESS DISORDER AND SUICIDE

Military stress, combat experience, and postservice are predictive of suicide. PTSD is a risk factor for suicide. PTSD is also a predictor of later medical morbidity and premature mortality (Boscarino, 2007). Thus, before I cite the essential research on the military, PTSD, and suicide, I will briefly make note of more general deaths. Does combat "friction" cause long-term health problems? Deaths?

Boscarino (2007) assessed the long-term health impacts of PTSD in the military. He and his team examined all-cause and cause-specific mortality among a national random sample of U.S. Army veterans with and without PTSD after military service. In the study, they

> examined the survival time and causes of death among 15,288 male U.S. Army veterans 16 years after completion of a telephone survey, approximately 30 years after their military service from the Vietnam War era. They adjusted for age, marital status, race, combat exposure, volunteer status, entry age, discharge status, illicit drug abuse, intelligence, and pack-years of cigarette smoking. (p 98)

Boscarino's (2007) findings

> indicated that the adjusted postwar mortality for all-cause, cardiovascular, cancer, and external causes of death (i.e., mortality due to suicides, homicides, drug overdoses, motor vehicle accidents and injuries of undetermined intent) was associated with PTSD among theater veterans with Vietnam. When level of combat exposure also was controlled among the theater veterans, PTSD was significant for cancer and external mortality, suggesting that PTSD is central in these outcomes. (p. 98)

The Centers for Disease Control (1988) has likewise concluded that soldiers exposed to combat service are at significantly increased risk for the long-term health impact of the trauma, and this is greater than in the general population. War causes death—even away from harm's way; even decades later. War causes death! Why?

Boscarino (1995, 2007) suggests that the trauma is PTSD itself.

> It is likely PTSD, not combat exposure per se, which is associated with increased mortality, . . . the specific cause-of-death classifications for external mortality suggested that PTSD-positive theater veterans were more likely to die from suicide, homicide, and from alcohol- and drug-related causes. Given these results and the lack of significance for combat status, our study suggests that it is PTSD that is associated with premature mortality, not combat exposure. (Boscarino, 2007, p. 110)

It is, however, interactional. There is an established "link between long-term exposure to severe psychological distress (including combat- and noncombat-related stress) and premature mortality from multiple causes" (Boscarino, 2007, p. 110). More research is needed to understand the link better.

Be that as it may, reports related to military personnel returning from Afghanistan and Iraq suggest that there are significant rates of psychological problems among these veterans. There is a higher risk for suicide (Farberow, Kang, & Bullman, 1990). Of course, TBIs are noted, but also are other physical injuries. Many of those wounded in Afghanistan and Iraq, as noted earlier, are surviving injuries that would have previously proven fatal (Gawande, 2004). There is hope; yet it is also a recipe for PTSD. There is an increase in the prevalence of psychopathology among wounded veterans (Jones, 1995a, 1995b). Sadly, we also know that returning veterans may not be availing themselves of mental health treatment (Hoge et al., 2004).

However, there are also many veterans, living with pain (mental and/or physical), fear, and hopelessness/helplessness, who continue to call for more acceptance, help, psychotherapy, financial stability, and *honor*. Too little is done, they state. Maybe it is not only about soldiers and veterans availing themselves of help, but also availability of service. This is especially so for veterans with mental illness; veterans are quoted as stating that "the criteria are arbitrary, the funds insufficient" for help. Many state that some times of the year, such as Christmas, are especially *unendurable*. "High expectations of joy and togetherness instead can turn to sorrow, isolation, and crisis" (Perreaux, 2011, p. A4). *I would add suicide!* Too many soldiers and veterans are left in the mental battlefield; the everlasting howling of the bombs and the deadly screams from seriously injured and dying never go away. Ask any veteran: there is shell shock! On Christmas or New Year's or Memorial Day, the traumatic aftershocks are more visible, but help is even less available; psychologists, psychiatrist, crisis workers, corpsmen are on holidays too. Veterans deserve more!

Finally, I can now discuss the classical studies of Norman Farberow. It may seem a long journey to this point; however, I believe that the general context of combat stress injury and PTSD is essential to understanding trauma and suicide.

## PTSD AND SUICIDE DURING/AFTER THE VIETNAM WAR

The long-term frictions of the Vietnam War on Vietnam veterans have been the subject of numerous U.S. studies. Studies suggest that Vietnam veterans are more likely than others to have drug- and alcohol-related problems, experience broken marriages and depression more often than others, suffer from PTSD as a result of their Vietnam service, and postservice suicide. Farberow et al. (1990) examined the potential risk factors for suicide among Vietnam veterans. The comparison group was veterans who died from motor vehicle accidents (MVA). The veterans were selected from the Los Angeles County Medical Examiner's file (1979–1982). Some 100 consecutive veteran deaths from suicide and 100 consecutive veteran deaths from MVAs were identified. A psychological autopsy was conducted. Demographic and military characteristics of Vietnam veteran suicide cases (VS) were similar to non-Vietnam veteran suicide cases (NS). Significantly higher proportions of the VS group than the MVA group indicated that they had PTSD symptoms. Men with PTSD were 12.3 times likely to have a manic-depressive disorder. Indeed, a substantial majority of the VS group had experienced depressive, manic, and psychotic symptoms. This resulted in questioning of the relationship between PTSD and suicide. Farberow et al. (1990) concluded,

> The characteristics of Vietnam veteran suicide cases were not substantially different from those of non-Vietnam veteran suicide cases with respect to known demographic risk factors and military factors. The psychological profile of Vietnam veteran suicide cases is also similar to that of non-Vietnam veteran suicide cases in most instances. It did not appear that the Vietnam veteran suicide cases were more involved in combat-related activities than Vietnam veterans who died from motor vehicle accidents. They did, however, seem to have experienced more PTSD-related symptoms than Vietnam veterans who died from motor vehicle accidents. (p. 37)

One fact is clear: veterans who died by suicide had a high incidence of PTSD. Did it add to the suicidal solution?

Hyer, McCraine, Woods, and Boudewyns (1990) studied suicidal behavior among chronic Vietnam veterans with PTSD. It is now believed that it is not simply about having a disorder. Many PTSD combat patients have demonstrated high levels of substance abuse, depression, unemployment, marital problems, traffic accidents, and aggression, as well as arrests and incarcerations. It also is known that the disorder PTSD is associated with increased combat exposure in a dose response relationship (Kulka, Schlenger, Fairbanks, Hough, Jordan, Marmar, & Weiss, 1988). The greater the exposure, the greater the number and extent of problems. Yet Hyer et al. (1990) note that there exist effects of various background and social variables of suicidal behavior on PTSD and suicide. But the process is unclear. There are multiple possible factors (see Chapter 2). Hyer et al write,

> To date, there are few data with regard to the influence of "high-risk" variables on suicidal behavior among PTSD combat veterans. A few research reports have addressed suicide as an adjustment complication of PTSD. . . . At the present time, the research literature has determined combat experience to be *the* prepotent factor in the etiology of PTSD. . . . Increasingly, however, the influence of selective premorbid variables, like family stability . . . is found to be important. In addition, moderator factors, such as social support have been found to be associated with psychopathology among these veterans. (p. 714)

To study the possible suicide risk factor in Vietnam veterans with PTSD, Hyer et al. (1990) studied 60 chronic PTSD veterans admitted to a Specialized PTSD Unit at the Augusta VA Medical Center. They were divided into two groups: 29 patients in a suicide group and 31 in a nonsuicide group. Hyer et al. found that

> results showed that the Suicide Group possessed problems in paternal child-rearing patterns, current adjustment difficulties, and the PTSD symptoms of survival guilt and crying. In a regression analysis, paternal inconsistency of love, survivor guilt, and tendency to cry, in addition to age and sex, accounted for the significant variance of suicidal behavior. (p. 714)

Hyer et al. (1990) further found that suicidal behavior among chronic PTSD Vietnam veterans is a distinctive behavior. Hyer et al. concluded that "taken as a whole, suicidal behavior among chronic PTSD patients is characterized by greater parental problems with the father, guilt feelings, and emotional liability, as well as lower levels of psychological adjustment, including paranoid and cognitively confused feelings" (p. 719). We need, thus, to know the mind better.

Adams, Barton, Mitchell, Moore, and Einagel (1998) looked into the hearts and minds of U.S. combat troops in Vietnam. They surveyed the epidemiology among U.S. ground troops in Vietnam from 1957 to 1973. The results suggest that certain types of combat troops were significantly more likely than others to commit suicide. They stated,

> The United States soldier's risk experience in Vietnam included a variety of factors that amplified its effect on participants. These elements included physical exhaustion, the constant drone of biting insects, unfamiliar tropical noises, an inhospitable climate, and unrelenting apprehension about one's personal safety, each of which heightened the combatants' level of physiological arousal. A person in this type of situation often develops specific explanations for his reactions to combat. Rationalization may include statements such as "I'm a coward" or "I'm losing my mind." Unable to observe this reaction in one's comrades, an individual may conclude erroneously that he—and he alone—is frightened while his comrades are brave. A mounting sense of group alienation soon begins to permeate his existence.
>
> The combatant may suffer in other ways as well. Unfavorable self-perception may spiral him into depression or anxiety, while group isolation, combined with the escalation of hostilities, may worsen his psychological state. The utility of this model lies not only in how the ubiquitously stressful experiences of a Vietnam-era combatant can not only lead him into significant psychological distress but, if nothing else, isolate him from his comrades. . . . Experiences of anxiety, depression and alienation are also related with self-cognitions that are strongly associated with suicide. Depressed individuals have low self-efficacy, i.e., they do not believe that they can alter their circumstances in any way. . . . Moreover, depression is often associated with a sense of learned hopelessness, a state in which an individual concludes that he or she cannot control unpleasant circumstances (Beck, Rush, Shaw, & Emery, 1979). Depressed or anxious persons typically make a series of cognitive errors that lead them to believe that they seldom do anything right. They selectively perceive their failures and view the future with intense pessimism. (p. 1688)

In their discussion, Adams et al. (1998) ask further questions: "Had the soldier recently suffered the psychological trauma of a battle-related death of a close friend? . . . Did he have a history of suicidal behavior or depression? Did genetic factors somehow influence the suicide?" (p. 1692). Did intrapsychic factors? Interpersonal factors? Community factors? Cultural factors? What? And how does the array of factors interact? Multiplicatively? Adams et al.

concluded with the now obvious: "The exigencies of warfare, however, demand the selection and training of young men (and women in many countries) to commit controlled violence in a rational, emotionally-detached manner. Is suicide under these circumstances, therefore, an altogether aberrant outcome?" (p. 1693). To answer such questions, we need a PA.

There is no question that the Vietnam War was friction. Not only PTSD, but also suicide, has been enduring consequence. The Vietnam War has been a shadowy presence—often a lethal one. Indeed, we now know that U.S. veterans are twice as likely to die by suicide as civilians (Kaplan et al., 2007). *That is a staggering risk factor.* Veterans are at risk! Is this true of other wars? Today, we can look at the Yugoslavian and Bosnian wars as well as the Iraq and Afghanistan wars.

## PTSD AND SUICIDE DURING/AFTER THE YUGOSLAVIAN WAR

In a Norwegian study of the aftershocks of the Yugoslavian War, Thoresen, and Mehlum (2008) noted that soldiers were often exposed to severe "friction," such as witnessing other soldiers being killed, witnessing atrocities toward civilians, firing at close range, being shot, and other threatening events. Thoresen and Mehlum, in fact, reported that 5%–15% of soldiers expressed high levels of posttraumatic stress reactions several years after deployment. Thoresen and Mehlum investigated, "whether traumatic stress exposure during peacekeeping service was associated with suicidal ideation in a representative sample of Norwegian peacekeepers, and whether such an association would be mediated by posttraumatic stress reactions or other mental health problems" (p. 815). The method called for study of 1,172 Norwegian male soldiers who had served as peacekeepers during the Yugoslavian and Bosnian wars. In this study, 6% of the cross-sectional sample of peacekeepers and 17% of the sample who were repatriated earlier reported suicidal ideation. What, however, does i mean? What does suicidal ideation mean? It can be defined in many ways. There is also a lack of reliable studies on the prevalence of suicidal ideation in both the general population and in military populations; thus, there is no basis for comparison. Is 6% high? There were further specific factors such as repatriation from peacekeeping service and service stress exposure. Why" Thoresen and Mehlum speculated that repatriation could be regarded as an expression of vulnerability to stress. Their main conclusion: "Our finding: indicate that the association between service stress and suicidal ideation was mediated not only by general mental health problems but also, more specifically, by PTSD" (p. 819). Thus, PTSD causes risk for suicide. The risk of negative health consequences increases with multitraumatization (the dose response). PTSD is one.

Of course, there are possible confounders. There is the possibility that their PTSD, when interpreting and reporting their stories influenced respondents. Maybe they remembered traumatic events more clearly. Maybe they never forgot the trauma, the nightmares, and the horror. The hopelessness and fear continue every day. Do they dissemble? Do they magnify? Or are they better reporters? There are studies that support the reliability of self-report measures. There are also studies that support the relationship between exposure and subsequent stress reactions. Thus, the soldiers' stories may be understandable and reliable. The association between military service stress and later suicidal ideation is likely mediated by both posttraumatic stress reactions and other mental health problems. This is even true if controlled for background factors, negative life events, and current social support. Thoresen and Mehlum (2008) concluded, "It is well established that there is an association between potentially traumatic events and PTSD, between PTSD and depression, and between depression and suicidal ideation and behavior" (p. 820). Yet questions remain regarding the relation between PTSD and suicidality. What about the wars in Iraq and Afghanistan?

## PTSD AND SUICIDE DURING/AFTER THE IRAQ AND AFGHANISTAN WARS

Guerra and Calhoun (2011), with the Mid-Atlantic Mental Illness Research, Education and Clinical Center Work group, examined the relation between posttraumatic stress disorder and suicidal ideation in a sample of veterans deployed during Operation Enduring Freedom and Operation Iraqi Freedom (OEF/OIF). Guerra and Calhoun studied the veterans of the U.S. Armed Forces who served in the military subsequent to September 11, 2001. Their report includes data from veterans recruited to the Registry between June 2005 and August 2008 who were deployed to a region of conflict, and administered measures, described below, of both suicidal ideation and pertinent DSM-IV psychiatric illnesses (e.g., PTSD). The application of these inclusion criteria resulted in a sample of 393 veterans, approximately 82% ($n = 322$) of who were male. These veterans ranged in age from 22 to 65 years, with an average age of 38.3. Guerra and Calhoun (2011) found evidence that PTSD is associated with an increased risk of suicidality. The finding was that even PTSD diagnosed persons, who are not dually diagnosed with other disorders, are at risk. PTSD predicts that this is important given the belief of some of the need for co-morbid disorders, such as depression or bipolar disorder. PTSD is enough. Indeed, consistent with other studies (Jakupcak, Cook, Imel, Fontana, Rosenheck, & McFall, 2009), dually diagnosed veterans were not significantly more likely to endorse suicidality than veterans carrying only a PTSD diagnosis. Furthermore, Guerra and Calhoun (2011) questioned what caused the greatest magnitude of suicide risk; was it numbing symptoms (e.g., feelings of detachment from others; restricted range of affect) or avoidance symptoms (e.g., efforts to avoid thoughts, feelings or conversations associated with the trauma). Regrettably, in DSM-IV nomenclature, numbing and avoidance symptoms are grouped together. These may need different attention in our soldiers, however. Avoidance and numbing phenomena are different. What else is most critical?

In the present research, suicidality was operationalized by the threshold of a self-report measure; yet it is a limitation common to much of the pertinent research that has previously been conducted (e.g., Jakupcak et al., 2009); and of course, the ever-present belief that we need more research among our soldiers at risk for PTSD and suicidality.

There are further studies on the obvious facts now. Koren et al. (2007) studied the interplay between combat, physical injury, and psychological trauma. They also asked some important questions on physical and psychological injuries.

Obviously, the relation between the severity of the injury and the severity of the posttraumatic reaction is not a linear one. Koren et al. (2007) concluded that "the growing mass of research in the field of physical injury and posttraumatic reactions indicates that physical injury during a traumatic event is probably a risk factor, rather than a protective factor, in the development of PTSD" (p. 124).

Koren et al.'s (2007) study began to answer some questions. They examined the unique contribution of physical injury during traumatic events to the development of posttraumatic symptoms. They employed a matched, injured-control design, which enabled them to compare warfighters injured during combat to their noninjured comrades who participated in the same events. They excluded soldiers who suffered TBIs or were treated for psychiatric disorders at the times of injury. They found that

> first, the prevalence of PTSD among the injured group (16.7%) was approximately 7 times higher [!] than the prevalence of PTSD among the noninjured group (2.5%). . . . Similarly, although somewhat less dramatically, the prevalence of other psychiatric disorders (such as depression, drug abuse, and adjustment disorder) that developed after the traumatic event was 2 times higher among the injured participants (10%) than the noninjured ones (5%). (p. 126)

Thus, and understandably, they found that physical injury is a risk factor for the development of PTSD. They asked, "How might the increased risk associated with injury be explained? The most straightforward answer is that traumatic injury increases the perceived threat to one's life or to one's physical integrity" (p. 128) That is, PTSD—traumatic injury exerts its effect on perceived threat to self.

The now obvious is not limited to American soldiers and veterans, of course. There are a few studies that examine the relationship between traumatic events and suicide in Canadian military personnel. Belik et al. (2009) examined the association, for example, between PTSD and suicide attempts in a Canadian sample. It is a potentially unique study. They asked, Do different types of traumatic events cause suicide risk? They looked into answers in military data, a limitation of this study. They examined the files from the Canadian Community Health Survey: Mental Health and Well-Being Canadian Forces Supplement (CCHS-CFS), a cross-sectional survey that provided a comprehensive examination of mental disorders, health, and the well-being of currently active Canadian military personnel ($n = 8441$; aged 16 to 54 years; response rate 81.1%). In the survey, soldiers were asked about exposure to 28 traumatic events that occurred during their lifetime (e.g., sexual trauma, interpersonal, accidents). Regrettably, suicide attempts were simply measured using a question about whether the person ever "attempted suicide or tried to take [his or her] own life." (This is a grave limitation.)

Belik et al. (2010) found that

> the prevalence of lifetime suicide attempts for currently active Canadian military men and women was 2.2% and 5.6%, respectively. Sexual and other interpersonal traumas (for example, rape, sexual assault, spousal abuse, child abuse) were significantly associated with suicide attempts in both men (adjusted odds ratios [AORs] ranging from 2.31 to 4.43) and women (AORs ranging from 1.73 to 3.71), even after adjusting for sociodemographics and mental disorders. Additionally, the number of traumatic events experienced was positively associated with increased risk of suicide attempts, indicating a dose-response effect of exposure to trauma. (p. 93)

The more traumas, the greater the risk (the dose-response effect). Belik et al.'s (2010) pilot study begins to ask the question whether sexual and other interpersonal traumatic events are associated with suicide. It may be more than combat stress injury and suicide risk; yet combat friction is enough!

## POSTTRAUMATIC STRESS DISORDER, TRAUMATIC BRAIN INJURIES, AND SUICIDE

In Chapter 2, I discussed the critical association of TBIs and suicide. TBIs are a large risk factor. The question can be raised, How do PTSD and TBIs interact? Is PTSD a single factor? Or does it combine with other factors to multiply the risk? There are only beginning answers to these questions; actually only one study exists, that of Lisa Brenner and her team.

In a unique study, "Posttraumatic Stress Disorder, Traumatic Brain Injury, and Suicide Attempt History Among Veterans Receiving Mental Health Services," published in *Suicide and Life-Threatening Behavior*, Brenner Betthauser, Villarreal, Harwood, Staves, and Huggins (2011) offer an answer to our question. As discussed, in addition to PTSD, history of traumatic brain injury is a frequently occurring condition among those serving in Iraq and Afghanistan (Brenner, 2010; Hoge, McGurk, Thomas, Cox, Engel, & Castro, 2008). There have been increasing efforts that have focused on identifying mental and physical health outcomes in those with PTSD and/or TBI (Brenner et al., 2010; Hoge et al., 2008;

Hoge, Terhakopian, Castro, Messer, & Engel, 2007). However, there has been, to my knowledge, no studies regarding suicide in veterans with both conditions. Brenner et al. suggest that this is probably in part related to challenges associated with retrospective identification of TBI. Like me, Brenner and her group were

> unable to identify any studies that explored nonfatal suicidal behavior among individuals with co-occurring PTSD and TBI. Given the research, which suggests that a history of PTSD or TBI increases risk for such behavior, the suicide attempt history among veterans with PTSD and/or TBI was explored. It was hypothesized that a history of PTSD, TBI, and both (TBI and PTSD) would increase risk of nonfatal suicidal behavior. (p. 417)

Thus, Lisa Brenner and colleagues (2011) undertook a most important study. In the study, "veterans from an archival clinical database at a large western Veterans Affairs Medical Center. Eighty-one veterans with a history of suicide attempts between October 2004 and February 2006 (target period) were identified" (p. 417). To allow for a randomized control trial, "two control patients per case ($N = 160$), matched on age and gender, were randomly identified from a mental health clinic database of 3,239 potential patients" (p. 417). Brenner and her team appropriately excluded veterans identified with histories of suicidality (attempt, death) in the control group. An internal retrospective archival dig of all the patients' electronic medical records (EMR) was undertaken (but not a PA); documents included patient notes, discharge summaries, and imaging reports. (Regrettably, no interviews with family, military buddies, or friends were undertaken.)

They searched for key words to identify potential suicidal behavior (attempts, deaths) that included "suic" and "suicide." The same key word search was conducted for control subjects. A second electronic search was undertaken for posttraumatic stress disorder, traumatic brain injury, and neurologic disease. Thus, there was an extensive word search review. Of course, searchers were limited by the data recorded, and there is no way of discerning validity or reliability. Despite this limitation, this is the best study possible to date. Statistical analyses were undertaken The distribution of neurological diagnosis, PTSD, TBI, and their combination was compared between cases and controls. The conclusions were most relevant to the military. Brenner et al. concluded,

> Results from this study suggest that a history of PTSD was associated with increased risk for a suicide attempt in veterans receiving mental health services. Specifically, the odds of a suicide attempt for those with PTSD were 2.8 times that of the odds for those without PTSD. (p. 420)

These findings supported the earlier work by Bullman and Kang (1994) who found that among Vietnam veterans, those with PTSD were more likely to die by suicide (relative risk = 3.97) and accidental poisoning (relative risk = 2.89) than those without PTSD. The results, thus, supported the now obvious fac that PTSD is a significant factor associated with suicide risk in veterans; this is especially so if these veterans seek military mental health services. However, Brenner and her team also discussed some new facts:

> The odds of a suicide attempt for those with both PTSD and TBI was 3.3 times the odds of an attempt for those with TBI alone. This finding suggests that psychiatric and emotional disturbance conveys an increased risk for suicidal behavior in individuals with a history of TBI. (p. 426)

They further found the following: "Findings from the current study seem to support the cumulative disadvantage theory in that veterans with co-occurring PTSD and TBI may be at increased risk for negative outcomes" (p. 421). Brenner and her team recognized the limitation of the retrospective nature of the study. A second and maybe major limitation is the use of VA data. Be that as it may, the study offers the first to answer to some key questions. Veterans with TBIs are at risk. Veterans with both PTSD and TBIs are at even greater risk. Was this true for General Upton? It is true of many of our wounded heroes. To use Brenner et al.'s own conclusion,

> Findings from this study suggest that in veterans seeking mental health services, a history of PTSD is associated with increased risk for a suicide attempt. This risk was present for those with and without a history of TBI. Results support including PTSD as a factor when assessing suicide risk, and those of TBI screening tools within mental health settings. (p. 422)

## COMBAT EXPOSURE AND INCREASED RISK

Do all types of combat exposure increase risk? There is very little information on the process through which military service influences suicide risk. It would seem, based on the discussion so far, that the greater the exposure to combat trauma, the greater the increased risk of suicide and suicidal behavior. The length of tour of duty may be relevant. Thus, the question: What is the association between types of combat exposure and the acquired capability for suicide? This is the very question that Bryan and Cukrowicz (2011) asked. In their method, they administered self-report questionnaires to 347 U.S. Air Force personnel who served in the Iraq War. They evaluated past suicidality, PTSD symptoms, acquired capability for suicide, and combat exposure. Combat experience was measured by a 23-item checklist, with a range of possible events (e.g., being attacked or ambushed, seeing dead or seriously injured babies or body parts, witnessing an accident, disarming civilians, patrolling areas with land mines, short on directing fire at the enemy, being taken hostage). Of course, the old problem of self-reports exists. Thus, which item(s) of combat experience predicted or loaded on suicide risk? Bryan and Cukrowicz found that, "*all* forms of combat exposure predict higher levels of capability" (p. 133). They also found that "combat events involving aggression and high levels of exposure to death and injury independently demonstrate stronger association with the capability for suicide relative to combat events that do not entail explicit exposure to death or aggression" (p. 133). The more violent, horrific, deadly, and so on, the more traumatization and increased suicide risk. Even members not directly involved in violent or aggressive combat were affected. Exposure is sufficient—not necessarily combat exposure; there is secondary traumatization in many armed forces personnel. Thus, all types of combat violence increase risk, but some events result in greater fear, horror, and helplessness. Some combat events may well be suicidogenic.

## WHY ARE THERE SO MANY CASUALTIES ALLOWED?

If we know that soldiers are at risk for PTSD and suicide, and we know that soldiers with PTSD are at greater risk for suicide, why are there so many casualties allowed? In Chapter 1, I argued, as have many, that the system, the culture, the secrecy, and so on—factors at the community and societal level—are key. Is this true? Are we allowing our heroes to die? We did after Vietnam. Are we to blame? Are we allowing suicide in our heroes? Is the system causing deaths? The ecological model will help us answer these questions. It is now time that we stop blaming the soldier for her PTSD and suicide! There have been some recent studies to answer the system questions. In 2011, Mills, Huber, Watts, and Bagian examined the national Veteran Affairs database, using a root cause analysis, looking for causes beyond the individual. They looked for common themes (commonalities) among veterans of Afghanistan and Iraq who died by suicide. They looked at system/organization problems. The most common root causes were lack of an adequate assessment of suicide risk, lack of coordination of the care, lack of timely access to care, and lack of communication among providers. These, of course, are not only in the military, but beyond (Hendin, Haas Maltsberger, Koestner, & Szanto, 2006). Yet they are evident in the military. Are there, thus, liability issues? Is there deliberate indifference? *There are huge system gaps*. Mills et al. (2011) state,

> One root cause . . . was unique: "special needs of OIF/OEF veterans." This theme was prominent in 10 of the 51 cases we reviewed and may reflect the difficulty some facilities had integrating these new veterans into their system. The RCA reports also reflected considerable uncertainty with identifying exactly what those special needs might be. Some providers felt they had missed opportunities to provide special services that were available for OIF/OEF returning veterans, while other reports reflected perceptions that OIF/OEF veterans may have a more difficult time accessing and staying involved in treatment especially with the demands of work and possible redeployment. (p. 30)

Is it stigma? Secrecy? Culture? Mills et al. (2011) note, "Returned veterans may feel increased stigma around disclosing psychiatric symptoms, may have experienced military sexual trauma, may be more likely to have severe physical wounds or amputations, or may have increased difficulty" (p. 30). These are problems. And there are more:

> A vexing problem for suicide prevention in any system is how to handle a patient who does not come to his appointments or declines services altogether. In this study, virtually all cases had previous mental health treatment either provided or offered. This suggests that the problem is not that patients are not known to the mental health system, but that they are not adequately helped, indicating a fertile area for action. (Mills et al., 2011, p. 30)

## TRAUMA, ALCOHOLISM, AND SUICIDE

Norman Farberow, in his classic book, *The Many Faces of Suicide* (1980), explicates the variety of self-destructive experiences. This includes self-harm, single-car road accidents, alcoholism, drug abuse, neglect of medical illness, criminal activity, and so on. It is an almost endless list. One, however, sticks out in the military: this is alcohol abuse. It is a pervasive problem. As a source of reference, I looked at the index of the three major books on the topic today. In Figley and Nash's (2007) book, *Combat Stress Injury*, it is cited: *Around 50% of people with PTSD abuse alcohol or drugs.* This is mammoth! In Freeman, Moore, and Freeman's (2009) book, *Living and Surviving in Harm's Way*, alcoholism is cited 15 times (and, if we add substance abuse, 19 times). In Everson and Figley's (2011) book, *Families Under Fire*, alcoholism is cited seven times. Also, for alcoholism, it cites prescription medication abuse (such as the destructive Oxycontin and Percocet). Alcoholism is related to war in the military in specific ways. There is no question that PTSD and alcoholism are associated, and so is suicide. It adds to the lethal mix.

The problem of alcoholism is not simply an individual soldier's problem or an interpersonal one; it is also a community and societal problem. To put it in a straightforward manner: the green culture and alcohol use are married. Although drug use is not tolerated, alcohol use very much is. The soldier often comes back from the battlefield to a beer. For most, this is okay, but for a few, and that includes many of those with PTSD, it is not. What do we know about suicide and alcoholism?

As the topic of alcoholism in suicide is so vast, I will provide here only very brief comment on the topic and discuss my study of alcoholism in suicide notes (Leenaars, Lester, & Wenckstern, 1999). Alcoholism is associated with a vast number of suicides (Murphy, 1992). The lifetime risk of suicide in alcoholics is about 2.5% (and 3.5% for those with a history of inpatient treatment). The most systematic study of suicide in alcoholism is the work of Eli Robins and George Murphy (Murphy, 1992). Their primary mode of study was the psychological autopsy. In the Murphy study of 50 cases (1992), 16 suicide notes are identified. Until Leenaars et al. (1999), the notes had remained unanalyzed.

Leenaars et al.'s (1999) multidimensional model for the study of suicide was utilized to analyze the notes. The purpose of the study was to assess whether the 16 suicide notes of alcoholics in the Murphy (1992) archive differ from matched suicide notes of nonalcoholics in the constructs provided in Chapter 2 (and the 35 specific protocol sentences comprised in these dimensions in the TGSP). The comparison sample was derived from my archive, matching for age and gender (see Leenaars, 1988). The notes of alcoholics met a strict criterion for alcoholism (Murphy, 1992) and were by 11 men and 5 women whose mean age was 45.1 years (age range 32–57). Comparisons of neither the eight clusters nor the 35 specified protocol sentences reached significance. The main conclusions about alcoholism in suicide are in support of the hypothesis that there may be more similarities than differences in suicide, regardless of whether one is an alcoholic or not. The underlying traumatization fuels alcoholism, substance abuse, and suicide. It is lethal self-destructive behavior.

Data from a descriptive point of view in Murphy's (1992) model suggest that alcoholism in suicide is often associated with several factors: suicide in alcoholics is a response to unbearable pain, often including the alcoholism itself (however, it could be physical injury, a lost relationship, etc.). There is a history of trauma; for example, a failing marriage or the inability to stop drinking. The person is mentally constricted in these events. There are only core lethal beliefs. The alcoholic person is hurt, injured, and angry. Although there is much more to the death, the alcoholic person in the end wants to escape, finding the alcohol an insufficient egression. This is true for our soldier in a green culture; alcohol use is accepted, so abuse can be easily hidden. Dissembling (deception) and alcoholism in the military are almost an order Soldiers adhere to it. Of course, there are limitations to my study; for example it had no soldiers. There needs to be study. Be that as it may, it is of methodo logical significance that the study of suicide notes yielded the same main conclusions as the more expansive Robins and Murphy psychological autopsy study. Much more needs to be learned about the suicide of soldiers with alcoholism and substance abuse. Yet, for the armed services, the importance of alcoholism in our suicidal heroes (and not suicidal PTSD soldiers) must be addressed at this time.

## SECONDARY TRAUMATIZATION: FAMILIES, SPOUSES, AND CHILDREN

The families of traumatized soldiers and veterans are indirect victims of their traumatic experience. This is called secondary traumatization (Dekel & Solomon, 2007). Secondary traumatization is one of several terms "that have been used to label the manifestations and processes of distress reported by persons in close proximity to victims of traumatic events that they themselves did not actually experience" (Dekel & Solomon, 2007, p. 138).

Of course, as Everson and Camp (2011) note, "Military families, with historically lengthy separations, frequent relocations, and constant strain associated with heightened occupational risk. Military families are just as likely to be rendered helpless by impinging stressors and vulnerabilities" (p. 21). Everson and Camp further state, "Families are often as unique as the branch of the military in which they serve, the rank the service members hold, the installation to which they are assigned, and the length of time they have served in a specific branch of service" (p. 26). Simply, the Marine family is different from the Air Force family, and the commanders are different from the enlisted men/women. There are different subcultures. Yet, as discussed in Chapter 1, there are great similarities. Furthermore, of course, individual families differ in resilience and vulnerability.

On resilience and vulnerability, Everson and Camp (2011) write,

> Families are often perceived in terms of strengths and weaknesses. The notions of resilience and vulnerability stem from attributions made . . . resilience is the ability to meet challenges and cope with changes that threaten the integrity of the system while remaining intact and creating a positive environment for fostering member well-being. Some families are more resilient than others. . . . Families also may be vulnerable as a result of both internal factors (e.g., chaotic structure and inadequate rulemaking) and external factors, including excessive periods of time apart, economic downturns, or unforeseen catastrophic events. . . .
>
> Resilient families are also less prone to the development of crisis responses to catastrophic events; but of course there are limits to which any family system can be taxed emotionally, and any given family's ability to adjust to change and adapt to its circumstances is crucial to the idea of resilience. . . . Inadequate adaptations rendering a family more helpless in the face of adversity and the presence of new stressor events may also lead to increased vulnerability as result of a phenomenon known as "pile-up." . . . The continued presence of unresolved demands (i.e., pile-up) eventually leads to emotional exhaustion under the tenets of disorders or psychosomatic physical maladies in some family members. (pp. 15–16)

Combat stress injury has many victims, and there are many faces of the aftershocks. They simply cannot bear to keep going. PTSD spreads. Wives, children, parents, and even friends can become victims and casualties. "Maloney (1988) describes six wives of Vietnam veterans with clear PTSD symptoms, including dreams of the war and panic attacks triggered by the same triggers as their husband's, such as the buzz of helicopters, sudden noises, gunfire, and the smell and sound of spring rain" (Dekel & Solomon, 2007, p. 138). One wife described surviving her husband's war stress, as *"walking on eggshells."*

Separation and divorce among couples is common. It is not, however, only divorce; PTSD veterans also reported less marital satisfaction, less intimacy, and less self-disclosure and expressiveness than non-PTSD veterans. There is, in fact, a great deal of hostility and physical violence (Dekel & Solomon, 2007). Dekel and Solomon note that "the chronic strain that PTSD places on both partners in the marriage and on the relationship between them might lead us to expect heightened divorce rates among such couples" (p. 142). Spouses often feel locked in their marriages. Threats are common, not only toward others but also toward the self. (It is all so unbearably painful.)

There is, of course, considerable variance in the adjustment of wives of PTSD veterans. PTSD severity is associated with multiple factors, many echo the ones that place soldiers at risk. Indeed, the soldier's PTSD predicts the secondary traumatization. "Studies show that the more severe the husband's PTSD, the more severe the wife's distress" (Dekel & Solomon, 2007, p. 144). Avoidance symptoms especially affect the quality of the marital relationship. (Many heroes do, indeed, deny and avoid talking about war trauma.) Of course, consistent with domestic violence literature, the more frequent the PTSD husband's violence toward his wife, the higher the wife's distress (Calhoun, Beckham, & Bosworth, 2002). These observations seem obvious, and there have been some theoretical explanations; the one most consistent with the evidence is identification and empathy. Charles Figley has offered some insights:

> Figley (1995, 1998) uses the term "empathy" rather than identification; but, in essence, his account seems to be an attempt to explain how and why these wives come to identify with their traumatized husbands. As Figley (1998) relates it, the process starts with the wife's efforts to emotionally support her troubled husband, which leads her to try to understand his feelings and experiences and, from there, to empathize with him. In the process of gathering information about his suffering, she takes on his feelings, experiences, and memories as her own—and hence his symptoms. (Dekel & Solomon, p. 147)

Janoff-Bulman (1992) suggests that, as with PTSD in soldiers and veterans, the spouses' and families' *world assumptions* (beliefs) *are* blown apart.

> Just as the basic world assumptions of direct victims of trauma are often upset (Janoff-Bulman, 1992), so too are the assumptions of indirect victims. The partner of a traumatized man learns, just as he had, that the world is unsafe and chaotic and that being a good person does not protect one from harm. Her basic assumptions about the relationship are also upset. (Dekel & Solomon, 2007, p. 149)

It is, therefore, understandable that secondary traumatization is a normal consequence; not crazy.

It is, of course, not only the spouse, but also the children who suffer from the combat stress injury. Indeed, the military culture affects the children, often negatively. Sometimes, these children become orphans of the war (Cohen, Goodman, Campbell, Carroll, & Campagne, 2009). The friction has consequences. "Stress, trauma, and loss are normative parts of the military culture and come in many forms via transitions, deployments, and permanent losses, as in the case of bereavement" (Cohen et al., 2009, p. 397).

The sudden death of a military parent is tragic, but it is not the only trauma. Traumatization puts them at risk for a stress or trauma reaction. There is secondary traumatization. Children of military parents, of course, encounter potential traumatic experiences as any child (e.g., fires, car accidents). Yet certain unique, traumatic events can pose risks to military children, such as a combat or noncombat (i.e., accidental) death. The incidence of certain forms of child abuse has been found to increase during parental deployment (Gibbs, Martin, Krupper, & Johnson, 2007). Suicide, of course, given the high rate, is one more increased traumatization.

Whether suicide or otherwise, "given the rise of military operations since 2001 and the danger inherent in military service, the risk of traumatic bereavement is very real for children of military personnel" (Cohen et al., 2009, p. 401). There is a smell of death in the air. PTSD would be expected. Traumatic aftershocks may follow, especially given a sudden and horrific death such as by improvised explosive device or accident. This is even more so if a suicide. The children too feel shame and disgrace. Yet a child may have a traumatic reaction from other types of death too. These children are stressed and overwhelmed by emotions and reactions, and sadly, some kill themselves. Thus, the children too die by suicide. There is an illegacy. *There are many more uncounted, wounded, and dead!*

Yet we must not forget: Children in military families are typically resilient. They most often persevere. They do not give up the fight. We just need to help them to do so.

## A TRAUMATIC CONCLUSION

War is friction. War is trauma. There is no question about that fact. The studies of our heroes after the Vietnam War demonstrated that painful reality. Combat stress injury is a normal reaction—not crazy or abnormal—to the reality of being in harm's way. Our warriors experience, witness, and/or are confronted with events that would horrify, repulse, disgust, and infuriate any sane person (Rudoffosi, 2006). Therefore, it is no wonder that many have aftershocks. There is a shadowy presence. PTSD in many—suicide among them—understandably occurs. How could it not be? It would evoke significant symptoms of distress in most people. It would in me. And, as Janoff-Bulman (1992) pointed out, PTSD is associated not only with suicide but also other traumatic events (such as serious crimes/incarceration, accidents, homicide, alcohol abuse, and so on). The aftershocks are multiple, lethal, violent, and enduring. We know this since the Vietnam War, and today in Afghanistan and Iraq the intentional traumatization is beyond the scope ever imagined.

PTSD is a stress-based disorder following a life-threatening event. Of course, given my beliefs—that is, no brain, no mind—there is also a biological, or more accurately, a neurocognitive connection (Vasterling & Verfaellie, 2009). This is, of course, beyond the scope here to discuss in detail, but I need to provide an example. Functional neuroanatomical models of PTSD, for example, have often focused on vulnerability to stress in the hippocampus (Vasterling & Verfaellie, 2009). There may be memory dysfunction, for example. In a unique study, Woodward, Kaloupek, Grande, Stegman, Kutter, Leskin et al. (2009) studied 95 U.S. veterans with and without combat-related PTSD. They studied multiple cognitive measures of verbal and visual declarative memory, multiple memory-relevant regions, and brain volumes that had been shown to exhibit effects of PTSD. The researchers, in fact, showed a pattern of small and moderate effects of PTSD on memory. There are function-structural relationships, of course (no mind, no brain). Yet, of course, much more study is needed. For example, in all probability, executive functioning problems may be evident. The executive functions are a collection of processes that are responsible for guiding, directing, and managing cognitive, emotional, and behavioral functions, particularly during stressful situations. We need to know more if we are going to treat our heroes, with or without PTSD and with or without suicide risk, *wisely*.

War causes unbearable psychological pain. We know this from General Emory Upton. It results in intrapsychic and interpersonal anguish—and, yes our communities and society suffer too. We know from the research, such as Dr. Farberow's, that a soldier's adjustment to a trauma is largely affected by the type of response provided (see Figley, 1985). Fortunately, there are programs now being developed, at least in the United States, whereas countries such as Canada lag far behind. (Why?) Scurfield (1985) has noted that sound intervention (and postvention) appears to have a positive effect in preventing and lessening the severity of PTSD. It will decrease suicide too. Figley (1985) suggested that the critical question is this: Is the environment supportive or not? Thus, we can ask, Is the military supportive or not?

A tragic fact: As I was writing this concluding remark on July 14, 2011, the headline in my city's newspaper, *The Windsor Star*, read, Family Blames Military After Ex-Soldier's Death (Wilhelm, 2011). Canadian Army Trooper Stefan Jankowski died by suicide on Sunday, July 10, 2011. Stefan Jankowski, a 25-year-old veteran, had served bravely in Afghanistan. Corporal Hunter Kersey, who had served with Stefan, described him as a hero. He stated that Trooper Jankowski "often volunteered for the most dangerous position." They served in southern Kandahar with the Royal Canadian Dragoon, doing reconnaissance and looking for improvised explosive devices.

Stefan Jankowski was a warrior. Yet, according to his mother, Gina Duguay, he was traumatized; for example, "At one point he told me about a boy, half a face missing." He had said, "God, mom you cannot believe the stuff you'd see over here." He witnessed, confronted, and experienced horror, deaths, and events unimaginable to most of us—but not to a soldier. His friend died. He was injured. He was haunted by having to shoot at children. It was horror, fear, and helplessness. Understandably, Trooper Jankowski developed PTSD. He was treated for his wounds, and like many, became addicted to pain medication (Oxycontin). According to family, it escalated. He went AWOL. He was arrested. He was incarcerated (more red flags!). The military knew that he was troubled, had PTSD, and was at risk. He needed help! He knew it. The Canadian Armed Forces' solution: They discharged him. The family stated,

> The military was good to him, as a trooper. But when it came to giving him professional help because of the things he saw and the things that happened, the military made mistakes. Their program is improper. The people that need help are not getting it fast enough. (Wilhelm, 2011, p. A1)

This sounds like the very findings of Mills et al. (2011) and is highly consistent with my beliefs. *There are system problems.* The military culture and secrecy are problems.

Stefan came home to Windsor; there too he sought help. His family and friends knew that he was in serious trouble and begged for help. He obtained help from a lawyer. Little, if anything, was done. The drug abuse increased. According to the family, "The military ignored him. They just said you're discharged." He felt useless. His father reported, "They don't need him anymore." Yet, sadly, professionals in our city did the same; Stefan was taken to psychiatrists, professionals, and hospitals, but little help was provided. Wilhelm (2011) reported, "Jankowski went to local hospitals in the days leading up to his death. Esco [his lawyer] took him to triage at . . . Hospital Thursday night. 'He told them he was suicidal,' said Esco. 'Apparently, they would not take him.'"

He died on Sunday. One more hero died needlessly. How many of our warriors with PTSD will we let die?

# CHAPTER 12
# Suicide Prevention in the Military

Karl Menninger taught that, "The patient is always right." Much of our treatment of suicidal soldiers and veterans is figuring out how they are right, even if they wear a green mask. The way to understand how the soldier is right is in the natural progression from understanding to application to practice. Once one knows the soldier or veteran, then treatment comes naturally. The proof of suicide *prevention* is in the "ventions"—as in prevention, intervention, and postvention. We are here most focused on suicide prevention and intervention in the military. We begin with the question, Can the military prevent suicide among soldiers and veterans?

## WHAT DO ARMED FORCES NEED TO DO?

Earlier, it was established that the main obstacle to help-seeking in our warriors is the negative attitude toward such. There are enormous green barriers. What are these green walls? Are they different than among civilians? Why has this remained the main obstacle? It was for General Upton during the great suicide epidemic of the Civil War in the United States. A fact (Lambert, 2004): A multimodal or multicomponent approach is needed to help suicidal soldiers; for example, psychotherapy, medication, hospitalization, and gun control.

Ultimately, the question remains, Does psychotherapy or environmental control or whatever help all soldiers? Are some warriors never helped? Why? The number one error: No communication. Every commander can resonate that one. General Colin Powell resonated this in the opening quote to Chapter 1:

> The day soldiers stop bringing you their problems is the day you have stopped leading them. They have either lost confidence that you can help them or concluded that you do not care. Either case is a failure of leadership.

How do we progress from conceptualization to understanding and then move to what we can do to prevent the events? The research says that we have to be person-centered or soldier-/veteran-centered (Leenaars, 2004). We also have to be system-centered. Ultimately, no treatment with suicidal soldiers can be isolated: Our treatment may involve the commanders, family, friends, fellow soldiers, minister, family doctor—all of us, not only the psychologist or psychiatrist. It takes the military community. Suicide prevention for our fellow soldiers and veterans is everyone's business, every military personnel's duty, beginning with the president and the prime minister.

What can the armed forces do? This chapter *begins* to answer this question. The reason for the word "begins" is simply the lack of military efforts to date. It is only of late that the military in the United States, after finally accepting the problem, has begun some prevention efforts. It is courageous of the U.S. Armed Forces to change this. Regrettably, in Canada, there is only a beginning official recognition of the problem. There is still denial about the problem, but then this is no different than about suicide in general (Leenaars et al., 1998). However, avoidance causes problems. Are they still labeling soldiers with PTSD; with personality disorders? The Canadian Forces need to accept *their* unacceptable. *The militaries have to be visible, and if they are visible, that's how the problem will become visible. Then we can do something about suicide.* Therefore, what is the evidence on suicide prevention in the military?

## THE FIRST STEP: ACCEPTING THE PROBLEM

The Department of Defense (2010) is attempting to meet the challenge. The department has called for action. They state,

> Institutions and entities must be organized structurally for suicide prevention, and leaders must be involved at every level. Consequently, the Task Force strongly believes that suicide prevention begins with coherent policy generated from OSD [Office of Secretary of Defense]. Command attention and review at the highest level will ensure that every other level of supervision and care pays careful attention to the well-being of members of the Armed Forces, that suicidal thoughts and behaviors are reduced, and that suicide risks are minimized. (p. 36)

The Department of Defense recognized the traumatization of war. They accept PTSD and its cousin, adjustment disorder, as military "friction." They state,

> PTSD and other traumatic stress disorders can be triggered by work-related traumatic events, particularly in occupations where exposure to such events is common. Workplace stress and conflict have been identified as common triggers for suicidal behaviour in service members. Thus, mitigation of work stress and strain through organization interventions such as training, policies, and programs might have suicide prevention effects. (p. 37)

Yet they recognize the main need, that of leadership. They argue that

> ordinary good leadership skills were likely to be far a more potent suicide prevention tool than specific suicide prevention skills. This general principle

applies when it comes to organizational policies: Those that effectively mitigate work stress are likely to be more powerful tools than suicide prevention policies *per se*. It should go without saying that preventing workplace stress and strain (e.g., through prevention of harassment) makes far more sense than helping employees cope with its consequences. And as was pointed out earlier, mitigation of work stress and strain will also offer innumerable benefits beyond suicide prevention. (Department of Defense, 2010, p. 37)

## SOME PRIORITIES FOR THE MILITARY

Regan, Outlaw, Hamer, and Wright (2005) note the now well-known fact that suicide is the second leading cause of death in the armed forces after unintentional injuries. (It is actually the first leading cause of death in the Canadian Forces.) Thus, Regan et al. argue that the immediate assessment of the suicide risk in a soldier is crucial to suicide prevention. However, military personnel are reluctant to seek mental health services even when feeling suicidal. This is *the* lethal fact. This reluctance or fear is for multiple reasons, but one reason is the very fact that being suicidal has a negative effect upon their military career goals. Simply, if you are suicidal, you lose your gun—and career. Regan et al. further note that, because of stigma, only 23% to 40% of those with mental health concerns sought assistance after serving in Iraq and Afghanistan. There are such high barriers.

This has been the same for veterans for decades, probably even during the Civil War. Few seek help. Basham, Denneson, Millet, Shen, Duckart, and Dobscha (2011) examined characteristics on VA healthcare utilization of U.S. veterans who died by suicide in Iraq and found that *very few* seek help. Suicidal warriors do not seek care. Basham et al. note that veterans who die by suicide access VA health care at similar rates as veterans who do not die by suicide. Thus, the researchers ask, are there veteran-specific risk factors in the suicide group? They found few significant demographic differences between suicide decedents who access VA care and those who do not. Is there no difference?

Basham et al. (2011) did find "that veterans who accessed care in the year prior to their death were more likely to have received service-connected disability benefits, consistent with prior evidence of the association between gaps in mental health treatment and not having a service-connected disability" (p. 293). There are, thus, a few clues to guide us. This is of importance, since the common belief is that psychiatric disorder is the best predictor. This may not be true. Basham et al. found that

> over half of veterans who received any care and completed suicide did not have a mental health diagnosis and over half did not see a mental health professional. Although psychiatric features are indeed an important risk factor for suicide, and may be under detected, it is perhaps equally important to understand patients with other, more general medical conditions may also be likely to complete suicide. (p. 294)

Regardless of whether this finding is due to underidentification, the low incidence of mental health disorder, or to stigma, we have to go beyond a diagnosis to prevent suicide in the military.

Do we really know how to identify a soldier at risk? The military is now also visible about this question. A major neglected priority is adequate training in suicide risk assessment and treatment. The Managing Suicidal Behavior Working Group (n.d.) in the *Air Force Guide for Managing Suicidal Behavior—Strategies, Resources and Tools*, Brooks AFB, TX: Air Force Medical Operations Agency (AFMOA) Population Health Support Division state, "Multiple surveys indicate that training on suicide assessment and intervention in mental health training programs is variable and often inadequate" (p. 6). The problem is being recognized; yet is anything really being done to change this lethal barrier? It can be done, maybe not perfectly, but at less cost.

General Eric Shineski, Army Chief of Staff, U.S. (2000), notes the same point on suicide prevention—"Could I have done more?"—published in *Hot Topics: Current Issues for Army Leaders*. He stated,

> We must understand the potential for suicides and increase awareness for recognizing individuals who are at risk or exhibiting self-destructive behavior. It is our responsibility to help our soldiers and civilians understand how to identify at-risk individuals, recognize warning signs, and know how to take direct action. Then we must act to provide immediate, active assistance and intervention. (p. 3)

This is most informative. It is for this very reason that General Shineski (2000) called for taking a proactive approach. General Shineski stated with courage, "*It is our responsibility.*"

This will help! Commanders need to own the problem. However, as Doctor Snow noted, we first have to know the problem. Research in the military is essential, including on mental health disorders (psychopathology), such as PTSD. The military needs to be research-centered. Hoge et al. (2003) outline some priorities for research in the U.S. military. They write,

> The first priority is to better define the burden of mental disorders in terms of incidence, prevalence, severity, risk factors, and health care use. The impact of mental disorders on occupational functioning, particularly among new recruits, needs to be better characterized. Suicide research should include efforts to validate mortality data, define the normal level of rate variability, and establish surveillance for clusters. (p. 182)

We should better define the burden of mental disorders in terms of incidence, prevalence, severity, and risk factors in the military population. This is surveillance. Yet, as we just noted, many soldiers and veterans at risk do not have a defined mental disorder. We need to go beyond this medical view. We need to adopt an ecological perspective and methodology. There are many things then that can be done. Public health can help. Hoge et al. (2003) offer the following tactics:

> The highly structured occupational environment in the military lends itself to studies of preventive interventions designed to reduce disability or occupational attrition resulting from mental/behavioral problems. Examples of such preventive intervention trials include evaluation of life skills training programs in basic training, anger management, stress reduction programs, cognitive-behavioral approaches, programs that are targeted at reducing specific health risk behaviors, and programs to destigmatize mental health treatment and improve earlier access to care. (p. 184)

Furthermore, there are also very unique efforts that can be put in place.

## SUICIDE PREVENTION IN THE MILITARY IS EVERYONE'S ORDER

Suicide prevention in the military will take everyone. Payne, Hill, and Johnson (2008) recommend the use of unit watch or command interest profile in the management of suicide risk. The same, they state, can be done for homicide risk. Military mental health professionals have, for decades, recommended that commanders implement a unit watch (now called a "command interest profile" at most Army posts) as a tool for enhancing the safety of personnel in the unit when a soldier presents risk. Yet the management of risk is difficult. Payne et al. note that unit watch is a convenient practice in a military setting. What does unit watch entail? Payne et al. clarify,

> A unit watch encompasses a variety of interventions initiated by a soldier's command team based on a recommendation from a clinician. These interventions typically include searching the soldier's belongings and living quarters for dangerous items, removing such items from the soldier's possession, prohibiting access to alcohol and drugs, minimizing contact with people that may negatively influence the soldier's mental health, continuously observing the soldier, and ensuring that the soldier returns for further evaluation and treatment. (p. 25)

A unit watch is an excellent example, they note, of the military clinician working together with the command team to address the soldier's needs. Unit watch protocols first and foremost introduce environmental control, namely, limiting the access to lethal means for suicide (or homicide). This means removing the gun. Of course, soldiers may gain access to items such as medications, ropes, or knives. Yet it is highly important to limit access to firearms, medications, and other means. We know that restricting a means decreases risk. Restriction of drugs can be effective. Payne et al. (2008) note, "the potential for access to alcohol and drugs can also be substantially reduced through a unit watch" (p. 26).

There is more that can be done environmentally in the military, often not possible in civilian life. We can limit contact with toxic individuals who may exacerbate the soldier's suicidal risk. Harassment and sexual harassment, for example, has been a toxin. Sexual attacks and rape are all too prevalent (Department of Defense, 2010). The incidences, of course, do not occur in the U.S. military alone. They occur, for example, in the Canadian military (Minsky, 2010b). There were 163 sexual attacks reported in the Canadian military in 2007; this is similar to incidence in 2009 ($N = 266$). This is troubling. Yet the Canadian military reported that these numbers are not necessarily accurate (Minsky, 2010b). There are probably more.

One benefit of many military programs, such as unit watch, is averting hospitalization. There is a benefit, not only for the soldier but also for the forces. Payne et al. (2008) state, "One reason to avoid hospitalization is that the soldier maintains occupational functioning at some level, which may help the soldier preserve a sense of self-worth and belonging" (p. 26). Stigma, discussed earlier, is huge, and "this stigma sometimes has a profound effect on the reintegration of soldiers" (Payne et al., 2008, p. 26). Suicidal soldiers are often called "psycho" or having been "locked in a rubber room." This has to stop! Commanders are responsible for addressing this need. Even on unit watch, *several episodes* of ridicule and verbal harassment by unit members have occurred. The stigma is one of the greatest barriers to wellness in our warriors. Tearing down the walls begins at the top, and that means with the president or prime minister. "For all of these reasons, it is vital to convince the command team to set a supportive tone for the soldier during a unit watch" (Payne et al., 2008, p. 27), or any mental health care. They are good soldiers, not cowards or manipulators, with a personality disorder.

However, is unit watch, as any care, helpful? Regrettably, "there exists essentially no research that directly addresses the safety and efficacy of a unit watch as an intervention" (Payne et al., 2008, p. 28). There is, in fact, very little research on any prevention in the military. This needs to change. Fortunately, this is slowly beginning in the United States—but only beginning; not so in Canada.

The main reason that unit watch has worked is because of a buddy system. A fellow soldier (read: friend) is attached to the suicidal soldier—soldiers helping soldiers. Greden, Valenstein, Spinner, Blow, Gorman, Dalack, Marcus, and Kees (2010) offer an excellent overview of a unique suicide prevention strategy in harm's way. The paper, "Buddy-to-Buddy, a Citizen Soldier Peer Support Program to Counteract Stigma, PTSD, Depression, and Suicide," was published in the *Annals of the New York Academy of Sciences*. Greden et al. note that 40% or more of approximately 2 million armed troops deployed to Afghanistan and Iraq were citizen soldiers (National Guard and Reserves). Out of the group, 25% to 40% have developed PTSD, clinical depression, sleep disturbances, or suicidal thoughts. Upon returning home (transition), these National Guard and Reserve soldiers encounter additional stresses and barriers to obtaining care:

Specifically, many civilian communities lack military medical/psychiatric facilities; financial, job, home, and relationship stresses have evolved or have been exacerbated during deployment; uncertainty has increased related to future deployment; there is loss of contact with military peers; and there is reluctance to recognize and acknowledge mental health needs that interfere with treatment entry and adherence. (Greden et al., 2010, p. 90)

Greden et al. (2010) estimate that half of those needing help are not receiving it. There are white walls. The question raised is, What can be done to remove these barriers? To address this question, one effort developed was a unique green-culture-friendly peer-to-peer program for soldiers called Buddy-to-Buddy.

Like other authors and military personnel have pointed out, Greden et al. (2010) note that the most important barrier is the inexorable stigma associated with seeking health care. Buddy-to-Buddy was a program specifically designed for returning citizen soldiers to counteract stigma and treatment entry. They were ingeniously "using culture to change culture." One has to remember a reality that many soldiers convey: "If you haven't been there, you don't get it." They also strongly hold some core beliefs: "We believe in taking care of our own" and "Other veterans can be trusted" and "Doctors cannot be trusted." Cultural barriers impede treatment entry or adherence; thus, peer-to-peer influences were seen to be "a crucial cultural starting point in overcoming them" (p. 93). For these and many reasons, Buddy-to-Buddy was developed. "Buddy-to-Buddy ensures contact with every returning soldier by using soldier peers" (p. 93). That will help.

Consistent with military cultural traditions, Greden et al. (2010) note, "Buddies, families, and resiliency" became constant messages, accompanied by the messages of "you are not alone, treatment works, it has helped many of your buddies, and pursuing help is a sign of strength" (p. 93). Buddy-to-Buddy programs made suicide prevention everybody's order. They have strong potential to augment suicide prevention programs by applying the use of culture to change the culture of treatment avoidance. There are many possible tactics. Strategies would emphasize earlier identification of suicide risk, threatening or self-injurious suicidal behaviors, knowing about referral sources, "helping show the way," and like all good support systems, long-term support of adherence to treatment once started (Greden et al., 2010). There is hope and help. Soldiers helping soldiers.

## CONFIDENTIALITY AND THE MILITARY

Not only is there, thus, a lack of study on practice, but there are also unique ethical and legal issues facing mental health professionals in the military. What is the legal and ethical meaning for a doctor or psychiatrist labeling a traumatized soldier as having a borderline personality disorder, because that is the order from the very top command?

Jeffrey, Rankin, and Jeffrey (1992), in their article, "In Service of Two Masters: The Ethical-Legal Dilemma Faced by Military Psychologists," highlight *the* major dilemma for healthcare personnel in the military. They state, "Military clinical psychologists may find themselves caught between apparently contradictory requirements of the Department of Defense (DOD) and the American Psychological Association (APA)" (p. 91). This is equally true for psychiatrists, nurses, and all professionals. This needs to be still addressed. To date, there is no guidance. Thus, I offer at least a prolegomenon on the topic of confidentiality.

Treatment of soldiers cannot work without an atmosphere of trust. The foundation of trust is confidentiality. Doctors, psychiatrists, psychologists, corpsman, nurses, and so on must adhere to a code of confidentiality. However, experience and research show that soldiers and veterans are reluctant to see a mental health professional. Many fear what they say in psychotherapy, for example, will not be protected. Some choose death over talking. Warner, Appenzeller, Grieger, Belenkiy, Breitbach, Parker, Warner, and Hoge (2011), for example, assessed "the influence of the anonymity of the screening process on the willingness of soldiers to report mental health problems after combat deployment" (p. 1065). The reporting of mental health problems after deployment is routine. Warner et al. undertook anonymous and unanonymous surveys ($N = 3,502$ U.S. Army soldiers). They found a twofold to fourfold higher reporting of PTSD, depression, suicide risk, and interest in receiving help in the confidential surveys. Soldiers believe that they cannot be honest! Confidentiality was/is a legitimate issue. Is knowledge of a soldier's PTSD, for example, kept secret? Regrettably, the answer is all too frequently "no." Many soldiers want help; yet there are barriers!

Allow me to illustrate with the Canadian case of Gulf War veteran Captain Sean Bruyea. He was an intelligence officer. After service, Captain Bruyea, like many veterans, had complained about his health care from Veterans Affairs Canada (Curry, 2010). He complained to the federal government. That is understandable; he wanted better care. However, because he complained, Veterans Affairs and other government personnel/staffers saw him as a problem and obtained Sean Bruyea's mental health records. Captain Bruyea became aware of this fact. Using privacy laws, Bruyea obtained 14,000 pages of government documents related to him. The documents showed that after making complaints for better care/compensation in 2005, his health records were widely shared. The Ministry of Veterans also saw the records, both from the Liberal (read: Democrat) and Conservative (read: Republican) parties. A circulated e-mail reads, "Folks it's time to take the gloves off here" (Curry, 2010). Government personnel saw Bruyea's behavior as harassment, in fact.

Canada's Privacy Commissioner investigated the case and concluded that Sean Bruyea's complaints of invasion of privacy were well founded (Curry 2010). Retired Colonel Michael Drapeau, who supported Captain Bruyea said, "I've never seen anything like this. It's outrageous." Yet the recommendations to address this problem in the Canadian military have been, Colonel

Drapeau said, "banal." Of course, in the United States, it is no different. Will it ever change? And on one more point: Is it any wonder that soldiers and veterans are reluctant to seek help?

There will be barriers, beyond confidentiality, but invasion of privacy is a major obstacle to the effective implementation of mental health services in the military. The docs are not trusted. "A major concern is confidentiality and the worry among veterans of being stigmatized or experiencing other negative consequences for future military participation or other activities if identified as being "high risk" for suicide" (Bruce, 2010, p. 102). It is, thus, totally understandable that soldiers and veterans have concerns about confidentiality. We need to change this.

## CHALLENGES AND CONSIDERATIONS FOR MANAGING SUICIDE RISK IN HARM'S WAY

There are many challenges and considerations for managing suicide risk in combat zones. Bryan et al. (2010) present probably the best paper in the literature on this topic: "Challenges and Considerations for Managing Suicide Risk in Combat Zones," published in *Military Medicine*. Having served as healthcare providers in harm's way, they are in the best position to write on this topic. Bryan, Kanzler, et al. state,

> Military mental health professionals deployed to combat zones face a number of challenges and barriers for effective risk management that are unique to the deployed setting. To date, there exists no body of literature identifying areas in which suicide risk management differs between garrison and combat settings to guide mental health professionals in improving clinical decision making with respect to managing suicidal service members in combat zones. (p. 251)

One of the main obstacles is the gun. This is no surprise by now. "Among deployed soldier suicides and suicide attempts, firearms are used with much greater frequency when compared to nondeployed soldiers (e.g., 93% vs. 52% of suicides)" (Bryan, Kanzler, et al., 2010, p. 714). The availability of lethal weapons is a problem; there is no gun control. Bryan, Kanzler, et al. wrote,

> Availability of lethal means for suicide—notably firearms—is perhaps the single most salient factor that distinguishes deployed from nondeployed settings. Simply put, "everyone's packing heat" in a combat zone. In fact, in many areas, military personnel are required to carry their weapon at all times and will be denied access to basic facilities (e.g., dining halls) if unarmed. (p. 714)

They advised that, "deployed clinicians should therefore give considerable weight to the widespread availability of highly lethal means in combat zones when assessing suicide risk, which may necessitate recommendations to medically evacuate service members from the combat zone to maximize safety" (p. 714).

Controlling the environment is paramount (Leenaars, 2009b; Leenaars Cantor, et al., 2002).

A second major barrier is the treated soldier feeling estranged. The suicidal soldier feels cut off and lost. He believes that he is no longer a good soldier. She believes that she is a burden. As discussed in Chapter 1, the wounded suicidal soldier often feels loss of the group. The collectivism fades. She is disenfranchised. This system factor is not to be underestimated in the military—but also, it is a potential protective factor. Bryan, Kanzler, et al. (2010) learned the following in combat zones: "The sense that one belongs to a group and feels connected with others is an important buffer against suicide" (p. 715). However, combat service can decrease feelings of collectivism in a variety of ways. Bryan, Kanzler, et al. offer the following example: While deployed, soldiers are separated from families and friends. Family and friends are often primary sources of support, and that loss can be traumatizing. Indeed, it is well known that soldiers "commonly describe difficulty talking about their experiences to friends and family, a sense of being 'out of place,' or feeling 'emotionally numb'" (p. 715). This is estrangement and can feel like loss/rejection and even abandonment. It may be a narcissistic injury. However, "military service can have a considerable positive influence on social cohesiveness as well" (p. 715). There is a "brothers-in-arms" identification (attachment). Unit watch is an example, but "it can be easily misconstrued as punitive action and can foster mental health stigma" (p. 715).

There are further walls. A third major barrier is limited access to mental health services. Bryan, Kanzler, and colleagues (2010) state,

> In a combat zone, the full spectrum of mental health care is unavailable. Specialized services such as substance abuse treatment, inpatient psychiatric units, and family therapy simply do not exist. This substantially constrains the nature and types of services that can be reasonably offered to deployed service members. . . . Even in locations with available mental health resources, psychiatrically trained providers are scarce, resulting in a preponderance of general medical providers managing psychotropic medications. (p. 716)

Traditional models of "once-a-week check-ins" for treatment and ongoing risk monitoring are often, thus, an unrealistic strategy within harm's way.

Bryan, Kanzler, et al. (2010) thus argue, "It is critical to note that these forward-deployed medical facilities are not inpatient psychiatric units, and they should not be mistaken as such" (p. 716). Bryan, Kanzler, et al. put it simply: "Forward-deployed medical facilities do not have the resources or manning to meet adequate standards for inpatient psychiatric care. For acutely suicidal service members requiring inpatient care, deployed providers should therefore facilitate rapid medical evacuation" (p. 716).

Like me, Bryan, Kanzler, et al. (2010) note the green walls. Dissembling, even in a combat zone, is a frequent problem. They note, "Arguably one of the most frustrating issues in suicide risk management is accuracy of self-disclosure by the service member" (p. 716). There may be a few soldiers suspected of overreporting or "faking bad"—stating that he is suicidal when really he is not. There may be a number of reasons, such as to avoid an undesired task or responsibility (e.g., wanting to go home instead of remaining deployed). However, it is even more common that the green walls, as discussed, create fear of confidentiality, being perceived as a coward, and even being labeled with a personality disorder. As Bryan, Kanzler, et al. note, a

> much more prevalent and hazardous issue is that of under-reporting or "faking good," in which service members with suicide ideation, intent, and plans minimize or outright deny the presence of these symptoms, often due to fears of repercussions secondary to self-disclosure (e.g., stigma, duty limitations, negative career impact). (p. 717)

Bryan, Kanzler, et al. (2010) offer a summary and some lifesaving considerations for suicide risk management in combat zones. They are worthy of study by all militaries.

## PROTECTIVE FACTORS AND RESILIENCE

Of course, the armed forces need to go beyond suicide. They need to focus on the larger context. They need to address PTSD, for example. McDowell et al. (1994) note the need for multiple, diverse programs. They write,

> It might make more sense to offer programs to help people deal with failed relationships and financial, substance-abuse, and work related problems. These people do not kill themselves because they want to die; they kill themselves because they cannot cope with their problems, and suicide is a vehicle for making the problems go away. Programs that deal with those kinds of problems may have the indirect consequence of reducing suicides. (p. 112)

One can also assist in developing protective factors—one is resilience (or hardiness). Resilience is a personality characteristic or individual-level variable that protects wellness under stressful situations. Donald Meichenbaum (2012) has noted that 70% of military personnel evidence resilience; 30% manifest persistent adjustment problems. In a Canadian study, Lee, Sudom, and Rammsayer (2011) argue for the development of resilience, thus, in soldiers in the military. McDowell and his team (1994) agree and so does the U.S. military. As a U.S. example, the U.S. Air Forces (2004), in a landmark program, have shown that by using a communitywide suicide prevention program, suicide is a preventable public health problem in the military. The factors isolated include decreasing stigma, building social networks and help-seeking behaviors, and enhancing understanding of mental health in the community. Not only this program, but also others in the military show beginning evidence that suicide is indeed a preventable health problem in the military. Only if we have the will to do something, does the soldier not have to die needlessly. Suicide in the U.S. military can be and is being prevented. Regrettably, there are few programs in Canada and almost everywhere else in the world. Has nothing changed since the time of the U.S. Civil War?

## A SYSTEMATIC REVIEW OF MILITARY SUICIDE PREVENTION PROGRAMS

Is there a systematic review of suicide prevention programs for the military? For veterans? Bagley, Munjas, and Shekelle (2010) have undertaken this large empirical task. Bagley and his team undertook a systematic review of the literature on suicide prevention. They examined 3,406 titles and reviewed 261 articles. Out of the sample, they identified seven studies involving military personnel; this is a small number. This calls for caution in drawing definitive conclusions. According to Bagley et al., one of the best articles located was by Knox, Litts, Talcott, Feig, and Caine (2003), who described

> a multifactorial suicide prevention program implemented in the U.S. Air Force (USAF), comprising over 5,000,000 active duty personnel.
> 
> The intervention was designed to reduce stigma and risk factors, and strengthen protective factors in a population-based approach. The program had 11 components. There was training for squadron commanders, addition of suicide prevention into required training, use of guidelines for mental health referral, addition of staff to support community-based preventative services at mental health centers, and required training for nonprofessionals in suicide risks and referral procedures. The program also assessed for suicide risk those under investigation for legal problems, established teams to respond to traumatic events including suicides, integrated the delivery system for human services prevention activities, established patient privilege in psychotherapy, conducted a behavior health survey, and established a suicide event surveillance system for tracking risk factors. To evaluate the program the USAF population from 1990–1996 was the control cohort, and the 1997–2002 population was the treatment cohort. No differences in demographic characteristics, changes in which would be expected to modify suicide rates, or in rates for mental health disability were found between the two groups. There was a statistically significant trend for decline in suicide rate over time, with a 33% reduction of risk for completed suicide compared to the baseline rate. The average rate in the pre-intervention period was 13.5 per 100,00, and 9.2 in the post-intervention period. (p. 261)

From the systematic review, Bagley et al. (2010) concluded that multicomponent interventions in military personnel probably reduce the risk of suicide. What about veterans? Bagley et al. located only three studies on veterans. Caution is, thus, even more warranted. There is an urgent need for systematic study. Will it occur? If history is predictive, then the answer is no.

In their review, Bagley et al. (2010) thus identified studies of suicide prevention for military personnel. They write,

> Most used a conceptual model of risk factor identification, based on review of suicides in the population under study, augmented with factors previously identified by others, followed by educational and organizational changes to reduce those factors or increase education and awareness about them. All reported declines in suicides or suicide attempts. However, the reporting of sufficient data to make proper comparison was incomplete, and the quality of the analysis that was reported was generally poor. The largest studies were deployed for the U.S. Navy and Marine Corps and for the USAF. The methodologically strongest study was that of Knox, Litts, Talcott, et al. (2003) for the USAF. Its strengths include: very large study population, explicit description of 11 component initiatives in prevention program, use of linear time-trend analysis, and formal consideration of potential confounds from changes in the demographic makeup of the population over time. (p. 263)

Bagley et al. (2010) concluded, "In summary, multicomponent interventions in military personnel are consistent in reporting reductions in suicide. The largest and best described such study is by Knox et al., and it provides the most convincing evidence of effectiveness" (p. 260). There are still numerous questions.

## SOME COMMON FACTORS IN TREATMENT

The main payoff of all our surveillance, research, and training activities lies primarily in making our clinical and public efforts more effective. That is what counts in everyday military society. The military needs evidence-based application. In the military's attempt to intervene in order to predict and control the occurrence of trauma and suicide, we have to determine what works. We have to be multicomponent and multidisciplinary. To present the knowledge is beyond the scope of this book (see Leenaars, 2004, 2006). I will, however, note seven challenges to implement a strategy for the broader armed forces population.

First, the military needs some basics for crisis intervention. What do you do in an acute traumatic suicide risk situation? There is a need (Bryan, Kanzler, et al., 2010; Figley & Nash, 2007; Leenaars, 2004). I illustrated this earlier with a sad case of a highly suicidal female armed forces personnel who was raped in Iraq.

Next, psychotherapy is needed. The basis of evidence-based psychotherapy is the therapeutic relationship. It is a common factor, if not core (Lambert, 2004). There is today substantial empirical evidence on what makes psychotherapy effective and for that matter, harmful (Hendin et al., 2006). What is the best evidence-based psychotherapy? Third, although not the only standard, we need to look at randomized control trials research on what is effective psychotherapy with suicidal people. A systematic literature review on the question, What is the *best* treatment? reveals that there is not *the* effective therapy (Leenaars, 2010b; Rudd,

2000). There are many effective psychotherapies; cognitive, cognitive-behavioral, psychodynamic, dialectical-behavioral, and so on. There are *common* factors! Fourth, what are some common implications for response? If we understand suicide, then we can generate some implications for how to treat suicidal soldiers better (Leenaars, 2004). There are powerful therapy techniques, especially if used in a collaborative manner. Response can be designed to allow change. Research shows that there is a long history of what we know is effective, and not (Beck, 1976; Leenaars, 2004). This is based on the very understanding of suicide.

Fifth, the armed forces need to look at the basics of medication and what is effective medication (Stahl 2000). Sixth, hospitalization of suicidal soldiers in the military may be necessary; however, this should always be implemented with caution. Seventh, environmental control is needed in military strategies. The WHO (2006) has stated that the best evidence for preventing suicide comes from the research on environmental control (Hahn, Bilukha, Crosby, Thompson, Liberman, & Mosscicki, 2003; Leenaars, 2009b; Leenaars, Cantor, Connolly, EchoHawk, Gailiene, He, et al., 2000). The U.S. military agrees. Soldiers do not need to die. There are many antisuicide tactics.

## THE HEALTH COSTS OF WAR

Military's healthcare costs are rising; in the U.S. military, healthcare costs rose twice as rapidly as civilian healthcare costs (Zoroya, 2010). Of course, this is due to war-related injuries, but also mental health traumatization. Defense budgets in the United States, Canada, and other countries are rising. For example, in the United States in 2001, the defense healthcare cost was $19 billion; in 2011, it is projected to be $50.7 billion, a 167% increase (Zoroya, 2010). Of course, veterans of Iraq and Afghanistan are not surprised that the injuries, such as TBIs, exceed those for the veterans in World War II, for example. The injuries are taking their toll. This is especially true for mental health wounds, and this is not only for veterans but also their families. There has been secondary traumatization. This is not only true in the United States, but also Canada (Bailey, 2011). PTSD is most common, but also alcohol abuse. Lieutenant-Colonel Rakesh Jetly, Canada Forces, stated, "We're dealing with trauma, we're dealing with stress, and we're dealing with loss." (Bailey, 2011). This is true.

Canadian Defence Minister Peter MacKay echoes the U.S. view, and recognizes the cost; he stated, "There is no higher standard of civilian responsibility for a government than to treat those veterans who have put it all on the line" (Taber, 2010). Yet veterans in the United States and Canada are wary asking if they will say to injured traumatized soldiers, "I'm sorry, your wounds aren't serious enough to warrant this [help]" (Taber, 2010)? Is the military health system broke? Broken? There is cost associated to war. U.S. Secretary of Defense Robert Gates stated that the costs are "beginning to eat us alive " (Zoroya, 2010). War is costly. Yet will we continue to pay our veterans for saving our lives? Or let them die? Veterans did during and after the Civil War.

## CONCLUDING REMARKS

In suicide prevention in the military, no soldier should fight alone (Milburn, 2009). Will there be help? Will they be labeled as having a personality disorder? Will the health care be confidential? Will the United States and Canada just allow the soldier to fight alone, or will there be real help?

President Barack Obama stated that it is "a solemn responsibility to provide our veterans and wounded warriors with the care and benefits they've earned when they come home" (Pace, 2010). PTSD is recognized. President Obama further noted that PTSD is true for soldiers and veterans suffering secondary traumatization. No longer will veterans have to prove that the illness was caused on the battlefield. For example, doctors and hospital corpsman at health clinics are recognized as suffering PTSD too. This is also true for the family members, such as the spouses, the parents, and the children. War is trauma for all. The president said, "I don't think our troops on battlefield should have to take notes to keep for a claims application," adding, "I've met enough veterans to know that you don't have to engage in a firefight to endure the trauma of war" (Pace, 2010).

PTSD is real. At least 300,000 U.S. veterans have symptoms of PTSD or major depression (Pace, 2010). Is it a normal reaction to the "friction?" Finally, a president is accepting the responsibility. Will care/benefits now change?

What about Canada? Will veterans like Captain Sean Bruyea finally receive the help that they deserve? Maybe. Canadian Defence Minister Peter MacKay stated that he is "outraged" at the care of soldiers/veterans (Minsky, 2010a). We will see.

Minsky (2010a) tells a soldier's story. Corporal Stuart Langridge, a model soldier and veteran of the Yugoslavian (Bosnia) War and Afghanistan War, suffered from PTSD. After returning from the Afghanistan War, Stuart Langridge died by suicide. His parents questioned his care and benefits. There are many more parents of our fallen heroes that do the same. Mike and Helene Purcell, who you will meet, are fighting for their son Chris' rights. Many families are fighting this battle. Will the great green walls of the military come down? Important matters need to be addressed. Will they?

Stuart and Chris did not need to die. Sean has a right to privacy. Our warriors have earned the right for evidence-based care. The governments in the United States and Canada have said that they will provide that care. Many lives can be saved, if the governments and militaries have a will to do so. Have we not lost enough warriors in battle? PTSD is no different; it is an unbearable wound; and for a few, suicide becomes the only solution to their "friction."

# PART SIX

# A Case Study

John Stuart Mill showed that both the general (nomothetic) and the unique (the idiographic) are critical for science's development. Dr. Gordon Allport, one of the great American psychologists in the early 1900s, provided us with a classical statement on the advantages of an idiographic approach. Psychology is committed to increasing man's understanding of man; soldiers in general and soldiers in particular. Like psychology, what concerns the military deeply is the individual soldier. The real concern in this area is to offer a rich, flexible, and precise approach that does justice to the fascinating individuality of each warrior. This section hopes to meet that challenge.

In this section, I present one chapter on a soldier's story told. This is the heart of this book, the death of a real hero in the U.S. Navy: Hospital Corpsman Third Class Chris Purcell. It is a PA. Through the courage of his survivors and his personal documents (e.g., his suicide notes), why a soldier kills him or herself comes alive.

## CHAPTER 13
# A Soldier's Story Told:
# A Psychological Autopsy

Children in military families are typically resilient. Military families, however, as we discussed earlier, also have unique stressors. They experience loss, trauma, and events that are outside the range of most nonmilitary families. Further, each branch of the military—Navy, Army, and so on—would result in the family being exposed to distinct traumatic events. This is true for the children. It, thus, would not be abnormal to respond with intense fear or helplessness.

Chris Purcell was born on December 19, 1986, in Keflavik, Iceland. His father, Mike Purcell, and his mother, Helene Purcell, were in the Navy. The Purcell family was a proud decorated military family. Chris joined the Navy in 2005. But before I tell Chris' story, I need to let you know who Mike, Helene, Chris, and the Navy are. I present what Mike Purcell told me.

Mike Purcell enlisted in the U.S. Navy in 1982 in Great Lakes Illinois, afterwards graduating from Radioman "A" school in San Diego, California. In his 27 years of naval service, he and his family were stationed in Naples, Italy; Keflavik, Iceland; Great Lakes, Illinois (3 tours); Charlotte, North Carolina; Sasebo, Japan; and San Diego, California. Purcell served as a Radioman [renamed Information System Technician in 1999] overseas in Naples, Keflavik, and Baghdad, in Great Lakes as an instructor, as well as two forward deployed ships and one ship on the West Coast. In addition to serving as a naval communicator, he served in various capacities as a Drill Instructor, Drug and Alcohol Program Advisory, Family Advocacy Representative, Fleet Quality Assurance Inspector, and Individual Augmentee Coordinator. He served during two wars, on dozens of shipboard deployments, as well as a 6 month tour "boots on ground" in Baghdad, Iraq, as a Communications Watch Officer in the Joint Operation Center at Camp Victory. He received numerous unit awards and citations; his personal medals include six good conduct medals, five Navy Achievement, one Navy Commendation, and one Joint Commendation. Purcell retired as a Chief Petty Officer in 2009 with 27 years of honorable service.

Helene Purcell enlisted in the U.S. Navy in 1981 in Orlando, Florida, afterwards graduating from Radioman "A" school in San Diego, California. Her first duty station was Naval Oceanographic Processing Facility in Dam Neck, Virginia, as a communications equipment operator. Her second duty station was Naval Communications Area Master Station Mediterranean in Naples, Italy, as a communications clerk. Her final duty station was Naval Communications Station in Keflavik, Iceland, where she served as a Communications Watch Supervisor. She was discharged in 1986 with 5 years of honorable service.

Christopher Lee Purcell enlisted in the U.S. Navy in 2005 in Great Lakes, Illinois, afterwards graduating from Hospital Corpsman "A" school. His first duty station was Naval Health Clinic in Brunswick, Maine. While serving at the health clinic, Purcell was handpicked to serve his Commander-in-Chief on the emergency medical response team during a presidential visit to the Kennebunkport compound. Purcell was posthumously promoted to Hospital Corpsman Petty Officer Third Class in 2008.

They are a proud family; however, like all military families, they had their traumatic events. The worst nightmarish tempest was on January 27, 2008. *Chris died by suicide.* The mode of death was unequivocally suicide. Here, I will answer the following questions: Why did Chris kill himself? How did Chris die and where; that is, why at that particular time? This does not mean that I will not address a third question, What is the probable mode of death? Yet, in this case, this was already known. This chapter will present Chris Purcell's story; it's a soldier's story. To answer the questions in the case of Hospital Corpsman Chris Purcell, I will follow, as I have done, the principle of *res ipsa loquitur*. The data convey the story. I assembled them here as the case unfolded to me, beginning with the suicide of Chris. Next, I present Chris' suicide notes, his suicide chat on MySpace, and the analysis of the last communications. The data, beyond the suicide note examined, include a psychological report, newspaper accounts, medical records, court documents, and other personal documents. In Chris' case, we also have his own reflections on life in a Navy family. A most important technique in the psychological autopsy is the third-party interviews. I plan to present abbreviations of the interviews of Helene Purcell, mother; Mike Purcell, father; Kristin Purcell, sister; and Derek Ozawa, friend. Let me begin with the death.

## THE DEATH

There are, of course, almost always different perspectives on a death. Death is a perception. I here present verbatim a court account. I am sure the Navy, the Purcells, and so on, may have different views. I begin with the account from Circuit Judges Bauer, Flaum, and Evans as it is presented in Purcell v. U.S. (2011)

> Christopher Lee Purcell ("Purcell") committed suicide in his barracks at the Brunswick Naval Air Station, where he was serving on active duty in the Navy. Navy and Department of Defense ("DOD") personnel were called

to the scene after being informed that Purcell planned to kill himself. They arrived at his residence before he attempted suicide, but did not find the gun they were told he had. Later, they permitted Purcell to go to the bathroom accompanied by his friend. Upon entering, he pulled a gun from his waistband and committed suicide by shooting himself in the chest.

Purcell was twenty-one years old and working on active duty in the Navy as a hospital corpsman at the Brunswick Naval Air Station when he committed suicide. The brief submitted by Purcell's father, Michael Purcell, notes that shortly after enlisting, at the age of eighteen, Purcell began experiencing social and emotional problems. It also mentions that the Navy intervened on several occasions by providing substance abuse treatment and mental health care.

On January 27, 2008, someone contacted the base at around 8:30 PM to inform them that Purcell had a gun in his room and was threatening suicide. In response to the call, Junior Corpsman Stephen Lollis told base security that Purcell had a gun and was about to kill himself, and provided Purcell's address. DOD Police Officers Shawn Goding and Matthew Newcomb were the among the first local law enforcement officers to arrive at Purcell's apartment, followed by DOD Patrolman Francis Harrigan and Petty Officer First Class David Rodriguez. Each was aware that Purcell had a gun and was suicidal.

Purcell was alive when the investigating officers arrived at his on-base residence. They searched his residence and found evidence indicating that he had a firearm, including an empty gun case and bullets on top of a television stand, but they did not find a weapon, and they never searched Purcell's person.

Rodriguez spoke to Purcell and suggested they go outside to talk. Purcell responded calmly. Outside, Petty Officer First Class Mitchell Tafel approached Rodriguez and stated that they needed to get Purcell into custody to protect him and local law enforcement. Purcell became irate and non-compliant when told he would have to be put in restraints. A struggle with Rodriguez, Tafel, Harrigan, Goding, and Thomas Robinson, also with DOD, ensued. The five eventually subdued Purcell, handcuffed him, and escorted him back to his room. Once upstairs, Tafel permitted Purcell to use the bathroom and instructed Robinson to remove one of Purcell's handcuffs. Purcell went to the bathroom accompanied by his friend, Nathan Mutschler. After entering the bathroom, Purcell pulled his gun from his waistband and committed suicide by shooting himself in the chest.

## THE AUTOPSY REPORT

I will abbreviate the Final Autopsy Report authored by Major M., Deputy Medical Examiner, dated February 12, 2008.

> Circumstances of Death: This 21 year-old active duty Sailor was, by report, in his housing at Brunswick NAS when he called a friend stating that he was going to kill himself. Authorities were notified and responded to his

residence. He was restrained by the Master-of-Arms with handcuffs. At some point, he requested to use the bathroom and one of the handcuffs were undone. A corpsman was helping him steady himself when the corpsman noted that he had a gun. The deceased shot himself in the chest and all resuscitative efforts were unsuccessful. After examination of the scene by NCIS, the body was transported to the Maine State Medical Examiner's facility in Augusta, Maine. Dr. G. signed the manner of death as gunshot wound to the chest and the cause of death as suicide. The body was released to the jurisdiction of the Armed Forces Medical Examiner.

Authorization for Autopsy: Office of the Armed Forces Medical Examiner, IAW 10 USC 1471.

Identification: Positive identification by ante-mortem and post-mortem fingerprint and dental comparisons.

CAUSE OF DEATH: Gunshot wound to the chest

MANNER OF DEATH Suicide.

## OPINION

This 21 year-old male, Christopher Lee Purcell, died of a gunshot wound to the chest. The bullet injured the heart, liver, and aorta (major blood vessel). A deformed, jacketed bullet and fragment of jacket were recovered. There was evidence of close range discharge of a firearm (soot deposition on the skin surrounding the entrance wound and soot deposition and searing of the clothing). The abrasions of the face and small subarachnoid hemorrhage of the brain are consistent with either the history of being restrained or terminal collapse. The contusions of the wrists are consistent with being restrained with handcuffs. The toxicology screen was positive for ethanol in the blood and vitreous fluid. Autopsy findings and investigation are consistent with a self-inflicted injury. The manner of death is suicide.

M.
Maj, USAF, MC
Deputy Medical Examiner

Captain P., Chief Deputy Medical Examiner, undertook a toxicological examination. The report, dated January 31, 2008, reads in part as follows:

VOLATILES: The BLOOD AND VITREOUS FLUID were examined for the presence of ethanol (cutoff of 20 mg/dL), acetaldehyde, acetone, 2-propanol, 1-propanol, t-butanol, 2-butanol, iso-butanol and 1-butanol by headspace gas chromatography. The following volatiles were detected: (concentration(s) in mg/dL)

|  | Ethanol |
|---|---|
| BLOOD | 140 |
| VITREOUS FLUID | 180 |

DRUGS: The URINE was screened for acetaminophen, amphetamine, antidepressants, antihistamines, barbiturates, benzodiazepines, cannabinoids, chloroquine, cocaine, dextromethorphan, lidocaine, narcotic analgesics, opiates, phencyclidine, phenothiazines, salicylates, sympathomimetic amines and verapamil by gas chromatography, color test or immunoassay. The

following drugs were detected:
None were found.
No mefloquine was detected in the blood at a limit of quantification of 0.01 mg/L using liquid chromatography/mass spectrometry.
    P., Ph.D.
    CAPT, MSC, USN
    Chief Deputy Medical Examiner
    Office of the Armed Forces Medical Examiner

A note: The information in the military documents may contain information exempt from mandatory disclosure under the Freedom of Information Act; however, the reports were released to Mike Purcell and he, with informed consent, released them to me to have his son's story told.

## PERSONAL DOCUMENTS

Like Allport, Shneidman, and many others, I believe that the best documents are personal ones. I hereby present Chris' very personal documents: an autobiography, suicide notes, and his last chat. I also report here on his psychological assessment at age 14; it seems to fit best among his own documents. I begin with his autobiography at age 14, the beginning of his tempest.

### An Autobiography

<u>Chris Purcell</u>
<u>Mr. O.</u>
<u>2-10-00</u>
<u>LA 5-6</u>

### Life Blows

Is change for the good? I would say not. Ever since I moved to Northbrook last March, I have hated it here ever since. Everybody thinks that change is so good, but why? I rather things always just be the same, forever and ever.

I remember the first day I moved to Charlotte. I have faint memories of how long the trip took to get there. I was little and have brief dreams of when we moved to there, but I can recall it well enough. I was constantly moving; always walking around catching other buses that we needed to take in order to get to our cousins. I remember when we finally got to my aunts and uncles house. It was in the middle of winter, and we spent the night at my cousin's house (my mom's sister). I some what have the memory standing there looking at my cousins that I had never really met them that much. We stood there gawking at each other, I remember I was standing behind my dad grasping his leg, just staring for like ten minutes or so. But when I got over all the shyness that was shown on the outside of my face, and showed my smile underneath the shyness, we had a blast, we played so many games and it was so fun!

Then I have memories of when I was older. Behind my cousins house, they were building a highway, and were under constructions, so . . . there was a lot of mud, and my cousins (I had a lot there were 7 kids in their family) and I would go back to the construction site when no one was around. And we would play around there for fun, and we would have so, so, much fun in the winter. They had these little mud puddles, not really little, but HUGE! They would freeze up, so we would take like these huge poles and just toss them into the frozen puddles and watch them crack open. I did some pretty weird things with my cousins, and we had a great time.

I remember one time my two cousins and I were walking and talking and all of sudden my older cousin (James, he's now 16) just dropped, we looked over at him and his leg was in a huge pile of mud, we had to pull him out. My cousin and I were laughing at him, he was mad, but afterwards, he too found it funny. I remember going to my even older cousins (John, he's 18) baseball and basketball games. I remember my cousin would wake me up in the morning and ask if I wanted to go to his games, so then I would wake my mom up and ask her, she would usually say yes, then I would tell my cousin yes and his dad would come and pick me up. After that I would usually end up spending the night at my cousins house for a couple of days.

I remember my uncle real well; he was kind of carefree, last minute decision guy. For example, I remember when the movie "Mortal Kombat" had just came out, and it was like eleven at night and my uncle couldn't sleep so he came downstairs (where my cousin and I were playing or something) and said "do you guys wanna go to a movie or something" and we said "yeah." And we went to see Mortal Kombat and got back at like two in the morning. That was awesome. I remember when it was a Friday morning and my cousins were going to the beach, and they asked if I wanted to come with them. So, I got to skip school that day and go to the beach with them. It was so, so, so FUN! I have great memories of my cousins and I and I hope that I will never lose or forget them.

Then there was my school Piedmont Middle, I loved it there also. It wasn't the richest school or the best, but I loved it. I remember going outside after lunch and playing tackle football then getting in trouble by the other teachers. I remember all my old friends that I had there. When I hear certain songs, I remember Charlotte and everything about it. Even today when I hear songs I have memories of Charlotte. At my old school, it wasn't the best because there were always a lot of fights and threats and stuff like that. But, I still loved it there. I really remember my best friend Chris, I remember going to his house all the time. We would talk about anything and everything. I had the greatest memories of Chris and me.

Then, I remember those dreadful words that my mom said to me, they were "Chris, w... we're going to be moving sometime". It was around the middle of the school year. And by that I thought that my meant maybe in a couple of years. But then we were expecting it sometime in the beginning of next year. Then we were moving in two weeks, and I was like, "You're kidding right"? And my mom said, "no I'm not." And I just sat down to take a breather.

I remember saying goodbye to all of my friends. But, I was telling lies, it wasn't a goodbye at all, it was a badbye. I was saying "see-ya-later" when I would not see most of them for the rest of my life. That last day went by slowly, it all felt like a dream, that last day of school. But, I got to go to my friends for the weekend. I got to go home with him at least. I was saying goodbye to everybody, then I had to go. My friends sister was honking the horn, and I had to go. I remember seeing my friends faces dissolving over the horizon as we drove to his house.

I still had a little while to have fun before I was dragged into hell. I got to spend the week at my friends house. Then we got to go to the Hornets game. We got V.I.P. tickets, so we got to go in the media entrance. So that was pretty cool. We didn't see any basketball players, but we did see scouts for other teams and stuff, and that was pretty cool. I remember playing all week at Chris's house, we rented video games, and we stayed up all night long. Playing video games. Watching videos, and watching T.V.

Then I remember when my parents came and it was time for me to go. I still keep in touch with some of them, but that is not enough for me. Well, what can I do? My mom says were never going to move back to Charlotte. I guess I'll just have to get over it.

I remember my first day here, it was my sister and I, and I remember when I opened the doors to Wood Oaks, I remember seeing the big Wood Oaks banner in the hall to the side of the gym. I remember the smell of Wood Oaks. It smelled pretty good for being opened in 1972. It looked like it was built recently. I remember Mr. Louis showing me around. I was so nervous coming into all these classrooms with all these people I didn't know. I felt like hiding, and going home. I didn't want to be at Wood Oaks, I was totally silent like the first week of class. I didn't do well on any of tests or quizzes. I felt like an outcast. After awhile I started to hate Wood Oaks. And teachers think that I have no friends at all, I have a lot of friends, I just don't like Wood Oaks or most of the people, and I probably never will like it here.

Coming from a school where it was half whites, and half blacks, then coming to Wood Oaks where it's a lot whites, a lot of Asians, and a couple blacks is kind of a big change (I got nothing against them too). And then I remember how many Jewish people there were. (I got nothing against them). I don't think I have ever met a Jewish person in my life until I came to Wood Oaks. I remember thinking that Wood Oaks was going to be just like my old school. But I was way off. I would pinch myself and try to wake up from this hellish nightmare. But, I wouldn't wake up. I remember thinking that this is all just a dream and that one day I would wake up from it all. But, I probably never will. I know that there are things I cant change, I have to change myself. But, I will never change myself. I will always hate it here no matter what anyone says, or no matter what anyone does. I know I don't show it much that I hate it here but I do. Sometimes I feel like I don't care about school, so I will slack off, but that's getting me nowhere, so I try my hardest but sometimes my hardest just ain't good enough. In Charlotte I was getting A's and B's, here I'm getting B's and C's. I just wish everything would change back to normal, the way it used to be.

My dad keeps on saying that it will get better, back to normal, better than it was in Charlotte, but I'm seeing no light at the end of the tunnel. Now, my parents are already talking about moving again in a couple of years to Indiana, where my dad grew up. *IT IS SO BORING THERE.* Well, I guess my life lesson learned is that everything happens for a reason, I just got to find my reason. And I know that life will take me on journeys, places that I don't want to be, things that I don't want to do, but I will have to learn to deal with them. And life will throw a lot of crap at me, but I will have to keep on going on. For me that's hard, but I know I will have to get over it, and go on with my life. I hate Wood Oaks and always will. That's why life totally blows.

## PSYCHOLOGICAL REPORT

Chris Purcell's autobiography, written on October 2, 2000, got the attention of his teacher. He was in grade 8 at White Oaks Junior High School, Northbrook, Illinois. He had moved to this school in March 1999. Mrs. Purcell had reported that Chris was "depressed." Chris was exhibiting a poor self-image and low self-esteem, for example, calling himself "stupid." Teachers were concerned about significant anxiety and depression. There was also a concern about suicide risk. Chris was overheard to have stated to another boy about getting a "stun gun" and shooting himself with it. Mr. Purcell had reported that Chris had spoken about killing himself before when they lived in North Carolina. Chris Purcell was reported to be very upset about the frequent moves, due to being in a Navy family and a plan to move to Japan. He dreaded that fact.

On December 21 and 23, 2000, Chris Purcell, at the age of 14, was seen for a psycho-educational assessment. A report, authored by Dr. N., offers some early insightful observations. The relevant background information indicated that Chris did well in Kindergarten and was a "joy to teach." During his early education, he received good grades, abundant acceptable scores on standardized tests, and exhibited "excellent" conduct. In sixth grade, he received A's and B's; however, math was problematic (he received an F). In grade 8, we read a different story, and I quote verbatim:

> His first quarter grades were: B's in applied studies, language arts, and PE; C's in art, reading, social studies, and math; and an F in science. Teachers note that the content of his work fluctuates. . . . he is not focused. He appears not to care whether he is successful.

Dr. N. saw Chris for an interview. Aside from academic concerns, when Chris was asked, Dr. N. recorded the following problems:

> Chris reported adamantly that he is "not good at school." His biggest problem at school is "keeping organized." He forgets to do or misplaces assignments. Wood Oaks is much harder than his other school. He admitted readily to feeling depressed, saying, "<u>Everything</u>" is bad, not just school." Chris does not like most of his fellow students, perceiving them to be "stuck up."

However, he does enjoy meeting students from Highland Park, whom he perceives as being more diverse, less driven by high academic achievement, and "not so stuck up." Interestingly, despite his academic and behavior problems, Chris appeared to genuinely like all his teachers, with one exception, about whom he feels "neutral because I don't especially like the subject."

The procedures used for the assessment were standardized ones; for example, Interview with Chris; Wechsler Intelligence Scale for Children-III (WISC-III); Wechsler Individual Achievement Test (WIAT); Achenbach Teacher Report Form (TRF)—1991 version; Achenbach Youth Self Report Form (YSR)—1991 revision, and so on. The assessment findings were as follows: Based on the WISC-III, Chris earned a Full Scale IQ of 103 (58th percentile), which classifies his overall ability as falling in the Average range. Chris' nonverbal/perceptual skills were better than his verbal skills. The nonverbal/perceptual abilities (Perceptual Organization Index) were in the high average range (POI = 110, 75th percentile), whereas his verbal abilities (Verbal Comprehension Index) were in the average range (VCI = 98; 45th percentile). Such a difference is not significant. Verbal subtests ranged from 8 on Comprehension to 11 on Information; Performance subtests ranged from 9 on Block Design to 15 on Picture Completion. Thus, there did not appear to be a specific learning deficit.

The WIAT assesses reading recognition, reading comprehension, written spelling, and mathematics. Chris' level of academic skills, based on the WIAT, is as follows: Reading Composite (104 Standard Score [SS], 61st percentile, approximate grade equivalent [ge] 8); Reading Recognition (102 SS, 55th percentile, ge 8); Reading Comprehension (107 SS, 68th percentile, ge 8); Spelling (107 SS, 68th percentile, ge 8); Mathematics Reasoning (97 SS, 42nd percentile, ge 7); and Arithmetical Calculation (71 SS, 3rd percentile, ge 4). Thus, Chris achieved at least average, except in arithmetic. It was a significant weakness.

The Achenbach Child Behavior Checklist (1991 version) for ages 6–18 asks people to describe behavior of the child, now or within the past 6 months. The teacher completed the Teacher Report Form (TRF), and Chris completed the Youth Self-Report Form (YSR). I will quote verbatim Dr. N's observations:

> Several of Chris' teachers completed the Teacher Report Form, a questionnaire that lists problem behaviors and attitudes that teachers may observe in their students and compares the student's behavior to age and gender peers. Both teachers reported a large number of concerns in several areas. Specifically, they observed a significant number of behaviors found in depressed teenagers: sadness, sulking, tiredness, self-consciousness, apathy, lack of motivation, withdrawal from class activities, inattention, failure to complete work, and poor school work. Teachers also observed that Chris often appears tense and moody and tends to associate with other students who get in trouble. All of Chris' teachers observed that Chris has a very poor self-image as a student. . . . On the TRF, teachers reported highly

significant levels of anxious/depressed behaviors (> 99th percentile) and significant attention problems (98th percentile).

Chris willingly filled out questionnaires. . . . Chris reported that he often could not get his mind off his thoughts in order to concentrate on schoolwork. He hates school and wishes that he could attend school in Highland Park. Chris acknowledged disobeying his parents and teachers many times but had shared no insights in these behaviors. He does not want to move to Japan but is preparing for this eventuality by seeking out other teenagers and adults who have lived there and asking them about their experiences.

Chris feels confused, "in a fog," anxious, and fearful most of the time. He has many depressive, self-defeating thoughts such as "nothing will ever work out for me," and believes that he has never done well in school and that is he "retarded." Chris ruminates about "the meaning of life" and is fearful that there is none. He is very disappointed in social relationships, painting a broad picture that "People suck." Chris stated that he feels "terrible," is always tired, has trouble sleeping at night, and hardly ever has fun at school. He does not see a future but only a painful present. He worries about many aches and pains—headaches, stomachaches, and nausea. . . . Chris denied on questionnaires and in the interview ever having suicidal thoughts or suicide attempts. He dismissed his comment about the stun gun as "just goofing off." On the assessment's structured questionnaires, Chris' responses fell in the highly significant range in somatic complaints, withdrawal, attention, and anxiety/depression (YSR).

Conclusions. The main and final conclusion was: "Chris presented as an extremely depressed, demoralized youngster."

Thus, Chris was diagnosed with a mood disorder, depression, already at the age of 14. It is probable that he was depressed before that age. Not only was the depression evident, it was deemed that the symptoms caused clinically significant distress or impairment in social, school, and other important areas of functioning. It caused, in fact, in light of no cognitive deficits, significant academic problems. Chris was deemed to be experiencing trauma, most significantly, the frequent moves due to the Navy life. He was reacting with horror, fear, and helplessness to the next move to Japan. Chris may well have been suffering from PTSD at about age 14. There was predictable traumatization; in fact, many similar children in military families are also stressed and respond with suicidal reactions.

Chris was suicidal at age 14. This is unequivocal. Yet we do not know the lethality. Was it medium? High? Chris also exhibited characteristics, already then, of denial, avoidance and dissembling. I believe that he lied. This was to be an enduring characteristic of his all too short life.

In her report, Dr. N made some strong recommendations. She recommended that Chris' emotional and behavioral needs needed to be monitored and responded to in school. She recommended, "Counselling within the school setting be considered"; a referral for psychotherapy in the community for Chris, "as some of Chris' issues appear to be non-school related"; and a "psychiatric consultation regarding medication should also be considered because of the chronicity of

Chris' depression." She specifically noted the need for Chris to be provided "a safe place in which he can express his feelings and begin separating from friends, relatives and his country in an adaptive way." Japan was, she believed, to be traumatizing. *It was*. Chris, at age 14, was on his road to destruction.

## SUICIDE NOTES

Chris Purcell left two suicide notes. I hereby present the notes and an analysis.

Note 1)
Mom, Pop, Kristin, and Blair
I'm so sorry. . . .
Laugh as much as you
breathe
and love as long as you live
I love you
   Chris

Note 2)
Suicide Note
Mother, Father, Kristen, and Blair: I'm sorry . . . I don't know what else to say. You all mean so much to me . . . I just can't go on anymore . . . I love you all more than you could ever imagine.

I feel it is only right to state my reasons for what I am about to do. First and foremost, this has nothing to do with anyone or anything. I just no longer have the desire to live. Life bores me. Blame it on whatever you want, the season, depression, alcohol abuse . . . it makes no difference, because in the end it is just me. I've always felt this way. I have suffered in silence most of my life . . . and I don't want to suffer anymore. I could go off on a philosophical tangent but in the end it doesnt matter. I appreciate everyones help and guidance . . . I really do. I've learned a lot from everyone, people I like, people I loathe . . . it's been an experience. I also want to apologize to anyone I may have caused issues with . . . im not a bad guy . . . im really not . . . I have the purest intentions inside my heart . . . I think they just come out all wrong. I think saying I've felt pretty misunderstood my whole life would be an understatement. I don't expect anyone to understand why I did this . . . I'm sure most will see it as the easy way out . . . me being a coward, etc. I'm not escaping any problem, im not running from anything. I'm saving me from myself. I came from nothingness and I will return to nothingness. This is probably making a pretty shitty suicide note, but I don't care, I'm just trying to convey my reasoning to who ever might care. It is all just meaningless to me . . . wake up . . . work, eat, masturbate, get drunk, pointless conversation . . . rinse and repeat. It's fucking madness.

I'm done with it. I'm not sure what kind of emotion this will ellicit to people I know . . . anger, denial, sadness, guilt . . . but just know that there was absolutely nothing anyone could have done. Like I said, this is about no one. It is about me. You can see it however you want . . . this has been the hardest decision of my life but it has also been the most liberating one as well. At only 21 I realize I never will . . . meet the love of my life, hear my favorite song, play the best video game, have sex. . . . hear the best joke, have the best original idea . . . etc . . . I will also never disappoint myself or anyone else, never fail at anything again, never suffer alone, never watch life pass me by, never wonder what the fuck my problem is, never be rejected again, and ultimately . . . never live. Goodbye life, I have lost all interest in you.

## THE ANALYSIS OF THE NOTE

Chris' case, given the suicide note (#2), presents a unique opportunity: to postdict a communication to answer why and why at that time. In performing my analysis, I first looked at the note before any background information was known and interviews were available. In Chapter 2, I presented the complete text of the protocol sentences (TGSP): I here use abbreviations (see Chapter 2). The note had the following characteristics of suicide:

Intrapsychic Drama

I – *Unbearable Psychological Pain*: 1) Suicide as a relief, 2) Suicide as a flight from trauma, 3) Emotional states in suicidal trauma, 4) Loss of interest to endure, 5) Inability to meet life's challenges, 6) State of heightened disturbance.

II – *Cognitive Constriction*: 7) A history of trauma, 8) Overpowering emotions, 9) Focus only on grief topics

III – *Indirect Expressions:* 11) Aggression has turned inward, 12) Unconscious dynamics

IV – *Inability to Adjust:* 13) Feels weak to overcome difficulties, 14) Incompatible state of mind, 15) Serious disorder in adjustment (f & g)

V – *Ego:* 16) Weakness in constructive tendencies, 17) A "complex" or weakened ego, 18) Harsh conscience

Navy Stage

VI – *Interpersonal Relations:* 19) Problems determined by situations, 20) Weakened by unresolved problems, 21) Frustrated needs (e.g., abasement, autonomy, harm avoidance, infavoidance), 22) Frustration to a traumatic degree, 23) Positive development not forthcoming

VII – *Rejection-Aggression:* 25) Report of a traumatic event, 26) Narcissistic injury

VIII – *Identification-Egression:* 33) Identification with person/ideal, 34) Unwillingness to accept life, 35) Suicide as escape.

## THE LAST CHAT

On January 27, 2008, Chris had his last chat with a stranger on MySpace. Having removed identifying data, I here present his last chat:

what branch are you in?

navy

thank you for your service

whats this talk about taking yourlofe?

i see no end. nothing but agony of my mind. you know . . . i don't rwant to die i just want ito stop. . . i see no other options . . .

agony from being overseas?

No . . . it's a very fucking long story . . .

have you talked to a councilor or psychologist, or anymilitary support?

yes i have. Im not supposed to see them until next week . . . I just feel like I cant go on . . . nothigh will ever change . . .

don't' let these people in the chat room ruffle you r feathers

whats bothering you?

i understnad if you don't want to talk about it.

yes they do, a lot of stuff is ruffering my feathers, my whole fucking life has been nohtig but depression . . . im a 21 year old vigin . . . the best friend of my life doesnt see me the way i see her . . . im just tired of everything . . . im done . . .

but i have been told, God only gives you what you can handle. at times that is very difficult to believe but, its amazing to look back and see what all i have made it thru

first of all change chat rooms or get out of that one

im out

wait

there is NOTHING you can't overcome

do you have an emergency number for your counsilor?

if you don't call 911 and they will take you to the hospital for immediate help and attentoin

i don't help. im done with help. nothing can help me. i just want this suffering to end. the thought of death is comforting

i don't believe that because you are asking for help

you have been asking for help from people in a chat room you don't even know,

you do want help

i want it . . . from people who mean something to me . . . i don't want ot die . . . i don't know what else to do . . . i have ea loaded gun in my lap right now . . . im so scared . . . i don't want to live anymore

unload the gun

i am not your family and i don't mean anything to you, but i khnow what its like to be scared

scared of the unknown

but it will all unvail itself and we become better stronger people

im not even scared of the unknown . . . i embrace it . . . when we die we die . . . im just so tired of life . . . i cant even describe it . . .

is Kristin your sister?

yeah . . .

why

when was the last time you saw her

in November

thanksgiving?

yeah

y

were you happy to see her?

yes . . .

i was happy to see all of my family . . .

have you spoken to you mom lately?

within the past week . . . i don't tell her much though . . . it hurts too much . . . so i don't bring anything up . . . she thinks im fine . . .

i don't know what type of relationship you have with her but trying talking to her. Its ok to get upset and express hurt

i don't want to talk to much people right now . . . im so lost . . .

did you unload the gun?

fuck no. its sittong my desk

unload it

no

other than Kristin do you have any siblings

yeah, they don't have myspace though. i have a bunch of good friends . . . im just done . . . im fucking ugly . . . im over it . . . ill never meet anyone worthy . . . there are SO many people out there to meet, but you will not find them sitting in front of a tv

have you looked on line or going out to clubs?

i go to clubs onc ein awhile . . . im just fucking ugly . . . im done with it . . . i don't give a fuck anymore

what about your appearance do you not like?

fucking everything. my face especially. look at my pictures for christ sake

there is nothing wrong with you face, you have two eyes a nose and a mouth

so sorry i don't believe that

fuck you. don't lie to me. i don't give a fuck anymore. i know im ugly jsut tell me it

you are not ugly

but people can tell you that all day and yopu wont believe it, you wont believe it until you are ready to

what else about your appearance bothers you?

as if you couldnt see it . . .

my receding hairline, my bags under my eyes. my crooked teeth. my 50 year old face

what else

im hideous

get some,rest that will take care of the bags, try invisiline for your teeth, shaving your head to hide you hair line and working out

fuck it. you odnt understand. im done talking.

goodbye

im just trying to show you for the problems you have expressed, there are answers to

it may take time to heal and change things but it can all work out

hello?

who do you live with?

myself

bye

what do you want to change in your life

don't ever do that again

do what?

don't act stupid

so . . . youre g0oing to stop me?

just goes to show you people do care

## ANALYSIS OF THE CHAT

As with the note, I analyzed Chris' conversation. The scoring would be identical to the suicide note *except* (and this is important), under Rejection-Aggression, the aggression that was missing in the note comes out screaming. The protocols are: 25) Report of a traumatic event; 26) Narcissistic injury; 27) Preoccupation with person; 28) Ambivalent feelings toward a person; 29) Aggression as self-directed; 32) Revenge toward someone else. Chris, thus, was not only self-destructive, but also other-destructive—toward Sue, his would-be girlfriend who spurned his romantic intentions. There is no question that he was angry! Yet, akin to the note, he dissembled; he is explicit; no one is to blame. I believe that is not what Chris believed; yet being a good soldier, he kept secrets.

## GOOGLE

Like many of us, Chris Purcell used Google (www.google.com). He Googled in the weeks prior to his dying by suicide, compiled from his laptop computer by his father on June 7, 2008, among others, the following items: how to commit suicide, adjustment disorders, philosophy & suicide, navy & suicide, sgli & navy, sgli & suicide, .38 hollow caliber, albert camus, albert camus the stranger, effective suicide methods, ptsd symptoms, suicide guide forums, suicide guide, suicide message boards, best suicide scenes, suicide notes, effective suicide, used guns & marine, gun registration state laws, signs of alcoholism, fuck you, van morrison.

## MOVIES

Not only notes, e-mails, chats, diaries, and poems, but also movies are a window into a mind; in this case, Chris Purcell's. There is a growing interest in suicide movies. Steven Stack and Barbara Bowman (2011) recently published a book on the very topic, *Suicide Movies* (They ask, "How is suicide portrayed in the cinema and what does it mean for suicide prevention?"). Without question, Chris' favorite movie was *Donnie Darko*. It is a suicide film. I think that it helps to know Chris and why he killed himself. *Donnie Darko* is best described as a psychological suicide thriller. The main character is a teenager named Donnie Darko. He has a troubled history; he sees a psychiatrist. One day, a demonic-looking rabbit, Frank, appears, draws Donnie out of the house, and hypnotically tells Donnie the world will end. This will occur in 28 days, 6 hours, 42 minutes, and 12 seconds. Then, there is a trauma. Donnie shares his vision. At a Halloween party, Gretchen, his friend, arrives. It is implied that they had their first sexual encounter. At midnight, Donnie realizes 28 days have passed; he waits for the sixth hour. Donnie, Gretchen, and two friends go to see "Grandma Death." Bullies assault them. A car suddenly comes toward Grandma Death; the car avoids Grandma Death, but Gretchen is killed. It is a huge loss. The rabbit appears. There is an overwhelming feeling of pain. This is trauma. A tornado comes. Then, after 6 hours, 42 minutes, and 12 seconds, Donnie kills himself. (I will leave the plot's details a secret.)

A suicide note is a snapshot; this film is a full-length movie. It reveals the unbearable pain and why people kill themselves. Unlike Alice's looking glass, there are not wonderful things in this film. In a simple word, it is *dark*. Since I cannot show you the film in a book's printing (today), I will reproduce a few important lines. One can analyze the words, no different from a suicide note. Here are some quotes:

> Jim Cunningham: You are a fear prisoner. Yes, you are a product of fear.

> Dr. Lilian Thurman: What did Roberta Sparrow say to you?

> Donnie: She said, "Every living creature on earth dies alone."

> Donnie: You're right, actually. I am pretty- I'm, I'm pretty troubled and I'm pretty confused. But I . . . and I'm afraid. Really, really afraid. Really afraid. But I, I, I think you're the fucking Antichrist.

> Donnie: [reading poem in class] A storm is coming, Frank says / A storm that will swallow the children / And I will deliver them from the kingdom of pain / I will deliver the children back the their doorsteps / And send the monsters back to the underground / I'll send them back to a place where no-one else can see them / Except for me / Because I am Donnie Darko.

One other film: Chris' second favorite film was *Jacob's Ladder*. It is a psychological PTSD thriller. The main character, Jacob Singer, is a U.S. soldier, fighting in the Mekong Delta during the Vietnam War. The story begins with helicopters flying overhead. There is a Viet Cong offensive; solders are killed. During the *trauma,* Jacob is stabbed with a bayonet. The film next chronicles Jacob's life after the battle. There is no lack of traumatic aftershocks—*Jacob suffers from lethal PTSD*. This movie is a second mirror into Chris' mind. Did Chris suffer from PTSD? Do the movies show it?

## HEALTH RECORD

Chris Purcell's early medical record was unavailable, with the exception of the psychological report at age 14. When he was born in Iceland, his mother was in the Navy. It was a caesarean birth; there were no complications. His medical record as a child and teen are unremarkable. Chris Purcell's records, a bankers box full, weighing over 50 lbs., were made available to me. I hereby summarize Chris' medical records from the Naval Health Clinic, New England (Newport), beginning March 17, 2006. Before Chris joined the Navy, there is no prior medical record of impairment or treatment available. The first important health date available to me was March 17, 2006; Chris had an outpatient record review performed. There is nothing remarkable noted. Subsequent notes discuss an array of topics such as gastritis, chronic pain (the first digit of right hand), and dental. On October 3, 2006, there is the first yellow flag. Chris Purcell is noted to be drinking alcohol and smoking cigarettes. There was medium risk. Health promotion services were recommended, specifically tobacco cessation and nutrition counseling.

In a visit in December 2006, Chris denied use of alcohol. On January 31, 2007, he is diagnosed with "Nicotine Dependence" and prescribed varenicline (Chantax) to assist in the problem. On August 30, 2007, Chris was complaining of insomnia; he reported that he felt always fatigued. On September 7, 2007, Chris is prescribed zapidem (Ambien) for the insomnia. Yet, at these times, Chris denied increased stress and reported normal mood. He reported being healthy. This changed. On November 9, 2007, Chris received a suspension for a bottle of liquor in the freezer. On November 9, 2007, Chris, a self-referral, was seen for alcohol problems. He denied using drugs. He did admit to personal issues at home. He wanted to escape; he was using the alcohol to self-medicate. The reason for his self-referral was an incident when he had been drinking and caused damage to a friend's property. Chris admitted to alcohol abuse, for at least one year. He stated that he was "having trouble cutting back on alcohol." He reported that he started drinking in the Navy. He admitted to binge drinking; he once drank 10 liquor shots in 15 minutes.

Chris was asked about trauma. He wrote that his childhood was "good." However, he reported that his family moved a lot. He reported a positive history of alcohol abuse on his mother's side. Chris stated that he was having "emotional problems." He acknowledged being worried about his dying maternal grandmother. He was then admitting to some of his problems. His mental status was normal. However, was he dissembling? Keeping secrets? The doctor, Dr. L, diagnosed the following DSM impairments:

| | |
|---|---|
| Axis I | Alcohol Abuse |
| | Adjustment Disorder with depressed mood |
| Axis II | NIL |
| Axis III | Not recorded |
| Axis IV | Not recorded |
| GAF | = 81 |

The Global Adjustment Assessment of Function Scale (GAF) was 81. This suggests minimal symptoms (e.g., depression over grandmother's dying). Overall, it means good functioning in all areas of life. One would be seen as having everyday problems or concerns. However, Dr. L's note suggests Chris' motivation for change was not optimal. The risk for first alcohol abuse incident was rated as moderately high. He questioned the supportive note of his Navy environment; he reported that he did not feel support from his superior commanders. Chris was referred for treatment. It was recommended that he be referred to a psychologist.

On November 13, 2007, Chris reported that he first drank at age 15. At 17 to 18, he consumed little; he reported drinking one 12 oz. beer two to three times a year. Between the ages 19 and 20, he consumed more, but with social limits, although, of course, the drinking age is 21. On November 13, 2007, his work performance suffered. On December 7, 2007, he was prescribed alprazolam (Xanax), as needed, for anxiety. On December 10, 2007, one can read progressive regression: Chris was now reporting an array of many painful symptoms: feeling restless, feeling nervous, anxiety interfering with social activities, interfering with work, and initial insomnia. He reported an increase in anxiety over the past week with a few episodes of shaking and heart pounding. He reported experiencing panic attacks and that the attacks came for no known reason. He had been worrying. He reported being helpless and that he attempted to control the anxiety, but he never stopped them. He attempted strategies such as exercise/jogging and relaxation, but failed. His belief: Nothing will help! His mental state was not recorded as normal now. He was seen as tired and his mood was irritable. Chris was diagnosed with a new Axis I disorder: Anxiety Disorder, NOS (not otherwise specified). In the DSM, Anxiety Disorder NOS includes disorders with prominent anxiety or phobic avoidance that do not meet criteria for another specific disorder. His impairments, thus, grew beyond an adjustment disorder. Things got worse!

On December 10, 2007, Chris is known to be attending AA. Not only counseling, but also educational opportunities were provided to him. On December 13, 2007, he evaluated the group work as not beneficial. He felt like he was talking to strangers. He felt estranged, cut-off, withdrawn. On December 13, 2007, he had a conflict with Navy authorities; he is reported to fear his commander. He fears that the commander would find out his secret and what he says. Chris was worried about confidentiality. On December 13, 2007, he denied abusing alcohol since initiating treatment. The same on December 14, December 17, December 19, and so on. This was not so on January 7th. Chris was drinking. On January 8, 2008, he admits to some abuse of alcohol. On January 14, 2008, he was in military trouble. He had brought a bottle of Jagermeister to the clinic on the weekend and proceeded to drink to intoxication. This would mean an immediate military disciplinary issue; he was in legal trouble in the Navy.

By this time, Chris' medical record suggests two diagnoses, Alcohol Abuse and Mood Disorder. Obviously, the health professionals were questioning the initial diagnosis. Chris presented more and more symptoms, and now shame and disgrace. He would be ordered to go in front of the command; he would face military court consequences—more loss of honor. His problems were also growing interpersonally. The alcohol use was causing problems with his peers. He was becoming socially isolated. He reported few sober friends. He had few healthy recreational habits. Chris also began to present a very different childhood. He reported now that his childhood was "lonely and unstable." The problems were associated with being in a military family. Overall, however, he was seen as having historical and current situational problems. From a DSM perspective, he was seen as having a co-morbid disorder: alcohol abuse and a mood disorder. It appears that more and more, health professionals were wondering if Chris had a mood disorder. Yet nothing is specific. Was it Mood Disorder NOS? Chris was seen at risk for further substance abuse. The primary reason for the assessed risk was an underlying mood disorder. The mood disorder was seen as contributing to his misery. It was also seen as resulting in a poor (green) attitude and poor work performance. Military disciplinary court was likely.

Chris was ongoingly questioned about suicide risk. He denied risk. He denied intent, plan, and so on. There is no suggestion of risk. At what point was he intentionally dissembling, keeping a mask about his deep howling battle? Dr. C, a licensed clinical psychologist, offered the following DSM diagnosis:

Axis I   Alcohol Abuse
         Mood Disorder, NOS
Axis II  Deferred
Axis III None noted
Axis IV  Problems with primary support system

His GAF was now 55–60. That is consistent with moderate symptoms in functioning (e.g., depressed mood, panic attacks), and moderate difficulty in social and occupational functioning (e.g., few friends, conflict with peers or commanders).

I should also offer a further comment on Mood Disorder NOS. Mood Disorder NOS includes a wide array of disorders, from bipolar (manic-depressive) disorder to depressive disorder. It is used with mood symptoms that do not meet criteria for any specific mood disorder and when it is difficult to choose between Depressive Disorder NOS and Bipolar Disorder NOS. To classify children to young adults whose symptoms will likely crystallize into a bipolar disorder, health professionals also commonly use it. A manic depressive disorder, early in life, is a developmental disorder. There is a positive family history of bipolar disorder. Greater intervention was recommended:

Recommendation #1: Chris was referred to individual counseling to address not only alcohol use, but also behavioral issues. An appointment was made.

Recommendation #2: Chris was referred for psychological consultation to further assess and clarify the mood related problems. Further assessment for mental health needs was planned. An appointment was made.

Recommendation #3: Chris was a referred to a medical doctor to determine whether he would benefit from medication to assist in mood regulation.

A day later, January 15, 2008, Dr. C saw Chris again. Chris was seen as co-operative. Dr. C evaluated Chris' mood, problems, and so on, and this included questions about suicide risk. Chris denied suicide ideation and risk. He also denied homicidal ideation, intention, or plan. In a nutshell, he was also denying any significant stressors at that time. He was treated; he was attending a group. He was receiving individual counseling for substance abuse. He was not deemed at risk. On January 15, 2008, Chris and Dr. C. discussed Chris' mother and her bipolar disorder. His mother's sibling, who also has a mental illness, was discussed. The helpers, I believe, were becoming more and more aware of Chris' underlying pain. He suffered from a developing mental disorder (psychopathology). Yet he denied any current or previous suicide ideation, intent, or plan. Was he wearing a mask? On January 15, Chris noted that his alcohol abuse and his recent binge drinking was a "stupid mistake."

On January 17, 2008, there was a record review by Dr. L, a medical officer, a counselor, and another clinical staff; the group recommended follow-up counseling, an After Care program supported by his commander. Only a member of a soldier's command office could order the plan. The Navy had concerns, and Chris was aware of the meaning of the concern in a green culture. Did he fear even incarceration? Dishonorable discharge? Involuntary repatriation? Being disenfranchised? Military humiliation? What were Chris' stressors? On January 22, 2008, Dr. C saw Chris for 60 minutes. This was to be their last visit. There was no suggestion of suicide risk. Chris reported some mood and adjustment related difficulties within the past 2 months. A comprehensive psychiatric evaluation was planned. The plan was to continue treatment; of course, sadly, it never occurred. He needed comprehensive care, which would have included psychotherapy, medication, hospitalization, and environmental (gun) control.

Therefore, mental health professionals, just prior to the tragedy, saw Chris Purcell, and those professionals did not document him to be at risk for suicide (or violence). Although he attended appointments with professionals, overall it appears that he was noncompliant with treatment. He was seen as having a mood disorder; he was seen as depressed, anxious, and abusing alcohol. However, he was not seen as a threat to himself. The noncompliance reference is in relation to the fact that he was attending counseling but perhaps not discussing the depth of his problems. Was he taking the medication as prescribed? He was known to be concerned about side effects. Did he follow through with the prescribed treatment? He had a history of noncompliance with prescribed medication. Albeit, based on the health record, this tragedy was not predictable. Was it preventable? That is a very different question, and the questions were raised for the military. We may never perfectly know why Chris killed himself; however, the personal documents are most revealing. Secrets come alive. In a PA, we can do more. From the interviews, we can get a pretty good idea why Chris died by suicide and why at that particular time.

## THE INTERVIEWS

Let us next present the interviews, utilizing Shneidman's basic outline, although one has to let the interview flow. Be that as it may, I routinely conduct the autopsy in a standardized fashion as Dr. Shneidman taught me. It is an open-ended function. It is not like a physical autopsy or forensic investigation (see Chapter 3).

For the psychological autopsy, the following people were interviewed: Mike Purcell, father; Helene Purcell, mother; Kristin Purcell, sister; and Derek Ozawa, best friend. We had hoped to interview some of Chris' Navy buddies; however, they were not allowed to participate. Thus, a question: Are there secrets that we don't know? Before I present the verbatim answers as recorded, I want to mention travel. Helene and Mike Purcell live now in Algoma, Wisconsin. It is an old fishing village on the shores of Lake Michigan, a beautiful rural village. On August 26, 2011, I flew to Green Bay, Wisconsin; rented a car; traveled through dairy country; and stayed two days in Algoma. Helene and Mike invited me into their home; we did the interviews; we talked; we went to a Friday fish fry; and we shared. The setting was quite juxtaposed to the reason for being there: Death.

After information identifying Chris, I proceeded with the PA.

Question 1: Tell me about Chris.

Helene Purcell: Chris was born on December 19, 1986, in Keflavik, Iceland. Mike and I were in the Navy. The birth was a caesarean; there were no complications. However, I had postpartum depression. What was he like as a baby? Chris was a good kid; he was sensitive. He was very attached to me. We attempted to get a babysitter; however, after six attempts, she didn't want to care for Chris. Chris cried all the time. We were very attached. His death is hard for me. Chris was my heart.

Derek Ozawa: It is hard to say. He was quiet. If you got to know him, he talked. He had to be comfortable. He was a really good kid; super funny.

AL: You were friends?

Derek Ozawa: Like grade 8, I met him. We skateboarded. That was huge then. That is how we met, skateboarding.

He moved a lot. He moved to Japan, then San Diego. We visited; I stayed in California. We got close. I remember in the summer, he would come back. He stayed with me and my family. Nothing was really different. It was nice to have a friend like this. Later, he moved back; he moved to the house next door. It wasn't planned. It was great. Chris didn't like high school. He went then to night school. We hung out. We skateboarded. After graduation, he wanted to go into the Navy. We were surprised. It didn't seem like what he would do. He didn't like structure, like in school. He wasn't that kind of person. It was a surprise! Chris went to boot camp and then Maine. We lost touch for a bit. We'd e-mail. He came home, but he was a little bit different.

AL: Different?

Derek Ozawa: He seemed depressed; he seemed to have fallen in a pit. There was something.

AL: ?

Derek Ozawa: I don't know how to explain it. I recall calling him on his 21st birthday. He never returned the call.

AL: Anything else?

Derek Ozawa: He was drinking. He came here on Thanksgiving; I picked him up; he was drinking. At 11 a.m. he was drinking. He would drink before breakfast. It was weird. He talked about what he was doing. We went to the skate park; but he didn't have his board. I didn't know if he was still skating. After that, I didn't hear from him.

Question 2: Tell me about the death.

Mike Purcell: The evening that Chris died, I had duty that night (Great Lakes Base, Chicago), from 8 p.m. to 8 a.m. We had eaten, gone for a walk, and were playing cards, when Kristin, our oldest daughter called. She said, "Chris has a gun; he wants to kill himself. He has been drinking." Kristin had called him; they had talked. Therefore, I called his base in Brunswick, Maine. I had told them that my son was suicidal and had a gun. They told me, "We know" and it was "being taken care of."

After the call, I went to work. I thought that I could find out more there. It was a bad night, just waiting. At midnight, the Captain and the Chaplain came in; they said what happened. Chris had killed himself. I had a forewarning. What I learned from the reports and from people that Chris had been drinking. He was on MySpace. He had a chat about being suicidal. Some were urging him on; some said, "Do it." There was a girl on line; they talked. She called security. Security arrived; Chris was drunk. They found a gun case and bullets. However, they were supposed to search him but they did not search him. They were there for 45 minutes; the tension escalated. Twelve people had come; there were nine when he died. Someone said, "We need to handcuff him." They did and went outside. They even put him in a head lock. Yet they did not search him.

Chris wanted to go to the bathroom; they took one handcuff off. There was no security. There should be. A friend went into the bathroom with him. There he pulled out the gun in his waistband and shot himself.

AL: Anything else?

**Mike Purcell:** I was upset! They did not follow standard operating procedure. I accept that as a suicide; Helene will never get past it. However, I want a proper accounting. They could have done something. Chris did not need to die. They did nothing. They should have taken the gun. That is the biggest issue! I can't change it; maybe for the next sailor, we can.

**Helene Purcell:** The night that he died, I was taking care of my mother. Kristin came and said Chris was on the phone. She said that he had a gun. He wanted to kill himself. So I told Mike. I tried to call Chris; he would not answer. I called the Quarter Deck; they said that they knew and said, "We have everything under control." Back then, I was taking care of my mother 24/7. I had no Facebook. I didn't text. I didn't know.

**AL:** How did you learn that Chris died?

**Helene Purcell:** Mike came home at 1 a.m. with the Chaplain and he told me. Mike was on duty at the Great Lakes Base. Chris was in Brunswick, Maine.

**AL:** What was your reaction?

**Helene Purcell:** Shock! Sadness! Confusion! I was confused because they said that they were handling it. I was shocked. They said that he was okay. I was saddened. They said everything was fine. It was shock!

**Kristin Purcell:** It is so vague. That night, I was on my computer. Someone messaged me on MySpace. I thought that it was spam; but it was a girl, _____. She said that, "Your brother wants to commit suicide." She got my attention. I then called Chris. We talked. I had no idea that he was suicidal. We were talking for about an hour. I asked if he was okay. He said, "Yes." Then the girl messaged me again; she pointed me to a conversation in a chat room. I read it. He said that he had a gun; I had no idea. I would not have hung up the phone. When I read the chat on MySpace, I called my boyfriend. We discussed telling my parents, so I talked to them. I would regret if not. So I told them. They contacted the base in Maine. We were in Chicago. Dad contacted base security. My parents and my sister Blair tried to call Chris. I was the last person to talk to him.

**AL:** What happened next?

**Kristin Purcell:** Around 1 o'clock, I heard my mother scream. I knew something wasn't right. There were two men in the living room; one was a chaplain. I had heard the scream and crying; I knew!

**Derek Ozawa:** His sister called; that is how I learned about his death. I didn't get to talk to him.

**AL:** Were you surprised?

**Derek Ozawa:** Definitely. It was a shock. A friend had been in that situation; he was a superb help to the friend. He was angry towards the friend. Chris was never suicidal. It was very surprising. He [Chris] said, "Such an idiot." He was always there for the friend and me.

Question 3: Outlining the victim's history.

Mike Purcell: Chris was born in Iceland. There were so many moves. It was my Navy life. Here is where Chris lived: Keflavik, Iceland 1986–1989; Great Lakes, Illinois, 1989–1992; Charlotte, North Carolina, 1992–1998; Northbrook, Illinois, 1998–2000; Sasebo, Japan, 2000–2001; San Diego, California, 2001–2004; Northbrook, Illinois, 2004–2005; and Brunswick, Maine, 2006–2008. Chris had problems with the life of a Navy family, the moves. He had problems adjusting to five high schools. He had problems socially. He had friends, and then we would have to move. He was not outgoing; he would make one or two friends. For Chris, it was traumatic. And Helene had a bipolar disorder. As he grew up, Helene would have problems. If she was depressed, it was bad. It was more disruption in the family. There were hospitalizations. It was all hard on the kids. Helene's illness and the moves had a big effect on Chris. In the back of my mind, I never wanted it to be this way. But this is Navy life.

Helene Purcell: One thing about Chris: he didn't academically like school. He always struggled. We moved a lot; we left Iceland when Chris was 2. Between 12 to 17, he went to five high schools. It was stressful. In retrospect, Navy life was hard. Chris loved sports, especially baseball and basketball. At 13, he got a skateboard. That was a big thing. He was good. He got a lot of praise.

The Navy life also resulted in problems in socialization. It is hard. At 18, Chris wanted to go into the military. I didn't want him to go. However, because all our moves, and the lack of jobs, it was a job, so he went into the Navy. It isn't easy. I got over that he wanted to join the Navy. He was first at Great Lakes for 6 months, so we got to see him. Then he went to Maine. The separation was difficult.

AL: Anything else about his history?

Helene Purcell: No, not really. About him being depressed, he never said, "I am depressed." At 15 or 16, I wanted Chris to see a counselor. He had trouble with all the moves. I told him to see a counselor; however, he said no.

AL: Medication?

Helene Purcell: No. He once saw a psychiatrist and prescribed Adderall. It did not help. When at Great Lakes, he was prescribed Zoloft, but he took it for only one month.

AL: Alcohol abuse?

Helene Purcell: I never saw it. I never saw him drinking. He didn't abuse substances. He started drinking in the Navy; they drink a lot in the Navy. He was clearly having liquor. He drank Jaegermeister and Wild Turkey. They found 20 bottles in his room. We had no idea.

AL: Can you tell me about his siblings?

Helene Purcell: He had two sisters; Kristin, she is 26; and Blair, she is 19.

AL: How did they react to Chris' death?

Helene Purcell: Kristin was devastated. She talked. Blair was devastated; however, she never talked about it.

AL: Medical illnesses?

Helene Purcell: He had a normal medical history; the usual cuts and bruises. Nothing else.

AL: Hospitalizations?

Helene Purcell: No.

AL: Accidents?

Helene Purcell: No. Well, with skateboarding, nothing major.

AL: Family?

Helene Purcell: Navy life is difficult. The moves are very difficult.

AL: Therapy?

Helene Purcell: No.

AL: Suicide attempts?

Helene Purcell: No.

Kristin Purcell: Chris always had a hard time adjusting. When he went to Japan, he had problems, he was stuck in the previous place, Illinois. It was really hard for Chris. Letting go was difficult. Yet, for me, I was the exact opposite. I was excited about the moves.

AL: As a brother?

Kristin Purcell: He was a good brother. He was very protecting. He was always looking out for me and Blair.

Derek Ozawa: His parents were in the Navy. I knew it was hard. He never said anything; I asked. He never said anything bad about his father or mother. It was just a Navy family? When I asked, Chris said, "It is the Navy."

AL: Parents?

Derek Ozawa: They are typical. He said, "Oh, my mom is so amazing." There was nothing unusual. I knew his sister.

AL: Friends?

Derek Ozawa: The thing about Chris was skateboarding. Everything was skateboarding. And that is what we did.

AL: Girlfriends?

Derek Ozawa: In grade 8, he hung out with a girl. She was a girlfriend. But later, I knew of no one. No one really.

AL: Did he talk about a girlfriend?

Derek Ozawa: Not much. He didn't talk really.

AL: School?

Derek Ozawa: He graduated from night school. We didn't talk about it. I knew he hated school.

AL: Navy?

Derek Ozawa: Not much. I asked.

AL: Depressed?

Derek Ozawa: Hard to say. Yes, I think that he was a lot of the time.

AL: About what?

Derek Ozawa: About school. If you hate school, and you have to go, it sucks. Chris was not meant to go to school. He would say, "I'm not going anywhere." I tried to tell him, "Bear through it." But he was down about it. But he was smart.

AL: Anything else?

Derek Ozawa: Well, Chris came to visit us before his death. When he was leaving, he was upset. He was crying. He would give everyone, my mom and sister, a hug. It was strange; he never cried. He was stone faced. It was weird.

Question 4: Details of victim's family.

Mike Purcell: Helene is bipolar. Other members of her family have similar problems; a sister is depressed, maybe bipolar. But they deny it; that has an effect. And a brother is an alcoholic. My grandfather was an alcoholic.

AL: Any suicides?

Mike Purcell: No. However, shortly after Chris, another distant relative shot himself.

AL: Did he know about Chris' suicide?

Mike Purcell: No.

AL: Illnesses?

Mike Purcell: Most died of old age. One had cancer.

Helene Purcell: Yes, I'm bipolar. My sister has clinical depression.

AL: Substance abuse?

Helene Purcell: Alcohol use.

AL: Illness?

Helene Purcell: My mother died from cancer and a brother died from cancer at 18. However, my family are old school; they don't talk about problems and illness.

AL: Anything else about your family?

Helene Purcell: I think Chris lacked self-esteem. Who am I? I never realized that he had problems.

Kristin Purcell: Just the moves. The military life was hard for Chris. He longed for Illinois. Oh, substance abuse runs in the family.

Question 5: Description of personality and lifestyle of the victim.

Mike Purcell: As a teen, he was competitive. He would play video games. He would stay up all night. He had friends in the neighborhood. He loved skateboarding. He was a smart person; however, he did not care for school. He was quiet, always respectful, and pleasant. We never had the normal teen issues, like sex or drugs. He was the best of kids. He gave us no issues. No trouble with the law, except once in Japan. He was skateboarding on a railing, so he had to repaint it. He was the best.

AL: Anything about his personality?

Mike Purcell: He did not like to talk to people. He would keep things inside, bottled up. Maybe every 6 months or so, he would get emotional and talk about it. However, typically, he kept to himself. He kept stuff to himself.

AL: Anything else?

Mike Purcell: He was very antimedicine.

Helene Purcell: Chris had a really good sense of humor. He was always joking. He would make light of things. He was easygoing. In hindsight, he was hard to get motivated. He was bright, not academic. But he was super smart. He loved music.

AL: ?

Helene Purcell: Anything. Van Morrison.

AL: Anything else?

Helene Purcell: Just that he loved skateboarding. It calmed him down. However, he stopped when he went into the Navy. It wasn't acceptable.

AL: Girls?

Helene Purcell: Chris never dated. He was shy. He had a lot of female friends; one was special. It was hard for him.

Kristin Purcell: As he grew up, he was very active in sports; especially baseball and basketball. He loved skateboarding. It was his thing. But the Navy stopped the skateboarding. He was less active. In the Navy, Chris was drinking. He started smoking cigarettes.

AL: ?

Kristin Purcell: About a half pack per day. Not before, maybe to control his weight.

Derek Ozawa: Chris was closed off if he didn't know you. However, if he knew you and trusted you, he would be open, but not pour his heart out. Other than that, he was a rebellious kid. He got in trouble at school. He got detentions. He had a really cool personality.

AL: Lifestyle?

Derek Ozawa: Skateboarding was #1. He was competitive at it. I would say, "Man, no more competitiveness." He didn't go out much, unless skateboarding. It bummed me out that he stopped skateboarding. He stopped doing that.

Question 6: Victim's typical pattern of reaction to stress.

Mike Purcell: He was a lot like me, on an even keel. He would see the other side of a thing. Not overreact. Not freak out. He would accept the stress. Analyze it. And do it.

Helene Purcell: Chris would get angry. He would have little bouts of anger. He would yell. Sometimes he would sit and play video games. I think the main stress was the moves. He was afraid. He kept a lot in.

AL: Anything else stressful?

Helene Purcell: The moves. New situations were stressful, like a new school all the time. For example, he never knew Japanese people, and then suddenly he was in Japan. It was difficult. He wasn't prejudiced; it was a new situation. That is why I didn't want him in the Navy.

Kristin Purcell: Chris was very emotional. Small things could send him off. He would close up and cry.

AL: Sometimes?

Kristin Purcell: Yes, very sensitive.

AL: Drink?

Kristin Purcell: After he joined the Navy. I would say the Navy was stressful.

Derek Ozawa: It seemed that he didn't talk about it. He would get frustrated, but not that stressed. He would brush stress off. If things bugged him, he brushed it off. He wouldn't stew; he would skateboard or go for a walk.

Question 7: Any recent upsets, pressures?

Mike Purcell: Yes, one of the things that started his downward spiral was while I was in Baghdad. It started the alcohol abuse. He grew close to a girl, Sue. She had been married and was separated. He was close to her. Chris wanted to move the relationship forward; she just wanted to be friends. Then, a friend, also a Hospital Corpsman, who had been in Iraq, returned. He started a relationship with Sue. Chris was very upset; he said, "Fuck it." Then he started drinking more. That got him into trouble. One time, while off duty, he arrived at the clinic drunk. He was arrested and had to go to an alcohol rehabilitation program.

In November, the doctor diagnosed Chris with borderline personality disorder. That really upset him. Chris was prescribed medication. Then another medication. Chris graduated from the program the day that he turned 21. But he started drinking again. On December 19, a few weeks later, he bought a gun. He continued to drink. Sue continued to reject him. He was charged for drinking. He had to go in front of the Discipline Review Board. He was disgraced and upset. But that is the routine of the review board. He felt pressure and was upset and overwhelmed.

AL: Do you think that having to go in front of the review board traumatized him?

Mike Purcell: Yes.

Helene Purcell: Yes, he said his friend Sue had a new boyfriend. She just got divorced. He said that, "I love her." She had a daughter, and he loved her daughter too. I didn't realize that he was in love. She rejected him. She said she did not know. He wanted a family. He said, "I lost my dream." Sue was his best friend. She started dating someone else. They were sneaky. He was very devastated. It was a huge loss. It was very important. That was the upset.

Kristin Purcell: From what I gather, he was close to a girl, Sue. He and Sue were best friends. He was more interested; not her. That was big. That led to the suicide. It was a loss!

AL: Anything else in the last 12 months?

Kristin Purcell: No. I didn't know Sue. We talked a little on MySpace. Chris and I, when he joined the Navy in 2005, didn't talk so much. When he joined MySpace, the last 6 months, we talked more. Sue was the loss!

Derek Ozawa: Chris was more comfortable here. The Navy was stressful for him. It was again moves in the Navy. The Navy was stressful.

AL: Any recent upsets?

Derek Ozawa: His mom talked about this girl, but he didn't say anything to me. What I know is that the last Thanksgiving, he was drinking, a lot and early. He was also smoking cigarettes.

Question 8: Role of alcohol or drugs.

Mike Purcell: Yes.

AL: What about guns?

Mike Purcell: Ever since Chris was young, we had guns. I had to shoot guns. They were in the house. We would go to the shooting range. We did in November, when we last saw Chris. We always had guns.

Helene Purcell: Yes, he also started smoking cigarettes. He started drinking in the Navy, not before.

AL: Guns?

Helene Purcell: We always had guns. We were in the Navy. We didn't think anything was wrong.

Kristin Purcell: Yes, but I didn't know too much about Chris. After, I learned.

AL: Guns?

Kristin Purcell: We always had guns in the house.

Derek Ozawa: Chris used to be a social drinker; we would have a drink. When he last visited, he was abusing alcohol.

AL: Prescription drugs?

Derek Ozawa: No.

AL: Drugs?

Derek Ozawa: No.

AL: Guns?

Derek Ozawa: I never talked about it.

Question 9 (a): Nature of interpersonal relationships.

Mike Purcell: He had a very close friend, Derek. He seemed to make close friends. There was Sue. However, she didn't see it as boyfriend-girlfriend.

[We then discussed borderline personality disorder. We discussed what it meant. I discussed the DSM definition. Mike asked, "Can they have friends?"]

Helene Purcell: Chris had a few friends. Derek for years. And Sue. He shared music. Laughed. Was funny. However, he was a deep person.

Derek Ozawa: He had a few friends. He had my family. He had skateboarding, but that goes only so far.

Question 9 (b): Nature of ego (Can you give me three adjectives to describe Chris?)

Mike Purcell: Intelligent. Impatient. Empathic.

Helene Purcell: Funny. Kind. Sweet.

Kristin Purcell: Introspective. Emotional. Loyal.

Derek Ozawa: Quiet. Caring as a friend. Felt comfortable with us—within the circle.

Question 10: Fantasies, dreams, thoughts.

Mike Purcell: He wanted to be a skateboard pro. He was very good, but not at a pro level. After high school, he said that he wanted to be a fireman. I said, "Great." I took him to a fire station. They showed him around, and he learned that he had to take a college course. Somehow he got sidetracked and decided to join the Navy. He went there to get trained. So he decided to join. That was part of the plan; he could get the college course. That was his goal.

AL: Death or suicide?

Mike Purcell: He liked odd music. He liked *Donnie Darko* and *Jacob's Ladder*.

AL: His friend's suicide attempt?

Mike Purcell: He didn't discuss it. He could not believe it. He was already in the Navy; he was not there. He said that he could not believe it. It was over a girl problem. He said, "Why would he?"

Helene Purcell: Dreams? He always wanted to move on his own. He wanted to get an apartment. He had a motorcycle and he wanted to get a car. Finally, he got the money and wanted to buy a car. He wanted to take care of himself. Be an adult. Be in the Navy. He wanted to be a firefighter. But we didn't discuss it much.

AL: Death or suicide?

Helene Purcell: Chris thought suicide was not a big deal. He saw a movie, *Donnie Darko*. It is about suicide. I couldn't watch it. He thought suicide was normal. I believe Chris didn't want to die. He wanted help. That is why he called. He wanted the relationship with Sue.

AL: His friend's suicide attempt?

Helene Purcell: He said that he believed that his friend swallowed pills over a girlfriend. He said, "That is so dumb." He didn't get it. So what happened at 21? Why?

Kristin Purcell: He wanted to be a firefighter or EMT. After the Navy, that is what he planned.

AL: Death?

Kristin Purcell: No.

AL: Suicide?

Kristin Purcell: Not at all; only his friend's suicide attempt. We talked about it. He brushed it off.

AL: Relationship?

Kristin Purcell: Yes.

Derek Ozawa: Not really. I remember before the Navy, he wanted to be a fireman. And dreams of skateboarding.

Question 11: Changes in the victim before death.

Mike Purcell: The alcohol abuse. It started about 6 months to a year before his death. He was sad!

Helene Purcell: Chris died January 27, 2008. He came here in November, on November 23, 2007, for Thanksgiving. He was unusually quiet. I was then taking care of my mother. However, we talked. I looked at him; he was just lying, looking into space. I asked if he was okay. He said, "Yes, I'm fine." However, he was very quiet. He looked sad. I asked, but I should have said more. You ask, but you don't know.

AL: Alcohol use?

Helene Purcell: No. He had said in November that he had been drunk and trashed a room. I asked; he said, "I don't remember." Chris was sensitive. He needed protection. He was my buddy.

Kristin Purcell: I was not aware. After, I learned that he wasn't able to sleep. He had insomnia. He had been excessively drinking and smoking. I didn't know.

Derek Ozawa: The drinking at Thanksgiving. That is it. When he came to visit, he wasn't the same. He was, like, defeated. It was weird.

Question 12: Information relating to lifestyle of victim.

Mike Purcell: He planned to be a fireman.

AL: Anything else?

Mike Purcell: He tried to take care of himself before. He was a vegan. He exercised. He did good stuff. He tried to stop smoking.

Helene Purcell: Success. He was really proud about his skateboarding. He was proud.

AL: Navy?

Helene Purcell: Yes. He had told Kristin that he enjoyed the Navy. He felt good about the Navy. He never said otherwise.

Kristin Purcell: He had been chosen to be on the President Bush team. When President Bush was in Maine in 2007, he was chosen to be one of the people to protect him; as EMT. It was a high honor. There were only 14 or 15 or so. He was one!

Derek Ozawa: When he first went to boot camp, he said it was amazing. He was going to be like his dad. He was in the Navy.

AL: Anything else?

Derek Ozawa: The skateboarding.

Question 13: Assessment of intentions.

Mike Purcell: I think that he wanted help. The night that he killed himself, if the right people were there, if they did the right thing, he would be alive. But that night, he planned to do it. He was on MySpace; he wanted help. He wanted to be rescued.

Helene Purcell: Did he want to die? No. He called people; he wanted help.

He was going to be disciplined. That was stressful. He was going to the military board. It is a big deal.

AL: Was he aware of such consequences with the board?

Helene Purcell: Someone told him. He was really worried about it. Stressed. It was the last straw. Plus, he had money problems. He had a car, motorcycle, and cell. He never said anything about selling his motorcycle. And Sue. He tried to talk to people; no one listened.

Kristin Purcell: No. He was pushed into a corner.

AL: Who?

Kristin Purcell: The people who were there. The seven or nine people in the room. The military police.

Derek Ozawa: I think it was a suicide. Chris kept things to himself. He bottled things up. He needed help.

AL: A cry for help?

Derek Ozawa: He talked to someone online. He wrote a note. He had asked me on his last visit, "Why do I have to feel like this all the time?"

Question 14: Lethality.

Mike Purcell: The gun.

Helene Purcell: He used a gun.

Derek Ozawa: He used a gun, but he didn't want to die.

Question 15: Reaction to information of death.

Mike Purcell: I was dumbfounded. How did it happen? They said he died by suicide. We were mesmerized. It was in the newspaper that things were not done right.

Helene Purcell: I want to talk about it. I need to talk. There is such stigma.

Kristin Purcell: Shocked. Devastated. If I had known, I would have done something. It is traumatizing.

Derek Ozawa: First, it was unreal. I was unreal. Shock!

## WHY DID CHRIS DO IT?

Like all suicides—and military suicides are no different—Chris's suicide is complex. It is a myth, for example, that suicides are due to psychopathology alone, although many, not all, have an inability to adjust/psychopathological impairment and disability (WHO definition). Suicide is a multidetermined event. This was true for Chris. There are, as you have read, many reasons for Chris' death. It was not simple. One has to take the ecological perspective. I will next explain some core individual, relational, community, and societal factors of his death—some contributing factors. I will discuss the following: the military family and culture, some quantitative factors or characteristics in young adulthood, a comment on Chris' lethal cognitive style, and an assessment of his psychopathology. After that, I will turn to the question of why Chris died at that time, followed by some opinions and a theory of Hospital Corpsman Chris Purcell's suicide. I will end with a concluding thought—actually a survivor's.

## THE MILITARY FAMILY AND CULTURE

There is no doubt in my mind; Chris' Navy family life was stressful. For Chris, it was traumatic; he says so in his autobiography as a teen. It is probable that he even suffered from PTSD because of the nature and culture of Navy life. In Chapter 1, I highlighted this often-overlooked reality; I discussed Lynn Hall's (2011) most insightful exegesis. I offer here some thoughts about Chris and his Navy family.

One often perceives families in terms of strengths and weaknesses (Everson & Camp, 2011). Families foster well-being; this is especially demanding for military families. They have to be more resilient than families in the general population. It is simply a fact that they are exposed to and experience more trauma than the average family; it is a family under fire (Everson & Figley, 2011). Like in the general population, of course, some military families are more vulnerable than others. Some family members are more vulnerable. Military families can be vulnerable due to both internal factors (e.g., history of psychopathology, alcohol abuse, dysfunctional structure), and external factors (e.g., frequent moves, spouse working outside of the home, lower median income, high percentage of divorces). There are endless lists. However, the top three current military at-risk factors (Everson & Camp, 2011) are frequent relocation, previous long-term deployment, and longer separations during combat-related deployment. There is no question that these three traumas were true for Chris. He did not cope well with the Navy lifestyle. It disabled him. I truly believe that he suffered from PTSD. It is a fact in "why Chris died by suicide." The data speak for themselves; just listen to the words of his family and friend. Chris was under fire.

Mike and Helene Purcell attempted to provide a resilient family; however, as Everson and Camp (2011) note that there are limits to which any family can be stressed. Sressed pile up, and emotional burn out occurs. Fear and helplessness are increased. In the Purcell family, there was secondary traumatization; Chris suffered almost all his life so. As Hall (2011) noted, it is paramount that the readers understand the military worldview or mindset. This is so for the children. One must take a system or ecological view to understand Chris' death. Military families are more frequently rendered hopeless and helpless by stressors and vulnerabilities (Everson & Figley, 2011; WHO, 2002). Further, each branch of the military presents its own unique "friction." This is so in the Navy! The Navy culture was a factor (cause) in Chris' suicide. As Hall (2011) argued, unless you understand this fact, you will never understand a suicide in the military.

There is one more military system factor: Chris joined the Navy. In all probability, he identified with the heritage of a Navy life. He wanted to be like his father (identification). He wanted his father and mother to be proud of him. This is not to negate the fact that Chris saw the Navy as a job and as a way to get education; however, I believe that the deep attachment was more central to his choice. He wanted his family to be proud. *It was an honor to serve.*

The Navy satisfied Chris' needs; at the very least, identity and escape. Chris wanted to get away from what he experienced growing up, his anxiety, his moods, his pain. He wanted to flee from "a need for dependence" (Hall, 2011). However, Chris could not escape from his intrapsychic pain, what I believe was accurately diagnosed at the end as Mood Disorder NOS. This was a precursor, I believe, to a diagnosis of Bipolar (manic-depressive) Disorder. The attempts to flee his problems, however, did not end the problems. The Navy was not a solution; it became a deeper unbearable pain. There was shame and disgrace.

The military has barriers, green walls. This was true for Chris; just listen to the words of his survivors. Chris kept things inside. He kept secrets. Further, the Navy heritage put Chris in a double bind. As Hall (2011) pointed out, military personnel are restricted from seeking counseling. They are then also admonished by their chain of command for seeking help. Chris feared his commanders. It is the way of the green culture. Military families foster secrecy, stoicism, and denial. They do so automatically, not willingly. It is what a soldier and his/her family do. It is expected in the collective culture. This may have dire consequences, and one is suicide. Hall (2011) noted, "Instead of providing a supportive, nurturing, reality-based mirror, the parents may present a mirror that only reflects their needs, resulting in children who grow up feeling defective" (p. 47). This is military life. This has to change. This is the very aim of Mike, Helene, Kristin, and Derek sharing their stories. Furthermore, there was the alcohol abuse. There is a last straw incident at the clinic. He was charged. There was a military disciplinary issue, a common factor in military suicide. That was traumatic. It was more shame and disgrace.

There is a further military factor: Chris was to present himself to the Discipline Review Board. Being in a Navy family, he knew what that meant (see Chapter 10). It was a trauma, a Navy humiliation. It was dishonor. As Mike Purcell stated, "He was disgraced and upset." This was *a* factor. Indeed, there is no question that there were multiple Navy associations and patterns.

To conclude this discussion of *a* cause, I want you to reflect again on what Emile Durkheim (1951) wrote over 100 years ago: "When he puts on his uniform . . . it spreads like a trail of gunpowder among persons thus prepared to follow it" (see Chapter 4, pp. 110–111). This was true for Chris, as it was true for Emory Upton and Mike Boorda.

## YOUNG SOLDIER SUICIDES ARE PSYCHOLOGICALLY DIFFERENT FROM THOSE OF OLDER SOLDIERS

A suicide note is the *penultimate* act. This is true for Chris'. Like Shneidman and Farberow (1957), my core belief is that a theoretical-conceptual analysis offers the richest contribution in suicide note analysis. I presented an outline for such interpretation in Chapter 2. This psychological theory, with both intrapsychic and interpersonal factors/clusters, will allow us to better understand, among other heroes of our current wars, Chris' death.

The first controlled study of suicide notes indicated that the dynamics or reaction patterns vary with critical demographic variables, notably age (Shneidman & Farberow, 1957). A problem with the present-day age studies is that various researchers have offered various definitions of young, old, and so on. The following three groups with characteristics and age ranges are the most common: Young Adulthood (Intimacy vs. Isolation, age 18–25); Middle Adulthood (Generativity vs. Stagnation, 25–55); and Late Adulthood (Integrity vs. Despair, 55 and up) (Erikson, 1963, 1968). We can thus ask the following question: Are suicide notes by young adults like Chris Purcell and, by implication, suicide psychologically, different from other adults? This question is important in light of the fact that most soldiers are in young adulthood.

Research suggested that it is the age group from 18 to 25 that is most different from other adults in their suicide notes and, by implication, suicide (Leenaars, 1989b). Young adults differ from other adults on several essential patterns. This observation is one of more or less, not presence or absence, since these patterns occur across the life span. It is simply a matter of degree. It is a continuum; at some point, it is a significant quantitative difference, not qualitative difference. The following observations, nomothetic with the idiographic, have been made.

## THE INTRAPSYCHIC DRAMA

### Indirect Expressions

Complications, ambivalence, redirected aggression, and unconscious implications are often evident in suicide. There is dissembling; Chris did ("I have suffered in *silence* most of my life" [italics mine]). The person may even confuse what is real and not real, objective and subjective, and so on ("It is all meaningless"). The observation of unconscious implications is especially important here. Of note, what was Chris consciously aware of? This is an important phenomenon in suicide—repression, denial, secrecy and so on; that is, unconscious dynamics may well be the driving force for an individual who defines an issue for which suicide is perceived as the best solution. Such characteristics are more evident in the suicide notes and, by implication, suicide of young (and middle) adults than late adults. Very often, these adults are confused and likely unaware of significant causes that lead to their death. This is true for Chris, I believe.

### Inability to Adjust/Psychopathology

Bipolar disorders, anxiety disorders, PTSD, and other disorders have been related to some suicides. All these disorders can be seen to reflect a maladjustment; the suicidal person himself or herself states in the note that he/she is unable to adjust. Chris did. Even more that other adults, young adults are diagnosed as exhibiting a psychological disorder; this is because psychopathology begins to crystallize at this stage of life. Considering themselves too weak to function effectively, young adults, more than other adults, reject everything except death—they do not survive *life's* difficulties. Chris could not ("It is just all meaningless"; "I don't want to suffer anymore").

### Vulnerable Ego

A consistency in suicide is with lifelong coping patterns ("I've always felt this way"). Like other people, soldiers have enormous consistency in their ego, even during life's timelines. This was evident in Chris at a very early age. The suicidal person consistently does not function effectively ("I just can't go on anymore"). This is especially true of young adults; young adults (compared with middle and late adults) more frequently exhibit in their suicide notes a relative weakness in their capacity to develop constructive tendencies (i.e., attachment, love) and to overcome their personal difficulties. Their ego has been weakened by a steady toll of traumas (dose response). This was true for Chris (*"I just can't go on anymore"* [Italics mine]).

# INTERPERSONAL STAGE

## Interpersonal Relations

The suicidal person across the life span has problems in establishing or maintaining relationships (although it may be to another ideal, the Navy). The suicide is often related to unsatisfied or frustrated affiliation (attachment) needs. Even more than other adults, young adults frequently describe a disturbed, unbearable interpersonal situation in their suicide notes. Idiographically, this is abundantly true for Chris. Minimally, the following needs were frustrated: attachment, abasement, autonomy, harm avoidance, and infavoidance.

## Identification-Egression

Intense identification (or bond) to a lost or rejecting person or to any ideal (e.g., being in the Navy) is critical in understanding the suicidal person. If this emotional need is not met, the suicidal person experiences a deep discomfort (pain); he is estranged ("never suffer alone"); and he or she wants to be gone, to be elsewhere, to exit—to egress. Such a process is more evident in young adulthood than late adulthood and is clearly related to the crisis of intimacy vs. isolation. It was for Chris ("I'm not escaping any problem, I'm not running from anything. I'm saving me from me").

It is especially the pain of isolation (i.e., Erikson's crisis of *Intimacy vs. Isolation*) in these young adults, like Chris, that is critical in understanding their suicide. Erikson stated,

> Where a youth does not accomplish such intimate relationships with others—and I would add, with his own inner resources—in late adolescence or early adulthood, he may settle for highly stereotyped relations and come to retain a deep sense of isolation. (1968, p. 138)

What is critical during young adulthood is that these adults develop intimacy, attachment, affiliation, and partnership. One typically develops mutuality with a love partner, although the military, as discussed in Chapter 1, sometimes offers a *family*. The psychosocial strength or virtue or protective factor in young adults is *love*. It is "mutuality of devotion" that allows young adults, according to Erikson (1968), the increased capacity to function (or adjust) effectively. Without intimacy, one becomes isolated and unable to establish or maintain relationships. In the military, the soldier becomes cut off or disenfranchised. The individual, however, might not only turn away from others but may turn his/her anger to whatever approaches (Frager & Fadiman, 1984). Indeed, it appears that some young adults, and that includes Chris, may not just settle for a deep sense of isolation but choose death as a solution to their unbearable interpersonal pain. Chris died because of loss! "*Goodbye life, I have lost all interest in you.*" [Italics mine]

What is unique in young adults generally was true for Chris as a unique person. His life and suicide mirrored the lives of many young adults. Chris was not that unique; indeed, he is prototypical of suicides in such young heroes.

## A DEVELOPMENTAL VIEW OF YOUNG ADULTS' SUICIDES

A life-span developmental perspective is *essential* in understanding suicidal soldiers, and most are young adults. I believe that many of the nomothetic findings reported apply to Chris; I believe he died by suicide; and the nomothetic findings help us to understand the *why*. The idiographic is no different. There appear to be highly perturbed levels on such psychological issues as "I love you," "I am caving in," "I want to be gone," "I am weakened," and "This is the only thing I can do." Chris simply wrote, "It is just all meaningless." Development should not be construed as simply discontinuous—development across the life span is both *discontinuous* and *continuous*. Adult development is dynamic, ongoing, *and* serial. The suicidal person does not respond anew to each crisis—with a small "c"—in his/her adult life, but his/her reactions are consistent in many ways with that individual's previous reaction to loss, threat, impotence, and such. There is an *elliptical* nature to his/her development (Leenaars, 2004) as well as change. This was true for Chris. We need to understand each soldier uniquely and generally.

## A COMMENT ON CHRIS' COGNITIVE STYLE

The cognitive style of the suicidal person is unique and lethal. I earlier offered a comment on the topic and examples of General Upton and Admiral Boorda. What were Chris' major premises (core beliefs)? Why *his* "therefore"? Based on the PA, we can at least speculate on his cognitive style. He was perturbed. Overpowering emotions intoxicated him. He was drunk! Concurrently, there were constricted logic and distorted core beliefs. His logic can be symbolized by the following: "All people rejected by the girl whom they loved ought to be dead. I was rejected by Sue. Therefore, I ought to be dead." Or "All disgraced sailors ought to be dead. I feel shame and disgrace. Therefore, I ought to be dead." Or "All soldiers with a military disciplinary issue and are to go before military court ought to be dead." Or "All people with a fucking life ought to be dead." Or "All people looking hideous ought to be dead." He was putting a lot of things into a nutshell. His "therefores" were lethal.

## ASSESSMENT OF PSYCHOPATHOLOGY

I have never assessed Chris Purcell, therefore I cannot provide a diagnosis. Based on the above data/evidence, however, I can speculate a likely diagnosis for him, with numerous cautions about postdiction of emotional disturbances. There were, however, clinical records, including DSM diagnoses. Chris was assessed as depressed and at risk for suicide by age 14. Despite a need for an ecological perspective of violence (WHO, 2002), the worldview of an inability to adjust/psychopathology of individuals who died by suicide is *most* important. The percentage of mental disorders in young adults who die by suicide is very high. This is true for military suicide, especially given the very above-the-norm and increasing rate for suicide in soldiers. Yet it does not adequately explain the death. Suicide is complex; psychopathology is only *a* factor, not *the* factor. There *never* is. With this preamble, the question raised is, Did Chris have a mental disorder?

Based on the DSM-IV criteria (APA, 1994), my best evaluation is as follows:

| | |
|---|---|
| Axis I | Posttraumatic Stress Disorder, Chronic |
| | Mood Disorder NOS |
| | Alcohol Abuse |
| | Adjustment Disorder. With Anxiety. Acute |
| Axis II | NIL |
| Axis III | The first digit on right hand |
| Axis IV | Relational Problem NOS. Occupational Problem. Acculturation Problem. Phase of Life Problem Other psychosocial and environmental problems |

GAF = 10.

Some may read this as overly "psychiatric"; however, some others would say, "That makes sense." I am only offering some understanding, keeping in mind that there are different points of view. Many people who die by suicide have a mental disorder. We have to get past the stigma of psychopathology.

Based on the interviews conducted for this PA, we know that Chris had significant problems in adjusting to traumatic events as a child and as an adolescent. Navy life, especially the frequent moves, made it so. This raised concerns about PTSD. He showed symptoms of avoidance, dissembling, and so on. There were persistent symptoms of difficulty adjusting, including the following: sleep disturbance; irritability, and even outbursts of anger; difficulty concentrating; hypervigilance; and anxiety. He was anxious throughout his life and was very anxious toward the end of his life. Thus, I believe that he had PTSD, complex or otherwise, chronic and longstanding. We do know that a great deal of trauma occurred when he was young. It did at age 14. He said so. We know that he was acting out early as a young teenager, especially at school. Moreover, he likely then already dissembled (e.g. avoidance, numbing, lying). He was secretive. As a young adult, he felt *unbearably* rejected by Sue, the one girl whom he had hoped loved him. The loss was the final straw. There was probably more trauma; psychological autopsies answer some questions, but not all. We will never know 100%. This is true in almost all suicides.

Chris had anxiety and worry. It must have become unbearable; he could not continue that way. He was restless, irritable, and had sleep disturbance. He, like his counselor, identified anxiety, as his family and friends did in the end. His *angst* raged especially in the last days of his life. He was highly perturbed.

Chris was depressed; overall he had a mood disorder; his psychologist diagnosed such. After the loss of Sue, there is a depressed mood and anxiety. I would suggest Mood Disorder NOS would be most accurate. However, I would predict that a Manic-Depressive Disorder, or if you prefer the term (I do not) Bipolar Disorder would over time have been diagnosed. (There are construct validity issues, among other real problems, in the current DSM protocol for this disorder; this is being addressed in DSM-V, I hope). It is a developing disorder, and children to young adults' symptoms are not entirely consistent with a clear DSM diagnosis. Chris had a mood disorder; this was evident after the loss of Sue. Chris had such symptoms as insomnia, low self-esteem, irritability, mood fluctuations, and feelings of hopelessness and helplessness. The mood symptoms caused clinically significant distress in relationships and community levels. *It did in the Navy.*

Yet I think Chris also had an adjustment disorder, with anxiety and possibly depressive symptoms. This inability to adjust is associated with the development of emotional and behavioral symptoms in response to an identifiable stressor; this began possibly with anxiety and depression after the loss of Sue and was quickly followed by alcohol abuse. However, the most proximal identifiable stress was the alcohol abuse charge and the pending Military Review Board hearing. The nomothetic literature discussed in this book is entirely consistent with this belief. Of course, there were (distal) historical traumas, ever-increasing traumas. The symptoms were clinically significant, based on the family doctor's and the counselor of LEAC's notes and the reports of the informants. There was marked distress beyond what would be expected, including in the Navy culture, from exposure. There was impairment. The distress was not only due to bereavement, although that too was present, unresolved, and complex (his grandmother). There was a lot of unbearable pain; the common stimuli for suicide.

Chris Purcell abused alcohol; he was reported to be drinking one to two bottles of hard liquor per day. There was, like in so many military suicides, a maladaptive pattern of alcohol use. There was continual use in the end. He had an alcohol level well above the well-known level of intoxication of 20 mg/dL. The toxicological examination by Dr. P reported the following item/ result for Chris Purcell's level of ethanol in blood; "140" mg/dL (vitreous fluids was 180!). Chris was highly intoxicated (7 times the level) at the time of his death. (Could he be judged not responsible for his suicide at that moment due to intoxication? Could he be responsible in intent? What would Litman say about *"intention"* in this case?) The alcohol abuse started after he went to Maine and increased after the loss of Sue. The alcohol abuse was a regression;

yet, in the end, even the bottle did not numb his pain. He wanted to escape; he wanted to die by suicide.

Now for the controversial; I do not believe that Chris had borderline personality disorder (BPD) traits; he did *not* have sufficient traits to warrant a diagnosis of the disorder. Chris functioned well; yet, after the loss of Sue, he regressed. His ego weakened. There was a narcissistic injury(ies). He became estranged. He never recovered. The acting out/alcohol abuse began and forever increased. His fragile ego broke. It was *not* BDP!

Chris had beliefs that deviated markedly from the expectations of his green culture. It led clinically to significant distress and he, as the events revealed, could not adjust. All was colorless and unattractive. Chris was paralyzed. However, this is not equal to a borderline personality disorder. BPD is a pervasive pattern of instability of interpersonal relationships, self-image, and affect, and marked impulsivity by early adulthood. This is not true for Chris. Yet he believed that he was diagnosed with BPD; there is no question about this fact. He told, as you read in the interviews, his parents. However, and this is a key, he was never diagnosed with BPD; an adjustment disorder, yes. He sometimes jumped to conclusions and held disabling beliefs. He was aware of the meaning of the diagnosis, and as I noted before, he was aware of the military's deliberate and intentional use of the diagnoses (see Chapter 10). He faced the loss of his Navy career and his honor. He held a lethal belief, and it added to his demise. The question remains, did someone tell him he did? It was a practice in the military.

Chris had multiple psychosocial and environmental problems; the main ones were Relational Problem NOS; Occupational Problem; Acculturation Problem; and Phase of Life Problem. The first two are obvious. Acculturation Problem involving adjustment to a different culture (e.g., the green culture). The Phase of Life Problem is associated with a particular developmental phase or life circumstance, such as entering the Navy. Finally, one of the most lethal problems was Noncompliance with Treatment. This category is used when a key focus is noncompliance with important aspects of treatment for his reported emotional disturbances. Chris did not comply with his medical doctor's prescription for medication, such as Zoloft. The reasons may be his belief about medication (e.g., medication side effects), but not only. He also did not comply with his counselor. (This is not uncommon with suicidal patients).

The GAF is a global assessment of functioning. Chris Purcell had received a GAF of 55–60, which is consistent with serious impairment. However, a score of 10 is warranted in cases of persistent probability of severely hurting self; Chris killed himself.

Thus, it is easy to discern that Chris suffered so. It was multidetermined and complex; however, psychopathology, some longstanding, was an element in his demise. Indeed, Chris' problems are highly consistent with what we see in the military, whether trauma or suicide.

## HOW DID CHRIS DIE?

"How did Chris die, and when, that is, why at that particular time?" The answer to that question is even easier than the basic one, "What is the most probable mode of death?" *The gun, stupid.* With absolute certainty, the availability of the gun is why Chris killed himself that day, hour, moment, and second. It is predictable. You do not have to be an expert on gun control to know that obvious fact. Chris had a gun. Furthermore, if you know the research on means restriction, you see it is evidence-based. Gun control works! Mike, Helene, Kristin, and Derek know that.

A basic in SOP intervention: Get the gun! You always have to get the gun. It is difficult in harm's way, but that is why the experts in the battlefield (Bryan, Kanzler, et al., 2010) say, "Evacuate." You remove the gun; it is the most effective tactic—*environmental control*. Dr. Snow did the same; he removed the "bomb." It works. In the military, you always get the soldier, like Chris, out of harm's way.

## OPINIONS REACHED AND BASIS THEREFORE

Based on the foregoing, using the PA method of investigation, I have reached the following opinions:

1. Chris Purcell died by suicide. The mode of death is unequivocal.
2. Beyond the question of what, we can answer with little doubt the why Chris died by suicide and how Chris died (He shot himself).
3. The PA richly illustrates Chris Purcell life; there was a history of trauma. He was traumatized by the military culture/system; he was distressed, depressed, and suicidal by age 14, if not before. There is a suicidal history and an acute (proximal) suicidal crisis.
4. Further to Purcell's suicide notes, the content and themes in the note, chat on MySpace, and other personal documents illuminate his suicidal mind, and many details of his life are tragically illustrated by the contents of his notes and documents.
5. There is a history of a mental disorder/psychopathology.
6. There is a history of PTSD. He also suffered, among others, from Mood Disorder NOS and alcohol abuse.
7. There were proximal signs of unbearable pain, mood disturbance, and suicide risk. There were distal ones. There are abundant precursors to his lethality; yet it is only in the last month(s) that there is high suicide risk. We know that in his last communications that he was highly lethal and perturbed.
8. In addition to the mental, physical, and spiritual imbalance, he had a most traumatic interpersonal relationship—a deep attachment to Sue and (I believe) the Navy. They were lost! The loss/rejection of Sue, I believe, was a narcissistic injury, the straw that broke his heart.

9. Chris Purcell did experience a number of traumas in his life (e.g., Navy life; frequent moves when young; experiencing and witnessing injury and death of others as a Hospital Corpsman; and threat to his Navy career). The forthcoming Military Discipline Review Board narcissistically injured him. This should not be a surprise. He came from a Navy family; he wanted his parents to be proud of him. There was shame and disgrace.
10. In the green culture, there is absolute obedience. It is a collective value. There were military disciplinary issues and conflict(s) with Navy authorities. Chris was charged with violating the law. There was military humiliation. There was loss of honor.
11. Hospital Corpsman Purcell was treated for a number of painful mental impairments, such as severe anxiety and alcohol abuse. The treatments were consistent with community standard, but also unsuccessful. The main reason was noncompliance with treatment. *This is always lethal!*
12. However, on the night of his death, the first responders care was not consistent with community standard; there was dereliction of duty causing death! It may well be deliberate indifference!

Thus, there is no doubt in my mind that, at the time of his death, Chris did intend suicide. We knew that before. However, I have also answered why, I believe. It was multidetermined. I will next summarize an answer to the question, Why did Corpsman Chris Purcell kill himself?

## A PSYCHOLOGICAL THEORY OF CHRIS PURCELL'S SUICIDE

Psychologically, Chris Purcell's suicide was multifaceted. It was an intrapsychic drama on a Navy stage. As Durkheim (1951) predicted over a century ago, there was loss of honor. There was high perturbation and high lethality. Utilizing evidence available, and an analysis as discussed earlier, I propose the following psychological answer to the question, Why did Chris Purcell die by suicide?

### Intrapsychic Drama

#### Unbearable Psychological Pain

Chris Purcell's mind, based for example, on his suicide note, was permeated with pain. Shame and disgrace were paramount. He had lost everything. *All* his dreams were lost. Although I believe that his suicide was much more, his suicide was a flight from these specters—the loss of Sue, his flawed Navy career, and his disgraced military reputation. He was in great distress. Deprivation, distress, and grief are evident. He felt boxed in, hopeless, and helpless. He arrived at what he perceived to be the end; suicide became *the* solution, a relief from intolerable psychological pain.

## Cognitive Constriction

Chris Purcell's mind was a constricted mind. There was constriction of perception, logic, reason, and conscience. He put things in a nutshell. He reported only permutations of a history of trauma—loss of Sue, a charge, a painful life. He was overwhelmed and overpowered. There was a paucity of thought, focusing only on the traumatizing and grief-provoking topic.

## Indirect Expressions

Chris felt he was encircled; he was estranged. He submitted to his pain and irrational core belief. He concluded that he had to kill himself. He turned his anger inward. He was not ambivalent. (This was so for Emory Upton.) Although we know a lot about his psyche, I believe that there were unconscious dynamics in his death—even the Navy family/culture was a deadly one; some do not survive.

## Inability to Adjust (Psychopathology)

Hospital Corpsman Chris Purcell considered himself too weak to cope (or adjust); he gave up the fight. He, without a doubt, believed that his death was best for his reputation, and himself. Like his treating psychiatrist and psychologist, I believe that Chris had a mental disorder. He says, in fact, that he was depressed; he may also likely have suffered from PTSD, mood disorder, and alcohol abuse, among others. There was, without question, *a deep psychological pain* and an inability to adjust to his life's demands, like he could not as a child and teen in a Navy family.

## Vulnerable Ego

Chris' ego was weakened; he lacked constructive tendencies. With the loss of Sue (he talked about her a lot), he felt desolation. This was the final loss toward the end. Then, the core belief of the loss of his flawed Navy career was all he could bear. (What was the identification?) Further, there were unresolved problems, something that he could not change—himself. He saw himself as hideous. Chris became discouraged, to a lethal level. He thus felt that he must be punished with death (attribution inward). He was so isolated (even from his best friend, Derek), illogical, and harsh.

Navy Stage

*Interpersonal Relations*

Chris' problems were determined by his history (the Navy family) and the current situation that presented itself. He never did transitions well. He was weakened and felt defeated by the loss of Sue, his alcohol abuse, and military disciplinary issues. His needs were unsatisfied—at least attachment, abasement, autonomy, harm avoidance, and infavoidance. The frustration of these core needs became exceedingly stressful—he was not capable of healthy action. He drank. He had hoped for the acceptance of Sue, but such he saw as not forthcoming. He was hopeless about being able to survive (a core belief). Chris was, by all eyewitnesses, under constant strain and frustration; it was so for months. He responded with fear, helplessness, and horror.

*Rejection-Aggression*

In his suicide note, he wrote about a single traumatic event—his life. His ideal and dreams were lost. He was rejected. He feared the loss of his Navy reputation. There was Navy humiliation. He was so discouraged in the end and suffered a deep narcissistic injury. Yet his aggression was attributed inward; there was no aggression turned outward in his note; however, it was in his chat (toward Sue). However, he expressed no aggression toward his family. He was defeated and severely anxious and depressed, not outwardly angry.

*Identification-Egression*

Hospital Corpsman Purcell deeply identified with Sue, his father, and the Navy culture. *All was lost!* He felt rejected. His life, he now thought, was flawed and lethal. His reputation would be lost. He was unwilling to accept the loss of this ideal—his Navy career. (The same can be seen in General Emory Upton and Admiral Michael Boorda. It is common!) In similar illogical thinking, he reasoned, "Fuck it." His beliefs were traumatizing. Hospital Corpsman Purcell lost his integrity, and the only solution was suicide. He first escaped with his alcohol abuse and then by his death. He escaped!

## CONCLUDING THOUGHT

There are many thoughts with which I could conclude this chapter. None will provide the honor that Hospital Corpsman Third Class Chris Purcell deserves. Thus, I will let Mike Purcell, Chris' father, speak. Joe Gould (2011) on May 24, 2011, wrote a piece for *Military Times* entitled, "Are Suicides Considered Less Honorable?"

If military leaders want to stamp out the social stigma of psychological illnesses, it has to start with the commander-in-chief, writes vet Mike Purcell.

As suicides rise among service men and women, and the nation comes to grips with the reality of their psychological wounds, why should they be accorded any less honor than those who have died of physical injuries, Purcell asks.

"This Memorial Day please remember those we have lost on 'the other battlefield,'" Purcell writes. "Their service mattered greatly, as did they. Their families deserve to be recognized with dignity and respect, in their time of profound loss."

Purcell is far from alone.

# PART SEVEN

# Prevention and Policies

Colonel Elspeth Ritchie from the U.S. Army Surgeon General's Office has reported that we now know that we have a problem—suicide. We know the patterns and associations, and we need to look at what works to control and prevent the problem. We can ask, What policies and procedures are needed in the military? In this section, after some preliminary observations, especially about recommendations and challenges in managing suicide risk in the combat zone, I present recommendations from the U.S. Department of Defense, the U.S. Army, and the Canadian National Defence. I conclude with a very personal call to action. Thus, this section covers my concluding thoughts and presents some reflections from the Armed Forces on policies and prevention. I end with my battle cry: *Persevere. Don't give up the fight!*

## CHAPTER 14
# Military Suicide: Policies and Prevention

Almost one million people die by suicide worldwide every year; we do not know exactly how many in the armed services, but there are many. *In the United States, veterans are twice as likely to die by suicide than their civilian counterparts.* That is far too many. Could it be even higher? What about Canada? They do not know. In order to reduce suicide and suicidal behavior in soldiers and veterans, many policies and procedures have been proposed and implemented. Although many of the approaches may have been/are effective, often they are not sufficiently comprehensive to address the complexity of trauma and suicide in soldiers.

"No organization or government starts strategy development with a blank sheet of paper" writes a world leader in mental health policy, Rachel Jenkins and her group (2002). There are creative, rational, and organized ways to marshal such efforts. There are, for example, good management models of strategy development that are critically relevant to suicide prevention in soldiers. Armed forces do not need to, to use a common metaphor, "reinvent the wheel." There are suggestions already available in the literature, both from the military and professionals/academics. However, there will be key questions to ask in the establishment of policies and procedures. Here is a list of Jenkins' key questions, and I cite: What has worked? Can we do more of it? What has not worked? Can we fine tune? Can we invest the resources elsewhere? What elements are we uncertain about? How do we improve evaluation so that we will know whether the various elements of the strategy are working?

### SUICIDE PREVENTION IN THE MILITARY

*The proof of the suicidological pudding is in the "ventions"—as in prevention, intervention, and postvention.* In other words, the main payoff of all our efforts to prevent suicide among soldiers and veterans lies primarily in making our policies and practice efforts more effective. That is what counts in everyday military society. This is what Sara Upton (already in the 1800s) and many others have called for.

There will be ethical issues and dilemmas, some commonly known since the Civil War. General Emory Upton addressed some of them in his very own written policies. (This resulted in harassment and rejection by the military elite and Washington, which, without question, added to his lethal solution.) There will be solutions only if we have the courage and perseverance to talk about it, to break down the green walls of the military, and to save soldiers' lives. There will be challenges, but we can help Chris, Emory, Mike, and all suicidal soldiers and veterans. We first need the policies and procedures to do so, a topic that will meet great resistance. It has for centuries. Will the commander and government have the courage to change what they can?

The greatest obstacle, as we have learned, was/will be the green barriers. General Upton hit the wall. There is silence and indirect expressions. The fact that soldiers may express more indirectness may be expected from research to date across collective cultures (Leenaars, 2007; 2009a), and also from a broader green culture perspective (see Chapter 1). Some cultures call for deception and dissembling. Military collective culture encourages soldiers, and thus suicidal soldiers, to adhere to the "warrior" tradition, including in matters of suicide. There is great stigmatization regarding self-harm and psychopathology in soldiers. Officers and enlisted men/women, in fact, take pride in the military collective style, and they should. It is a proud heritage in many ways. Admiral Boorda was proud, and so was General Upton. It is what Emile Durkheim called honor. Maybe for suicidal people, however, the system fosters not communicating pain and suicidal intent, not even being conscious of individual pain. It reinforces non–help seeking. The indirectness may, thus, add to the lethal mix in a vulnerable and self-harsh soldier. It may well be that soldier beliefs foster indirectness of expressing intrapsychic pain, emotional disturbance, suicidality, and so on. PTSD is not even accepted—never was. Individualism is not fostered in the culture whereas green relatedness is. The suicidal state may, thus, be more veiled, clouded, or guarded. This is what we have called dissembling or masking and is a significant factor not only in military suicide risk but also in the many faces of violence. In soldiers and veterans, there are abundant unconscious dynamics, even self-deception; there are more reasons for the suicide than the soldier expresses and maybe is consciously aware of. Admiral Boorda, for example, chose a coffin over disgrace and dishonor. Further, based on the writings from and since General Upton, we can speculate that these dynamics in soldiers or veterans are highly associated with collectivistic military processes of being a warrior; but this speculation is only provisional. Yet, not only Upton, but also many others have stated so. Research is needed, but will it be allowed? Why did administration in the United States keep suicide a secret? It was an epidemic after the Civil War. Why was little done?

The best intervention is prevention, a basic belief in suicidology, since the very beginning of the science (Shneidman & Farberow, 1957). Surveillance is needed—reliable surveillance, not lies and deception. We need psychological autopsies. The very aim of psychological investigations, retrospective or otherwise, into military suicide is to turn the tragedy around. In the United States, veterans have asked since the Vietnam War. This is the request of every survivor. It is of Mike Purcell, the father of Chris Purcell. Many people, not just survivors, hope to do so. We can do so in different ways: here, I present in this chapter, some highlights from the military literature available. Does it help to predict and control? Once we know what trauma is and what suicide is, we can make better recommendations to prevent a future tragedy, whether PTSD or a suicide. Our aim, like all prevention efforts in this field, is to prevent further human misery. Our heroes deserve nothing less.

## CHALLENGES AND CONSIDERATIONS FOR MANAGING SUICIDE RISK IN HARM'S WAY: A REMINDER

There are many challenges and considerations for managing suicide risk in combat zones. I here remind the reader of a quotation of Bryan, Kanzler, et al. (2010). They served as healthcare providers on the battlefield and offer a reminder to all of us working to prevent suicide in a combat injury, as quoted earlier in Chapter 12, "Military mental health . . . in combat zones" (see page 251). We cannot lose sight of this reality. Craig Bryan and his team (Bryan, Jennings, Jobes, & Bradley, 2012) offer more extensive recommendations for suicide prevention in the military.

## POLICIES AND PROCEDURES FROM THE MILITARY

Understandably, there has been a host of people, including from the Armed Forces, the Department of Defense, veterans, friends, family members, and so on, both from the United States and Canada, who have presented some recommendations for prevention. Everyone wants to prevent another traumatic event. Of course, I cannot present all, so I decided to select the recommendations of some major efforts in the field: the U.S. Department of the Army (2009), the U.S. Department of Defense (2010), and National Defence (Canada) (2010).

However, before I present the recommendations, I want to state one final belief: armed forces need to develop a plan of action with clear goals and objectives; they need to develop a protocol for dealing with suicide among the armed forces. This appears to be especially true since forces have experienced an above average number of suicides since the Iraq and Afghanistan Wars. There were many deaths after Vietnam. However, probably the current numbers have only been surpassed by *the suicide epidemic of the U.S. Civil War*. I hope the thoughts from the military literature will help us not only to understand the following challenge, "We ought to know what we are preventing," but also the challenge, "We have to know who you are helping." These are great challenges. Yet the military knows how, only if they have the will do to so. Chris did not need to die. I hope that I have brought the reader, whether a commander, soldier, veteran, survivor, health provider, or many more, a little closer to that plan. Regardless of what the policies and procedures will be, they have to be grounded in evidence-based study. We have to be science-based. Ultimately, suicide can be prevented. As already seen in the Civil War, our greatest obstacles will be the many green walls. We need to get past the stigma of PTSD, suicide, alcohol abuse, incarceration emotional disturbance, and some specifically green problems—but so does the world. Soldiers have to own the problem, otherwise nothing will get done. We cannot wait for a president or prime minister to do something. We—soldiers and those of us who are not soldiers, but working with them on multidisciplinary teams; providing mental health services to the soldiers and their families; administrating to such soldiers; surviving a soldier's suicide; providing survivor bereavement services after the death by suicide; and many more—need to own the problem; nothing will get done otherwise. Suicide among soldiers and veterans has, and can be, prevented. Sara Upton believed that fact! Emory, Mike, and Chris did not. A core belief: Soldiers and veterans do not have to die.

## MILITARY SUICIDE: POLICIES AND PREVENTION

Back in the 1990s, Lande (1992) stated that he believed that the U.S. military was sensitive to suicide. He noted that there were military policies that direct the formation of active suicide prevention programs. The U.S. military, Lande stated, emphasized a humanitarian approach. Yet modern military law, especially if a soldier violates a law and is incarcerated or dishonorably discharged and involuntarily repatriated, may view suicidal behavior as deviant. Soldiers were not only labeled as having a personality disorder, such as (at least in Chris's mind) a borderline personality; they were prosecuted for having PTSD symptoms. Often the many faces of suicide were understandably behind a "don't tell" wall. Sometimes, the soldier was prosecuted for posttraumatic and/or suicidal behavior. Regrettably, this has occurred recently. This needs to change. The U.S. military has, in fact, convicted soldiers for attempted suicide (Lande, 1992). Revisions in the military law are needed. Lande was correct in 1992, as Upton was a century earlier. The military needs to be sensitive and humane. It is PTSD, not criminality.

With this call for humanity, given one of my limitations (I'm not a soldier), I present some military highlights on policies and procedures. There are many military documents on suicide prevention; however, to reduce the number to a manageable size, I decided to choose three of the best on policy and tactics. This should not be seen to mean that other publications are not of help, only that there are some we should consider. They are

> Department of Defense. (2010). *The challenge and promise: Strengthening the force, preventing suicide and saving lives.* Washington, DC: Author.
>
> Department of the Army. (2009). *Health promotion, risk reduction, and suicide prevention.* Washington, DC: Author.
>
> National Defence (Canada). (2010). *Report of the Canadian Forces Expert Panel on Suicide Prevention.* Ottawa: Author.

## Recommendations from the United States and Canada

> Department of Defense. (2010). *The Challenge and Promise: Strengthening the Force, Preventing Suicide and Saving Lives.*

As directed by Section 733 of the National Defense Authorization Act (NDAA) for fiscal year 2009, the Secretary of Defense established a Task Force "to examine matters relating to prevention of suicide by members of the Armed Forces." The Department of Defense (DoD) Task Force on the Prevention of Suicide by Members of the Armed Forces was created.

The Task Force, established in August 2009, has prepared the following report for the Secretary of Defense, detailing the research, results, and recommendations from a year-long review of data, studies, programs, and discussions with Service Members, their families, and their caregivers. The intent of this report is to provide the Secretary of Defense and DoD leadership with actionable and measurable recommendations for policy and programs designed to prevent suicide by members of the Armed Forces. The Task Force arrived at 76 recommendations.

The Task Force arrived at 49 findings and 76 targeted recommendations. The findings fall into four primary focus areas: Organization and Leadership; Wellness Enhancement and Training; Access to and Delivery of Quality Care; and Surveillance, Investigations, and Research.

### Focus Area 1: Organization and Leadership

> Suicide prevention begins with a comprehensive strategy that has the support of leaders at every level. Develop a coherent policy. Effective organizational structure is essential to develop enterprise-wide policy as well as procedural standardization and oversight. Suicide prevention is a leadership responsibility from the most senior leaders down to frontline supervisors (first-line leaders). . . .

### Focus Area 2: Wellness Enhancement and Training

> Military life, particularly in wartime, is inherently stressful on individuals and presents a unique challenge to maintaining wellness. Efforts to enhance well-being must include all areas of fitness: physical, psychological, family,

social, spiritual, financial, vocational, and emotional. Substantial effort to mitigate stressors by supporting programs that strengthen protective factors in these domains. Stress on the force *must* be reduced. Heightened operational tempo, repeated deployments, and insufficient quantity and quality of dwell time have had a cumulative fatiguing effect on Service Members, and a degradation of the overall fitness and readiness of the force. Service Members have been incredibly resilient. In addition, family members generally do not receive adequate education and training in suicide prevention and they, above all, are the best "detectors" of subtle behavioral changes associated with suicidal risk.

### Focus Area 3: Access to and Delivery of Quality Care

An effective, multifaceted suicide prevention program must provide access to high-quality professional services across the entire health and wellness continuum. These services include assessment, diagnosis, counseling, and treatment . . . strong lines of communication between service providers are essential.

### Focus Area 4: Surveillance, Investigations, and Research

The Task Force strongly believes that well-constructed surveillance is necessary to inform and shape future suicide prevention programs. For surveillance to be effective, it must be standardized, centrally driven, and reported from the Service-level to DoD in a timely, consistent, and reliable manner.

### Foundational Recommendations

1. Create a "Suicide Prevention Policy Division" at OSD within USD (P&R) to standardize policies and procedures with respect to resiliency, mental fitness, life skills, and suicide prevention. The office will provide standardization, integration of best practices, and general oversight, serve as a change agent, and establish an ongoing external review group of non-DoD experts to assess progress. Furthermore, this office will provide guidance from which the Services can design and implement their suicide prevention programs.
2. Keep suicide prevention programs in the leadership lane and hold leaders accountable at all levels for ensuring a positive command climate that promotes the well-being, total fitness, and "help seeking" of their Service Members. A significant focus on developing better tools to assist commanders in suicide prevention must be undertaken.
3. Reduce stress on the force. The pace of operations in today's military exceeds the ability of Service Members to be restored to their optimal state of readiness. There is a supply and demand mismatch that creates a cumulative negative impact on the force. Reduce stress by ensuring the quantity and quality of dwell time allows for individual restoration as the force is reconstituted over and over again. This will allow Service Members to reestablish relationships and connectedness. If necessary,

either grow the size of the force to ensure additional uniformed end-strength to meet the demand or reduce the mission demand.
4. Focus efforts on Service Member well-being, total fitness (of the mind, body, and spirit), and development of life skills and resiliency to increase protective factors and decrease risk factors. This is the pinnacle of primary prevention.
5. Develop a Comprehensive Stigma Reduction Campaign Plan that attacks the issue on multiple fronts to encourage help-seeking behavior and normalizes the care of the "hidden wounds" incurred by Service Members.
6. Strengthen strategic messaging to enhance positive communications that generate the behaviors and outcomes desired rather than highlighting the negative messaging about today's challenges. The focus of messaging must migrate from speaking solely about the "tragedy" of suicide and the "actions" being taken to messages that reduce stigma, encourage help-seeking, portray concerned leadership, and inspire hope by showing that help really works.
7. Develop skills-based training in all aspects of training regarding suicide prevention. The current awareness and education efforts about suicide prevention are adequate, but skills-based training is deficient, especially among buddies, family members, first-line supervisors, clergy, and behavioral health personnel.
8. Incorporate program evaluation in all suicide prevention programs to determine the effectiveness of each program in obtaining its intended outcome.
9. Coordinate and leverage the strengths of installation and local community support services for both Active and Reserve Component Service Members. Community health and access to quality, competent services are essential to suicide prevention.
10. Ensure continuity and the management of quality behavioral healthcare, especially while in transition periods, to facilitate a seamless transfer of awareness, management, and treatment as Service Members change locations. Transitions need to be actively managed and tools must be developed to actively manage them.
11. Mature and expand the DoDSER to serve as the main surveillance method to inform future suicide prevention efforts. Further standardize data collection processes. Robust surveillance will produce data that allow us to anticipate and avoid future occurrences of that event before the individual or population (or unit) reaches a crisis point.
12. Standardize suicide investigations and expand their focus to learn about the last hours, days, and weeks preceding a suicide or attempted suicide. Pattern suicide investigations on aviation accident safety investigation procedures and use the safety investigation process as a model to develop a standardized suicide investigation process.
13. Support and fund ongoing DoD suicide prevention research to enhance knowledge and inform future suicide prevention efforts, and to incorporate evidenced-based solutions. Focused research in suicide

prevention for Service Members is essential to identifying best practices, decreasing variation in prevention practices, and in achieving desired outcomes.

Department of the Army (2009). *Suicide Prevention. Health Promotion, Risk Reduction and Suicide Prevention.*

## Prevention

3-1. Suicide prevention

Suicide prevention is a continuum of awareness, intervention, and postvention to help save lives. Prevention refers to all efforts that build resilience, reduce stigma, and build awareness of suicide and related behaviors. Ultimately, the goal of prevention is to develop healthy, resilient Soldiers to the state that suicide is not an option. . . . [The military has no choice but to address the green walls. —AL] It is important to establish a culture that reinforces help-seeking behavior as an appropriate and generally accepted part of being responsible.

3-2. Factors contributing to suicide

Individuals may have difficulty coping with intense feelings or emotions and consider taking drastic measures to deal with the emotional pain. Strategies to address suicide should include both the mitigation of these intense emotions and the circumstances which lead to them. Most suicides and suicide attempts are reactions to one or more of the following intense feelings. . . . [The common stimulus in suicide is *unbearable pain*. These unendurable feelings include shame, disgrace, worthlessness, and especially, hopelessness and helplessness. —AL]

    *c.* Hopelessness is a strong sense of futility, due to the belief that the future holds no escape from current negative circumstances. The intensity of this emotion is fed by the belief that no resources exist to bring relief or change the current perception of reality.

    *d.* Helplessness is a condition or event where the Soldier thinks that they have no control over their situation and whatever they do is futile, such as repeated failures, to include failed relationships, and so on. . . .

3-3. Life skills and resiliency

    *a.* Resiliency-building programs help Soldiers and Families develop life skills and directly impact the success of suicide prevention efforts by enhancing protective factors and mitigating stressors at the earliest stages. Life skills classes are available on a wide variety of subjects to include couples communication, child rearing, money management, stress management, conflict resolution, anger management, and problem solving. Commanders at all levels are encouraged to work with ACS and local agencies to make these classes available to Soldiers and Families.

    *b.* Resiliency is the ability to recover and adapt despite adversity, trauma, illness, changes or misfortunes. Resiliency means "bouncing back" from difficult situations. Soldier resiliency is a combination of factors including a sense of belonging in the unit, having inner strength to face adversity and fears, connecting with buddies, maintaining caring and supportive relationships within and outside the Family, maintaining a positive view of self, having confidence in strengths and abilities to function as a Soldier, and managing strong feelings and impulses. . . .

3-4. Stigma reduction

One of the greatest barriers to preventing suicides is a culture that shames Soldiers into believing it is not safe to seek help. Stigma can render suicide prevention efforts ineffective unless elements are incorporated into the program to counter these destructive attitudes.

- *a.* Individuals may not seek help because they believe that their problems or behavioral health issues should remain a secret. Reasons for this may include shame and embarrassment, fear that their careers are affected, concern that personal issues are exposed, belief that seeking help is a sign of weakness, and a feeling of helplessness and hopelessness.
- *b.* Keeping personal problems or behavioral health issues a secret can result in the development of depression and anxiety, compounded stressors, degraded ability to think clearly, difficulty making decisions, thoughts of suicide, suicidal attempts, and completed suicides. . . .
- *d.* We must all reduce actual and perceived stigma of seeking help. Stigma is a cultural issue that will take a deliberate and focused effort to combat. The key to stigma reduction is leadership emphasis at all levels. Leaders can accomplish this by:
  (1) Eliminating policies that discriminate against Soldiers who receive mental health counseling.
  (2) Supporting confidentiality between the Soldier and his/her mental health care provider.
  (3) Reviewing policies and procedures that could preclude Soldiers from receiving all necessary and available assistance. . . .

3-5. Awareness

- *a.* An essential foundation to the suicide prevention program is communicating key suicide prevention messages to Soldiers, Leaders, DA civilians, and Families. As a result, the following goals may be achieved:
  (1) The subject of suicide is normalized. Soldiers and Families need to feel comfortable discussing suicide and asking those who are contemplating suicide the tough questions. Individuals need to be aware that they are not alone and do not need to suffer in isolation and silence.
  (2) The seriousness of the problem is highlighted, with specific emphasis on consequences and long-lasting effects of suicide on the Family members and loved ones who are directly affected.
  (3) Stigma is reduced and help-seeking behavior is encouraged. . . .

3-7. Intervention

- *a.* Intervention attempts to prevent a life crisis or mental disorder from leading to suicidal behavior, and includes managing suicidal thoughts that may arise. At its most basic level, intervention may simply include listening, showing empathy, and escorting a person to a helping agency. This is something that can be done by any Soldier, Family member, or DA civilian with minimal training at the unit level. . . .

b. Intervention may also include the use of more advanced skills by trained personnel who are capable of providing a greater level of crisis intervention, screening, care, and referral. . . .

c. The loss of a family member, especially the loss of a child due to suicide, is perhaps the most difficult form of death for survivors to accept. On top of their grief over the death of a loved one, families of suicide victims often experience shame, humiliation, and embarrassment. Other common reactions are fear, denial, anger, and guilt, all of which combine to produce one of the most difficult crisis a family will ever experience.

National Defence (Canada). (2010). *Report of the Canadian Forces Expert Panel on Suicide Prevention.*

Public interest in military suicides has been particularly acute over the past year or two, coincident with a rise in the suicide rates seen in the US Army and Marine Corps. Other militaries coping with the extraordinary demands of the conflicts in SW Asia are concerned that they, too, will see such increases in time. . . .

*Targets for Suicide Prevention in the CF* (Canadian Forces): The Panel began by reviewing a familiar model of targets for suicide prevention in civilians. This model included the following elements:

A. Education and awareness programs;
B. Screening and assessment;
C. Pharmacotherapy;
D. Psychotherapy;
E. Follow-up care for suicide attempters and high-risk patients;
F. Restriction of access to lethal means; and
G. Media engagement (to encourage responsible reporting of suicides).

The Panel then extended this model by adding additional prevention targets in military organizations, including:

H. Organizational interventions intended to mitigate work stress/strain (leadership training, policy, programs, etc.);
I. Selection, resilience training, and primary risk factor modification;
J. Interventions to overcome barriers to mental health care; and
K. Systematic efforts to improve the quality of mental health care. . . .

[They make some 59 recommendations.—AL]

## Prevention Strategies

A. Education and Awareness Programs
Recommendation 2: Routine suicide awareness and prevention training should be incorporated into the rest of the regular, coordinated mental

health training that will occur across each member's career and deployment cycle. . . .

Recommendation 8: The linkage between good leadership, work-related stress, and mental health problems should be covered in the CF's routine mental health education program, as should the leader's role in overcoming barriers to mental health care. . . .

**Assessment of Suicidality**
Recommendation 20: Assessment of suicidality should of course occur at the first encounter for evaluation of patients with symptoms of mental health problems in both mental health and primary care settings. Suicidality should be reassessed during care for patients who have deteriorated, have developed new symptoms or co-morbidities, have failed to improve as expected, or are experiencing a crisis or significant new stressors. . . .

Treatment of Suicidality
C. Pharmacotherapy
Recommendation 25: The CF should follow conventional, evidence-based guidelines for the drug treatment of mental disorders. . . .

D. Psychotherapy
Recommendation 27: Suicidality should be identified and addressed as a separate problem in mental health patients. Patients with suicidal ideation, intent, or behaviour should receive evidence-based psychotherapy specifically targeting the suicidality and the interpersonal problems that are driving it. . . .

E. Follow-up Care for Suicide Attempters/High-risk Patients
Recommendation 31: The CF should consider developing and implementing a single, system-wide process for assuring follow-up for patients receiving care for mental disorders in both primary care and specialty mental health care settings. . . .

F. Restriction of Access to Lethal Means ("Means Reduction")
Recommendation 33: The CF should ensure that suicide surveillance data captures the source of the firearm involved (service vs. personal), as well as enough information to determine whether policies and procedures for firearm access were followed. This surveillance data should inform any needed changes in firearm control policies. . . .

J. Systematic Efforts to Overcome Barriers to Mental Health Care
Recommendation 43: The CF should continue to identify the remaining barriers to mental health care (both in garrison and on deployment) and to make any needed changes to address these.

K. Systematic Clinical Quality Improvement Efforts
Recommendation 56: "Buddy aid" mental health skills should also be integrated into the CF's comprehensive mental health education program.

Department of the Army. (2009). *Suicide Postvention. Health Promotion, Risk Reduction, and Suicide Prevention.*

Postvention

4-1. General

Postvention consists of a sequence of planned support and interventions carried out with survivors in the aftermath of a completed suicide or suicide attempt. Postvention is prevention for survivors. The goal of suicide postvention is to support those affected by a suicide or attempt, promote healthy recovery, reduce the possibility of suicide contagion, strengthen unit cohesion, and promote continued mission readiness.

*a.* When implementing a Postvention program, commanders will do the following:
   (1) Provide long term support to Families, unit members, and co-workers who experience loss due to suicide. Care can be provided via external services and outreach programs including civilian services for grief and recovery (that is, Department of Veterans Affairs Bereavement Counseling, Tragedy Assistance Program for Survivors (TAPS), Survivor Outreach Services [SOS]). . . .
   (2) Postvention activities include unit-level interventions following an attempted or completed suicide in order to minimize psychological reactions to the event, prevent or minimize the potential for copy cat suicides, strengthen unit cohesion, and promote continued mission readiness. . . .
   (3) Provide care to the friends of someone who has attempted or completed suicide. The command must proactively address the situation and provide an outlet for those affected to express and process their emotions. . . .

4-2. Army suicide behavior surveillance

Army suicide behavior surveillance is a critical postvention activity which includes the collection of informational data about suicide behavior by all components. . . .

*b.* Psychological autopsies may be requested by the Armed Forces Medical Examiner (AFME) and/or the Criminal Investigation Command (CID) on Active Duty deaths under special circumstances, in accordance with AR 600–63, paragraph 4-4 Additionally, the senior commander may request a psychological autopsy through CID. The psychological autopsy is a forensic investigative tool that is used to confirm or refute the death of an individual by suicide. It is not to be confused with gathering of information for suicide event surveillance for epidemiological purposes. Specifically, psychological autopsies assist in ascertaining the manner of death; and will primarily be used to resolve cases where there is an equivocal cause of death; that is, death cannot be readily established as natural, accidental, a suicide, or a homicide.

## POLICIES AND PROCEDURES: RECOMMENDATIONS

Whether we need the U.S. Department of Defense's 76 recommendations or Canada's National Defence's 59 recommendations, I do not know. What I do know is that suicide can be prevented. If a military system says 50 or 70 tactics are needed to win a war, then they will need those tactics. The enemy is strong. It is in PTSD and suicide. There are barriers (walls). However, the green culture is not to be underestimated. They can win this war. We will need, as General Upton knew, a sound policy and tactics.

Of course, I could highlight more of the armed forces tactics. However, the ones above are some military beliefs about what the problem is and what needs to be done. To echo the military leadership, *the issue now* is the one that Lieutenant Commander Kenneth Cooper (Jetley & Cooper, 2010) stated at the end of his presentation on suicide prevention in the military at the conference of the Canadian Association for Suicide Prevention in Halifax, Nova Scotia: "We, the military, have made our recommendations, now we wait for the government to act on the recommendations." *There is a call to action!*

## THE METHOD OF DIFFERENCE: UNITED STATES AND CANADA

Like my mentor, Captain Edwin Shneidman, I strongly believe in John Mill's Method of Difference. I have highlighted this method throughout this volume, especially examining the U.S. and Canada's armed forces. Knowledge about oneself can often be gleaned from someone else (Leenaars, 1995b). This is true for people, systems, and even militaries. And the more similar the comparison, the more useful it is. I do not wish to imply that broader comparison is not useful. Yet comparison of similar units often allows for unique discoveries. As Seymour Lipset (1990), in his highly recommended book, *Continental Divide*, notes, "The more similar the units being compared, the more possible it should be to isolate the factors responsible for differences between them" (p. xiii).

Canada and the United States are obvious similar units. These two countries share not only close proximity but similar languages and cultural backgrounds. They have also been open to continued comparisons in art, literature, politics, religion, and such (Lipset, 1990)—and suicide (Leenaars, 1995b). Despite a vast number of similarities, divergence has been observed in such areas as recognition of authority (with Canadians fearing authority more), patterns of deviance (with Americans exhibiting deviance more), achievement motivation (with Canadians alienating their achievers and heroes more), and patterns of association (with Americans joining organizations more). Not only has such research aided in our understanding of diverse areas but also aided the relationship between Canada and the United States. My comparisons have been on suicide among the armed forces, a neglected area of contrast, even in encyclopedic volumes on the continental divide (e.g., Lipset, 1990).

## HISTORICAL SPECULATION ON THE CONTINENTAL DIVIDE

Max Weber (1949), a father of modern sociology, was one of the first to provide the speculation that to understand current events, one had to search history. Historic events explain current differences. Weber argued by analogy that as one rolls a die (i.e., a decision) a certain number (that same decision) becomes more likely to be repeated. A decision in the Canadian military reinforces that decision. Another in the U.S. Armed Forces reinforces that decision. The countries' military history reinforces their values and behavior.

As one examines both countries, one learns that indeed the nations differ considerably in values and behavior. The foremost Canadian scholar in the area was S. D. Clark (1962), who noted, "Whereas the American nation was a product of the revolutionary spirit, the Canadian nation grew mainly out of forces of a counter-revolution" (pp. 190–191). He speculated that differences in Canada and the United States can best be explained by their history; namely, how people in these countries settled their respective frontiers. The Canadians moved westward under the organization of the British Government, the influence of the military police, and the favor of the two predominant churches (the Church of England and the Roman Catholic Church). In contrast, America's frontier was settled independently, not as an "outpost of the empire." Canada remained loyal to Britain; the United States was established as a revolution against Britain. The revolutionary attitude in the United States is distinct from the more conservative attitude in Canada. It is this explanation, according to Clark and others (e.g., Lipset, 1990) that best accounts for the divergence in the two nations. It "threw the dice" to develop the continental divide, and that includes in the militaries.

As an example, Margaret Atwood (1972), one of Canada's best-known authors, suggests that the symbol for the United States is "the Frontier . . . a place that is *new*, where the old can be discarded." The American Revolution divided the new from the old. In contrast, in Canada the symbol is "Survival," a place where one is "hanging on, staying alive." Canadians hang on to Britain, the churches, the RCMP, for example. A quote from Atwood's intriguing book, *Survival* (1972), is most revealing, especially in its direct application to suicide prevention in the military. She notes,

> Canadians are forever taking the national pulse like doctors at a sickbed. The aim is not to see whether the patient will live well but simply whether he will live at all. Our central idea is one which generates, not the excitement and sense of adventure or danger which The Frontier holds out, not the smugness and/or sense of security, of everything in its place . . . but an almost intolerable anxiety. (p. 33)

I do not know whether these explanations help to understand the two different systems; however, I do know that Mill's method of difference allows us to see the two militaries better and what they are today doing about trauma and suicide in the armed forces. This, I hope, has made the siblings trauma and suicide among the armed forces more visible. I do not mean to suggest that there are not even more commonalities in the armed forces (Mill's method of agreement and difference); for example, both militaries use the "frontier circle-the-wagons tactic"; what I have called green walls. There is a continental divide in suicide prevention in the armed forces.

## CONCLUDING THOUGHTS: A CALL TO ACTION FOR SUICIDAL SOLDIERS

The Civil War General William Sherman famously stated, "War is hell." The nature of war is stressful. "War is friction" (von Clauzswitz, 1982). There is intent to cause actual death, or serious injury, or threat to physical and psychological integrity of others. General Sherman, for example, led his army to mass destruction of the city of Savannah and the country side. He intentionally killed and injured people, slaughtered the animals, torched the crops, burned the homes, ruined the railways, and so on. General Sherman, a hero among the Union troops, was accused of being a terrorist by Confederate soldiers and civilians. The Southern people endured and never forgot the trauma. His response to the charges of war crimes was his well known statement, *"War is hell."*

War also causes some soldiers and civilians to kill themselves. This has been known in the United States, but kept a secret, since the great suicide epidemic after the Civil War. It is known that trauma has a shadowy presence. Traumatic events would "horrify, repulse, disgust, and infuriate any sane person" (Rudofossi, 2006). After the Vietnam War, the now obvious question was asked, "Why shouldn't that be true for heroes and veterans?" In this book, I have attempted to break down the green wall of silence, which is exemplified by the statement, "Don't talk about it."

This book has asked some tough questions: Why does suicide occur? Was Durkheim (1951) correct when he suggested that the military system predisposes a soldier to make away with him/herself? Are some suicides altruistic? Military suicide can be altruistic, but also egoistic, anomic, or fatalistic. Ultimately, suicide takes place or occurs in the mind of a suicidal warrior; military suicide is an intrapsychic drama on a military stage. Minimally, the intrapsychic causes are unbearable psychological pain, cognitive constriction, indirectness and unconscious processes (e.g., dissembling, self-deception), psychopathology, and a deep (ego) vulnerability. The military stage minimally includes problematic relationship in the armed forces (or at home), loss/rejection-aggression (almost always self-directed; sometimes also other-directed), and the need to escape (egression).

Furthermore, whatever else suicide is, at the community and societal levels, within Durkheim's system, *the soldier's suicide is an altruistic suicide.*

What are the community and social factors that affect the suicidal history? Is individualism or collectivism a main factor? Does the military system (culture) promote suicides, especially among officers? It did for General Upton and Admiral Boorda. It is a closed system. Sometimes the military system saves lives; sometimes it causes deaths. Secrecy is our enemy. The availability of guns can be. The green walls have to come down!

The ecological model suggests that there is society, community, and relationship elements in every suicide, but also that there are deep individual ones (deeper than most are aware). The common stimulus for suicide is almost *always* enduring psychological pain. We need to ask, What was the unbearable pain? Is it shame and disgrace? Loss of honor? Hopelessness and helplessness? Whatever, *the traumatizing pain was a defeating enemy!*

What is the evidence-based "pudding"? Can the system be a protective factor? What are the evidence-based guides to treatment and intervention that might allow a suicidal soldier or veteran to courageously overcome not only the shocks but also the aftershocks? *Prevention* has to be multidimensional. Psychotherapy (counseling) is always necessary. What are the barriers (green walls) to wellness in the military? What can be done to more effectively educate and help our at-risk for suicide (and homicide, self-harm, motor vehicle accidents, etc.) warriors and veterans? This at-risk hero may even include the "tough" sergeant, the war-decorated General, and the Admiral of the U.S. Navy. And, of course, in light of current events in America and Canada, there is the dire system/service question: Will we continue to allow our soldiers to suffer the traumatization and die by their own hands? Will research, such as PAs, be allowed? I have maybe too repetitively called for more study among the armed forces, especially PA studies (Captain Shneidman did the same). This is not only for surveillance and developing evidence-based *preventions,* but also for meaning-making for the survivors (Dyregrov, Dieserud, Hjelmeland, Straiton, Rasmussen, Knizek, & Leenaars, 2011). The needless deaths affect all of us.

We need to understand first, then we can predict and control. That is evidence-based science and should be evidence-based military policy. I asked key questions using the Socratic method. Socrates knew that this form of exploration empowered the person, being asked the question, not the teacher (or therapist, or the sergeant). Our suicidal soldiers need to feel empowered, not the howling tempest of the suicidal mind. There were many questions throughout the book—about the heart and soul of soldiers and veterans. What were the shocks and aftershocks of the gruesome death of a civilian child, the rape of a nurse, a partner's suicide, an inflicted TBI, and simply war?

To our heroes and veterans, I state, You need your courage and hope. You are an intelligent, adept soldier. You have to *accept the unacceptable*—what you cannot change—and you have to have the courage to change what you can.

You need to get beyond the unbearable pain, mental constriction, vulnerability, and so on. You are a hero, and I believe in your ability to stop, pause, and reflect. You can be resilient. Healing is possible. The true warrior seeks *help*! Here I follow the wisdom of Jacob Bronowski (1973) in the famed book, *The Ascent of Man*. What makes a person a person—and a soldier a soldier—is the ability to wait, to think, to talk, to pause, to reflect, and so on, before the act. In the battlefield, the soldier does no different.

You, in the military, have worked hard as a soldier or pilot or Marine or sailor; now you can trust that *strength* to work hard on choosing life. You can have confidence that there is help available and that there are people—a Secretary of Defense, a four-star General, a sergeant, a psychologist, a fellow armed services member/buddy and so on—in the military that care! You have *green courage*. Courage is to change what you can. The anodynic experience, to somewhat quote Aldous Huxley, is not what happened to you; it is what you do with what happened to you. I offer some scripts: Don't give up the fight. Don't buy into the green walls. There is effective help. Choose life.

The soldier needs to trust her or his courage; despite all that has happened to you in harm's way and since, you have adjusted to stress, beyond what you imagined the first day of boot camp. I strongly believe that your life, and mine, is like that of the mythical Greek Sisyphus. Sisyphus lived in the heavens with the gods and on Earth with mortals. He saw the painful and depressing life of humans and knew what would help. The gods had an anodyne. Despite Zeus's orders, Sisyphus stole the gods' secrets and helped humankind. Zeus raged and banished him from the heavens. Sisyphus was doomed to the *human* condition; each day he had to push a boulder up a mountain, only to watch it tumble back down, causing the task to be repeated the next day. Your and my life is no different. Each day we must ceaselessly roll our distinctive rock to the top of our mountain, and the next day we must persevere and do the same. This is not to be condemned; this is life. We have to accept the *unacceptable*. Military life makes it even more so; the mountain is even higher. It is Mount Everest! The system and culture make it so. Yet, if you believe the Greek wisdomkeeper, Homer, Sisyphus was the wisest and most prudent of humans. As with our ancient Greek hero, I do not want to inoculate you against trauma, the common green culture approach; I have attempted in this book, from a suicidologist's perspective, to do something different. I have attempted to make suicide among the armed forces more visible. What I have learned what is *most* helpful is to *persevere*. I hope that this mantra will help you get in touch with your Sisyphean strength (what are called protective factors) that build the natural surviving of the aftershock; what St. Paul called the sting, of everyday military service, deep within the mind, heart, body, and soul. This book, I hope, will help you to heal your pain, to end your suicide risk. *You can survive!*

## Don't give up the fight!

# References

Aaron, D. (Ed.). (1985). *The Inman diary: A public and private confession.* Cambridge, MA: Harvard University Press.

Adams, D., Barton, C., Mitchell, G., Moore, A., & Einagel, V. (1998). Hearts and minds: Suicide among United States combat troops in Vietnam, 1957–1973. *Social Science & Medicine, 47,* 1687–1694.

Allen, J., Cross, G., & Swanner, J. (2005). Suicide in the army: A review of current information. *Military Medicine, 170,* 580–584.

Allen, N. (1980). *Homicide.* New York, NY: Human Sciences.

Allen, N. (2009, November 8). Fort Hood gunman had told U.S. military colleague that infidels should have their throats cut. *The Telegraph.* Cited in Wikipedia. Retrieved February 6, 2010 from http://en.wikipedia.org/wiki/Nidal_Malik_Hasan

Allport, G. (1942). *The use of personal documents in psychological science.* New York, NY: Social Science Research Council.

Allport, G. (1962). The general and the unique in psychological science. *Journal of Personality, 30,* 405–422.

American Psychiatric Association (APA). (1994). *DSM-IV: Diagnostic and statistical manual of mental disorders* (4th ed.). Washington, DC: Author.

Andreason, N., & Grove, W. (1982). The classification of depression: A comparison of traditional and mathematically derived approaches. *American Journal of Psychiatry, 139,* 45–52.

Apter, A., Bleich, A., King, R., Kron, S., Fluch, A., Kotler, M., & Cohen, D. (1993). Death without warning? A clinical postmortem study of suicide in 43 Israeli adolescent males. *Archives of General Psychiatry, 50,* 138–142.

Arlington National Cemetery Website. (2011). *Jeremy Michael Boorda.* Retrieved May 1, 2011 from http://www.arlingtoncemetery.net/boorda/htm

Asberg, M., Traskman, L., & Thorien, P. (1976). 5-H1AA in cerebrospinal fluid: A biochemical suicide prediction? *Archives of General Psychiatry, 33,* 1193–1197.

Atwood, M. (1972). *Survival.* Toronto, Canada: Anasi.

Ayer, A. (Ed.). (1959). *Logical positivism.* New York, NY: Free Press. (Original work published 1931)

Bagley, S., Munjas, B., & Shekelle, P. (2010). A systematic review of suicide prevention programs for military or veterans. *Suicide and Life-Threatening Behavior, 40,* 257–265.

Bailey, S. (2011, May 2). 450 troops sought help for mental health in 10 months. *The Globe and Mail,* p. A6.

Barak, A., & Miran, O. (2005). Writing characteristics of suicidal people on the Internet: A psychological investigation of emerging social environments. *Suicide & Life-Threatening Behavior, 35,* 507–524.

Barnes, J. (2007). Murder followed by suicide in Australia, 1973-1992: A research note. *Journal of Sociology, 36,* 1–14.

Barraclough, B. (1986). Illness and suicide. In J. Morgan (Ed.), *Suicide: Helping those at risk* (pp. 61–69). London, UK: King's College.

Barry, R. (1918, June 16,). Emory Upton, military genius. *The New York Times,* pp. 2, 12.

Basham, C., Denneson, L., Millet, L., Shen, X., Duckart, J., & Dobscha, S. (2011). Characteristics and VA health care utilization of U.S. veterans who completed suicide in Oregon between 2000 and 2005. *Suicide & Life-Threatening Behavior, 41,* 287–296.

Beck, A. (1963). Thinking and depression: I. Idiosyncratic content and cognitive distortions. *Archives of General Psychiatry, 9,* 324–335.

Beck, A. (1967). *Depression: Clinical, experimental and theoretical aspects.* New York, NY: Hoeber.

Beck, A. (1976). *Cognitive therapy and the emotional disorders.* New York, NY: International Universities Press.

Beck, A., & Greenberg, R. (1971). The nosology of suicidal phenomena: Past and future perspectives. *Bulletin of Suicidology, 8,* 10–17.

Beck, A., Kovacs, M., & Weissman, A. (1975). Hopelessness and suicidal behavior: An overview. *Journal of the American Medical Association, 234,* 1146–1149.

Beck, A., Resnick, H., & Lettieri, D. (Eds.). (1974). *The prediction of suicide.* Bowie, MD: Charles.

Beck, A., Rush, A., Shaw, B., & Emery, C. (1979). *Cognitive therapy of depression.* New York, NY: Guilford.

Belik, S., Stein, M., Asmundson, G., & Sareen, J. (2009). Relation between traumatic events and suicide attempts in Canadian military personnel. *The Canadian Journal of Psychiatry, 54,* 93–104.

Berg-Cross, L. (Ed.). (2000). *Basic concepts in family therapy: An introduction* (2nd ed.). New York, NY: Hawthorn.

Bodner, E., Ben-Artzi, E., & Kaplan, Z. (2006). Soldiers who kill themselves: The contribution of dispositional and situational factors. *Archives of Suicide Research, 10,* 29–43.

Boscarino, J. (1995). Post-traumatic stress and associated disorders among Vietnam veterans: The significance of combat exposure and social support. *Journal of Traumatic Stress, 8,* 317–336.

Boscarino, J. (2006). External-cause mortality after psychologic trauma: The effects of stress exposure and predisposition. *Comprehensive Psychiatry, 47,* 503–514.

Boscarino, J. (2007). The mortality impact of combat stress 30 years after exposure: Implication for prevention, treatment, and research. In C. Figley & W. Nash (Eds.), *Combat stress injury* (pp. 97–117). New York, NY: Routledge.

Bostwick, J. (2000). Affective disorders and suicide risk: A re-examination. *American Journal of Psychiatry, 157,* 1925–1932.

Brenner, L. (2010, April). *Traumatic brain injury and suicide in veteran and returning military personnel.* Plenary presented at the American Association of Suicidology conference, Orlando, FL.

Brenner, L., Betthauser, L., Homaifer, B., Villarreal, E., Harwood, J., Staves, P., & Huggins, J. (2011). Posttraumatic stress disorder, traumatic brain injury, and suicide attempt history among veterans receiving mental health services. *Suicide and Life-Threatening Behavior, 41,* 416–423.

Brenner, L., Ivins, B., Schwab, K., Warden, D., Nelson, L., Jaffee, M., & Terrio, H. (2010). Traumatic brain injury, post traumatic stress disorder, and post concussive symptom reporting among troops returning from Iraq. *Journal of Head Trauma Rehabilitation, 25,* 307–312.

Breshears, R., Brenner, L., Harwood, J., & Gutierrez, P. (2010). Predicting suicidal behavior in veterans with traumatic brain injury: The utility of the Personality Assessment Inventory. *Journal of Personality Assessment, 92,* 349–355.

Bronisch, T., Wolfersdorf, M., & Leenaars, A. (Eds.). (2005). Bipolar disorders & pharmacotherapy. *Archives of Suicide Research, 9,* 231–319.

Bronowski, J. (1973). *The ascent of man*. Boston, MA: Little, Brown, & Co.

Bruce, M. L. (2010). Suicide risk and prevention in veteran populations. *Annals of the New York Academy of Sciences, 1208,* 98–103.

Bryan, C., & Cukrowicz, K. (2011). Association between types of combat violence and the acquired capability for suicide. *Suicide & Life-Threatening Behavior, 41,* 126–136.

Bryan, C., Cukrowicz, K., West, C., & Morrow, C. (2010). Combat experience and the acquired capability for suicide. *Journal of Clinical Psychology, 66,* 1044–1056.

Bryan, C., Jennings, K., Jobes, D., & Bradley, J. (2012). Understanding and preventing military suicide. *Archives of Suicide Research, 16,* 95–110.

Bryan, C., Kanzler, K., Durham, T., West, C., & Greene, E. (2010). Challenges and considerations for managing suicide risk in combat zones. *Military Medicine, 175,* 713–718.

Bullman, T., & Kang, H. (1994). Posttraumatic stress disorder and the risk of traumatic deaths among Vietnam veterans. *Journal of Nervous and Mental Disease, 182,* 604–610.

Buteau, J., Lesage, A., & Kiely, M. (1993). Homicide followed by suicide: A Quebec case series, 1998–1990. *Canadian Journal of Psychiatry, 38,* 552–556.

Canetto, S., & Lester, D. (Eds.). (1995). *Women and suicidal behavior*. New York, NY: Springer.

Calhoun, P., Beckham, J., & Bosworth, H. (2002). Caregiver burden and psychological distress in partners of veterans with chronic posttraumatic stress disorder. *Journal of Traumatic Stress, 15,* 205–212.

Caplan, G. (1964). *Principles of preventive psychiatry*. New York, NY: Basic.

Carnap, R. (1959). Psychology in physical language. In A. Ayer (Ed.), *Logical positivism* (pp. 165–197). New York, NY: Free Press. (Original work published 1931)

Carr, J., Hoge, C., Gardner, J., & Potter, R. (2004). Suicide surveillance in the U.S. military—Reporting and classification: Biases in rate calculations. *Suicide and Life-Threatening Behavior, 34,* 233–241.

Carr, R. (2011). When a soldier commits suicide in Iraq: Impact on unit and caregivers. *Psychiatry, 74,* 95–106.

Cassimatis, E., & Rothberg, J. (1997). Suicide in the United States military. In A. Botsis, C. Soldatos, & C. Stefanis (Eds.), *Suicide: Biopsychosocial approaches* (pp. 23–32). Amsterdam, The Netherlands: Elsevier.

Castro, S., & McGurk, D. (2007). Suicide prevention down range: A program assessment. *Traumatology, 13,* 32–36.

Centers for Disease Control. (1988). Health status of Vietnam veterans: I. Psychosocial characteristics. *Journal of the American Medical Association, 259,* 2701–2707.

Centers for Disease Control. (1989). *Health status of Vietnam veterans: Volume IV. Psychological and neuropsychological evaluation.* Atlanta, GA: Author.

Centers for Disease Control. (1999). Suicide prevention among active Air Force personnel—United States. 1990–1999. *Morbidity and Mortality Weekly Report, 48,* 1053–1057.

Centers for Disease Control and Prevention. (2009, December 5). *WISQAR website and fatal injury reports.* Retrieved December 5, 2009 from http://www.cdc.gov/TraumaticBrainInjury/index.html

Chavez-Hernandez, A., Leenaars, A., Chavez-de Sanchez, M., & Leenaars, L. (2009). Suicide notes from Mexico and the United States: A thematic analysis. *Salud Publicda de Mexico, 51,* 314–320. [In Spanish]

Christian, J., Stivers, J., & Sammons, M. (2009). Training to the warrior ethos: Implications for clinicians treating military members and their families. In S. Freeman, B. Moore, & A. Freeman (Eds.), *Living and surviving in harm's way* (pp. 27–49). New York, NY: Routledge.

Clark, D., & Sutton, L. (1992). U.S. Army experiences with suicide in the Gulf War. In D. Lester (Ed.), *Suicide '92. Proceedings. Silver Conference. American Association of Suicidology.* Chicago, IL: AAS.

Clark, S. (1962). *The developing Canadian community.* Toronto, Canada: University of Toronto Press.

Cohen, J., Goodman, R., Campbell, C., Carroll, B., & Campagne, H. (2009). Military children: The sometimes orphans of war. In S. Freeman, B. Moore, & A. Freeman (Eds.), *Living and surviving in harm's way* (pp. 395–416). New York, NY: Routledge.

Cronbach, L., & Meehl, P. (1955). Construct validity in psychological tests. *Psychological Bulletin, 52,* 281–302.

Curphey, T. (1961). The role of the social scientist in the certification of death by suicide. In E. Shneidman & N. Farberow (Eds.), *The cry for help* (pp. 110–117). New York, NY: McGraw-Hill.

Curry, B. (2010, October 8). Ex-soldier's case sparks privacy audit. *The Globe and Mail,* p. A7.

Cutler, T. (2002). *The Bluejackets manual.* Annapolis, MD: Naval Institute Press.

Dahlberg, L., & Krug, E. (2002). Violence—A global public health problem. In WHO (Ed.), *World report on violence and health* (pp. 1–21). Geneva, Switzerland: WHO.

Datel, W. (1979). The reliability of mortality count and suicide count in the United States Army. *Military Medicine, 144,* 509–512.

Dekel, R., & Solomon, Z. (2007). Secondary traumatization among wives of war veterans with PTSD. In C. Figley & W. Nash (Eds.), *Combat stress injury* (pp. 137–157). New York, NY: Routledge.

Department of Defense. (1986, March). *Health promotion.* Directive 1010.10. Washington, DC: Author.

Department of Defense. (1996). *Worldwide U.S. active duty military personnel casualties, October 1, 1979 through March 31, 1993.* Washington Headquarters Services, DIOR/MO7-93/02. Washington, DC: Author.

Department of Defense. (2010). *The challenge and promise: Strengthening the force, preventing suicide and saving lives*. Washington, DC: Author.

Department of the Army. (2009). *Health promotion, risk reduction, and suicide prevention*. Washington, DC: Author.

Desjeux, G., Labarère, J., Galoisy-Guibal, L., & Ecochard, R. (2004). Suicide in the French Armed Forces. *European Journal of Epidemiology, 19*, 823–829.

Donaldson-Pressman, S., & Pressman, R. (1994). *The narcissistic family: Diagnosis and treatment*. San Francisco, CA: Jossey-Bass.

Durkheim, E. (1951). *Suicide: A study in sociology* (J. Spaulding & G. Simpson, Trans.). London, UK: Routledge & Kegan Paul. (Original published in 1897)

Dyregrov, K., Dieserud, G., Hjelmeland, H., Straiton, M., Rasmussen, M., Knizek, B., & Leenaars, A. (2011). Meaning-making through psychological autopsy interviews: The value of participating in qualitative research for those bereaved by suicide. *Death Studies, 35*, 685–710.

Eaton, K., Messer, S., Wilson, A., & Hoge, C. (2006). Strengthening the validity of population-based suicide rate comparisons: an illustration using U.S. military and civilian data. *Suicide and Life-Threatening Behavior, 36*, 182–191.

Erikson, E. (1963). *Childhood and society* (2nd ed.). New York, NY: Norton.

Erikson, E. (1968). *Identity: Youth and crisis*. New York, NY: Norton.

Erikson, E. (1980). *Identity and the life cycle*. New York, NY: Norton.

Everson, R., & Camp, T. (2011). Seeing systems: An introduction to systemic approaches with military families. In R. Everson & C. Figley (Eds.), *Families under fire* (pp. 3–29). New York, NY: Routledge.

Everson, R., & Figley, C. (Eds.). (2011). *Families under fire*. New York, NY: Routledge.

Farberow, N. (Ed.). (1980). *The many faces of suicide*. New York, NY: McGraw-Hill.

Farberow, N., Kang, H., & Bullman, T. (1990). Combat experience and postservice psychosocial status as predictors of suicide in Vietnam veterans. *The Journal of Nervous and Mental Disease, 178*, 32–37.

Fawcett, J. (1997). The detection and consequences of anxiety in clinical depression. *Journal of Clinical Psychiatry, 58*(Suppl. 8), 35–40.

Fear, N., Ward, V., Harrison, K., Davison, L., Williamson, S., & Blatchley, N. F. (2009). Suicide among male regular UK Armed Forces personnel, 1984–2007. *Occupational and Environmental Medicine, 66*, 438–441.

Ferri, E. (1917). *Criminal suicidology*. Boston, MA: Little Brown.

Feigl, H. (1970). The "orthodox" view of theories: Remarks in defense as well as critique. In M. Radner & S. Winakur (Eds.), *Minnesota studies in philosophical science* (Vol. 4, pp. 3–16). Minneapolis, MN: University of Minnesota.

Figley, C. (1983). Catastrophes: An overview of family reactions. In C. Figley & H. Cubbin (Eds.), *Stress and the family: Vol II. Coping with catastrophe* (p. 3220). Levitown, PA: Brunner/Mazel.

Figley, C. (Ed.). (1985). *Trauma and its wake*. New York, NY: Brunner/Mazel.

Figley, C. (Ed.). (1995). *Compassion fatigue: Coping with secondary traumatic stress disorder in those who treat the traumatized*. New York, NY: Brunner/Mazel.

Figley, C. (1998). Burnout as systematic traumatic stress: A model for helping traumatized family members. In C. Figley (Ed.), *Burnout in families: The systematic costs of caring* (pp. 15–28). Boca Raton, FL: CRC.

Figley, C., & Nash, W. (Eds.). (2007). *Combat stress injury: Theory, research, and management.* New York, NY: Routledge.

Flaherty, A. (2010). Hundreds of PTSD soldiers likely misdiagnosed. *Associated Press.* Retrieved August 15, 2010 from http://news.yahoo.com/s/ap/20100815/ap_on_go_ca_st_pe/us_soldiers_wrongly_discharged

Florkowski, A., Gruszcynski, W., & Wawrzynlak, Z. (2001). Evaluation of psychopathological factors and origins of suicides committed by soldiers, 1989 to 1998. *Military Medicine, 166,* 44–47.

Fragala, M., & McCaughey, B. (1991). Suicide following medical/physical evaluation boards: A complication unique to military psychiatry. *Military Medicine, 156,* 206–209.

Frager, R., & Fadiman, J. (1984). *Personality and personal growth* (2nd ed.). New York, NY: Harper & Row.

Frederick, C. (1969, March). Suicide notes: A survey and evaluation. *Bulletin of Suicidology, 27*–32.

Freeman, S., Moore, B., & Freeman, A. (Eds.). (2009). *Living and surviving in harm's way: A psychological treatment handbook for pre- and post-deployment of military personnel.* New York: Routledge.

Freud, S. (1974a). Psychopathology of everyday life. In J. Strachey (Ed. & Trans.), *The standard edition of the complete psychological works of Sigmund Freud* (Vol. VI, pp. 1–310). London, UK: Hogarth. (Original work published 1901)

Freud, S. (1974b). Mourning and melancholia. In J. Strachey (Ed.), *The standard edition of the complete psychological works of Sigmund Freud* (Vol. XIV, pp. 239–260). London, UK: Hogarth. (Original work published 1917)

Freud, S. (1974c). General theory of neurosis. In J. Strachey (Ed. & Trans.), *The standard edition of the complete psychological works of Sigmund Freud* (Vol. XVI, pp. 243–483). London, UK: Hogarth. (Original work published 1917)

Freud, S. (1974d). A case of homosexuality in a woman. In J. Strachey (Ed. & Trans.), *The standard edition of the complete psychological works of Sigmund Freud* (Vol. XVIII, pp. 147–172). London, UK: Hogarth. (Original work published 1920)

Freud, S. (1974e). Group psychology and the analysis of the ego. In J. Strachey (Ed. & Trans.), *The standard edition of the complete psychological works of Sigmund Freud*( Vol. XVIII, pp. 67–147). London, UK: Hogarth. (Original work published 1921)

Freud, S. (1974f). The ego and the id. In J. Strachey (Ed. & Trans.), *The standard edition of the complete psychological works of Sigmund Freud* (Vol. IXX, pp. 3–66). London, UK: Hogarth. (Original work published 1923)

Freud, S. (1974g). The economic problem of masochism. In J. Stachey (Ed. & Trans.), *The standard edition of the complete psychological works of Sigmund Freud* (Vol. XIX, pp. 157–170). London: Hogarth. (Original work published 1924)

Freud, S. (1974h). Civilization and its discontent. In J. Stachey (Ed. & Trans.), *The standard edition of the complete psychological works of Sigmund Freud* (Vol. XXI, pp. 64–145). London, UK: Hogarth. (Original work published 1930)

Freud, S. (1974i). New introductory lectures. In J. Strachey (Ed. & Trans.), *The standard edition of the complete psychological works of Sigmund Freud* (Vol. XXII, pp. 3–182). London, UK: Hogarth. (Original work published 1933)

Freud, S. (1974j). Moses and monotheism. In J. Strachey (Ed. & Trans.), *Standard edition of the complete psychological works of Sigmund Freud* (Vol. XXIII, pp. 3–137). London: Hogarth. (Original work published 1939)

Gaines, Jr., T., & Richmond, L. H. (1980). Assessing suicidal behavior in basic military trainees. *Military Medicine, 145*, 263–266.

Gawande, A. (2004). Casualties of war: Military care for the wounded from Iraq and Afghanistan. *New England Journal of Medicine, 351*, 2471–2475.

Gibbs, D., Martin, S., Kupper, L., & Johnson, R. (2007). Child maltreatment in enlisted soldier's families during combat-related deployments. *Journal of the American Medical Association, 298*, 528–535.

Gilligan, J. (1996). *Violence: Reflections on a national epidemic*. New York, NY: Random House.

Goffman, E. (1974). *Frame analysis*. New York, NY: Harper Colophon.

Goldblatt, M. (1992). *Richard Cory suicides: Diagnostic questions*. Paper presented at the annual conference of the American Association of Suicidology, Chicago, IL.

Gordana, D., & Milivoje, P. (2007). Suicide prevention program in the Army of Serbia and Montenegro. *Military Medicine, 172*, 551–555.

Gould, J. (2011, May 24). Are suicides considered less honorable? Retrieved August 28, 2011 from http://militarytimes.com/blogs/outside-the-wire/2011/05/24/are-suicides-considered less honorable?

Greden, J., Valenstein, M., Spinner, J., Blow, A., Gorman, L., Dalack, G., Marcus, S., & Kees, M. (2010). Buddy-to-buddy, a citizen soldier peer support program to counteract stigma, PTSD, depression, and suicide. *Annals of the New York Academy of Sciences, 1208*, 90–97.

Grossman, D. (1996). *On killing: The psychological cost of learning to kill in war and society*. New York, NY: Little, Brown.

Guerra, V., & Calhoun, P. (2011). Examining the relation between posttraumatic stress disorder and suicidal ideation in an OEF/OIF veteran sample. Mid-Atlantic Mental Illness Research, Education and Clinical Center Workgroup. *Journal of Anxiety Disorders, 21*, 12–18.

Hagmann, D. (2009). *Profile of Major Nidal Malik Hasan*. Retrieved from http://www.homelandsecurityus.com/archives/3262

Hahn, R., Bilukha, O., Crosby, A., Thompson Fullilove, M., Liberman, A., Moscicki, E. et al. (2003, October 3). First reports evaluating the effectiveness of strategies for preventing violence: Firearms laws. *MMWR. Morbidity and Mortality Weekly Report, 52*, 11–20.

Halbwachs, M. (1978). *The causes of suicide* (H. Goldblatt, Trans.). London, UK: Routledge & Kegan Paul.

Hall, L. (2008). *Counseling military families: What mental health professionals need to know*. New York, NY: Routledge.

Hall, L. (2011). The military culture, language, and lifestyle. In R. Everson & C. Figley (Eds.), *Families under fire* (pp. 31–52). New York, NY: Routledge.

Hawton, K., Harris, L., Casey, D., Simkin, S., Harrison, K., Bray, I., & Blatchley, N. (2009). Self-harm in UK armed forces personnel: Descriptive and case-control study of general hospital presentations. *The British Journal of Psychiatry, 194*, 266–272.

Hawton, K., & van Heeringen, C. (Eds.). (2000). *The international handbook of suicide and attempted suicide*. Chichester, UK: John Wiley & Sons.

Hearst, N., Newman, T., & Hulley, S. (1986). Delayed effects of the military draft on mortality: A randomized natural experiment. *New England Journal of Medicine, 314*, 620–624.

Helmak, J. C. (1996). Occupation and suicide among males in the US Armed Forces. *Annals of Epidemiology, 6,* 83–88.

Hempel, C. (1966). *Philosophy of natural science.* Englewood Cliffs, NJ: Prentice-Hall.

Hendin, H., Haas, A., Maltsberger, J., Koestner, B., & Szanto, K. (2006). Problems in psychotherapy with suicidal patients. *American Journal of Psychiatry, 163,* 67–72.

Henry, A., & Short, J. (1954). *Suicide and homicide.* New York, NY: Free Press.

Herrell, R., Henter, I., Mojtabai, R., Bartko, J. J., Venable, D., Susser, E., Merikangas, K., & Wyatt, R. (2006). First psychiatric hospitalizations in the US military: The National Collaborative Study of Early Psychosis and Suicide (NCSEPS). *Psychological Medicine, 36,* 1405–1415.

Hill, J., Johnson, R., & Barton, R. (2006). Suicidal and homicidal soldiers in deployment environments. *Military Medicine, 171,* 228–232.

Hitschman, M., & Yarrell, Z. (1943). Psychoses occurring in soldiers during the training period. *American Journal of Psychiatry, 100,* 301–305.

Ho, T., Yip, P., Chiu, C., & Halliday, P. (1998). Suicide notes: What do they tell us? *Acta Psychiatrica Scandinavica, 98,* 467–473.

Hoge, C., Auchterlonie, J., & Milliken, C. (2006). Mental health problems, use of mental health services, and attrition from military service after returning from deployment to Iraq or Afghanistan, *Journal of the American Medical Association, 295,* 1023–1032.

Hoge, C., Castro, C., Messer, S., McGurk, D., Cotting, D., & Koffman, R. (2004). Combat duty in Iraq and Afghanistan, mental health problems, and barriers to care. *New England Journal of Medicine, 351,* 1798–1800.

Hoge, C., McGurk, D., Thomas, J., Thomas, J., Cox A., Engel, C., & Castro, C. A. (2008). Mild traumatic brain injury in U.S. soldiers returning from Iraq. *New England Journal of Medicine, 358,* 453–463.

Hoge, C., Messer, S., Engel, C., Krauss, M., Amoroso, P., Ryan, M., & Orman, D. (2003). Priorities for psychiatric research in the U.S. military: An epidemiological approach. *Military Medicine, 168,* 182–185.

Hoge, C., Terhakopian, A., Castro, C., Messer, S., & Engel, C. (2007). Association of posttraumatic stress disorder with somatic symptoms, health care visits, and absenteeism among Iraq war veterans. *The American Journal of Psychiatry, 164,* 150–153.

Hourani, L., Coben, P., & Warrack, A. (1999). Suicide in the U.S. Marine Corps, 1990 to 1996. *Military Medicine, 164,* 551–555.

Husserl, E. (1973). *The idea of phenomenology* (W. Alston & G. Nakhnikian, Trans.). The Hague, The Netherlands: Martinus Nijhoff. (Original work published 1907)

Hyer, L., McCraine, E., Woods, M., & Boudewyns, P. (1990). Suicidal behavior among chronic Vietnam theatre veterans with PTSD. *Journal of Clinical Psychology, 46,* 713–720.

Hyson, J., Mosberg, W., Sanborn, G., & Whitehorne, J. (1990). The suicide of General Emory Upton: A case report. *Military Suicide, 155,* 445–452.

Jakupcak, M., Cook, J., Imel, Z., Fontana, A., Rosenheck, R., & McFall, M. (2009). Posttraumatic stress disorder as a risk factor for suicidal ideation in Iraq and Afghanistan war veterans. *Journal of Traumatic Stress, 22*, 303–306.

Janoff-Bulman, R. (1992). *Shattered assumptions: Towards a new psychology of trauma.* New York, NY: Free Press.

Jeffrey, T., Rankin, R., & Jeffrey, L. (1992). In service of two masters: The ethical-legal dilemma faced by military psychologists. *Professional Psychology: Research and Practice, 23*, 91–95.

Friedli, C., Jenkins, R., McCulloch, A., & Parker, C. (2002). *Developing a national mental health policy.* East Sussex, UK: Psychology Press.

Jetley, R., & Cooper, K. (2010, October) *The Canadian forces expert panel on suicide prevention.* Workshop presented at the Canadian Association for Suicide Prevention Conference, Halifax, Nova Scotia.

Johnson, L., & O'Kearney, R. (2009). Impact of cultural differences in self on cognitive appraisals in PTSD. *Behavioral and Cognitive Psychotherapy, 37,* 249–266.

Jones, F. (1995a). U.S. Air Force combat psychiatry. In R. Zajtchuk & R. Bellamy (Eds.), *Textbook of military medicine Part 1. War Psychiatry* (pp. 177–210). Washington, DC: Office of the Surgeon General.

Jones, F. (1995b). Psychiatric lessons of war. In K. Zajtchuk & R. F. Bellamy (Eds.), *Textbook of military medicine: War psychiatry* (pp. 10–34). Washington, DC: Office of the Surgeon General.

Jung, C. (1964). *Man and his symbols.* Garden City, NY: Doubleday.

Kagitçibasi, Ç. (1996). *Family and human development across cultures: A view from the other side.* Mahwah, NJ: Lawrence Erlbaum.

Kamerman, J. (1993). The illegacy of suicide. In A. Leenaars (Ed.), *Suicidology: Essays in honor of Edwin S. Shneidman* (pp. 346–355). Northvale, NJ: Jason Aronson.

Kang, H., & Bullman, T. (1996). Mortality among US veterans of the Persian Gulf War. *New England Journal of Medicine, 335,* 1498–1504.

Kang, H., & Bullman, T. (2008). Risk of suicide among US veterans after returning home from Iraq or Afghanistan war zone. *Journal of the American Medical Association, 300,* 652–653.

Kang, H., & Hyams, D. (2004). Mental health care needs among recent war veterans. *New England Journal of Medicine, 352,* 1289.

Kaplan, M., Bentson, P., McFarland, B., & Huguet, N. (2009). Firearm suicide among veterans in the general population: Findings from the National Violent Death Reporting System. *The Journal of Trauma Injury, Infection, and Critical Care, 67,* 503–507.

Kaplan, M., Huguet, N., McFarland, B., & Newsom, J. (2007). Suicide among male veterans: A prospective population-based study. *Journal of Epidemiology and Community Health, 61,* 619–624.

Keen, S. (1991). *Fire in the belly: On being a man.* New York, NY: Bantam.

Keith, D., & Whitaker, C. (1984). C'est la guerre: Military families and family therapy. In F. Kaslow & R. Ridenour (Eds.), *The military family: Dynamics and treatment* (pp. 147–166). New York, NY: Guilford.

Kelly, M., & Vogt, D. (2009). Military stress: Effects of acute, chronic, and traumatic stress on mental and physical health. In S. Freeman, B. Moore, & A. Freeman (Eds.), *Living and surviving in harm's way* (pp. 85–106). New York, NY: Routledge.

Kemmelmeier, M., Wieczorkowska, G., Erb, H., & Burnstein, E. (2002). Individualism, authoritarianism, and attitudes toward assisted death: Cross-cultural, cross-regional, and experimental evidence. *Journal of Applied Social Psychology, 32,* 60–85.

Kendler, K., & Zachar, P. (2008). The incredible insecurity of psychiatric nosology. In K. Kendler & J. Parnas (Eds.), *Philosophical issues in psychiatry: Explanation, phenomenology and nosology* (pp. 368–383). Baltimore, MD: Johns Hopkins University Press.

Kennedy, C., & Malone, R. (2009). Integration of women into the modern military. In S. Freeman, B. Moore, & A. Freeman (Eds.), *Living and surviving in harm's way* (pp. 67–81). New York, NY: Routledge.

Kerlinger, F. (1964). *Foundations of behavioral research.* New York, NY: Holt, Rinehart & Winston.

Kimmel, D. (1974). *Adulthood and aging.* New York, NY: Wiley.

King, L., King, D., Fairbank, J., Keane, T., & Adams, G. (1998). Resilience-recovery factors in post-traumatic stress disorder among female and male Vietnam veterans: Hardiness, postwar social support, and additional stressful life events. *Journal of Personality and Social Psychology, 74,* 420–434.

Klerman, G. (Ed.). (1986). *Suicide and depression among adolescents and young adults.* Washington, DC: American Psychiatric Press.

Knox, K., Litts, D., Talcott, G., Feig, J., & Caine, E. (2003). Risk of suicide and related adverse outcomes after exposure to a suicide prevention programme in the US Air Force: Cohort study. *British Medical Journal, 327,* 1376.

Koren, D., Hilel, Y., Idar, N., Hemel, E., & Klein, E. (2007). Combat stress management: The interplay between combat, physical injury, and psychological trauma. In C. Figley & W. Nash (Eds.), *Combat stress injury* (pp. 119–135). New York, NY: Routledge.

Kuehn, B. (2010). Military probes epidemic of suicide: Mental health issues remain prevalent. *Journal of the American Medical Association, 304,* 1427–1430.

Kuhn, T. (1962). *The structure of scientific revolutions.* Chicago, IL: University of Chicago Press.

Kulka, R., Schlenger, W., Fairbank, J., Hough, R., Jordan, B., Marmar, C., & Weiss, D. (1988). *Contractual report of findings from the National Vietnam Veterans Readjustment Study.* Research Triangle Park, NC: Research Triangle Institute.

Lambert, M. (Ed.). (2004). *Bergin and Garfield's handbook of psychotherapy and behavior change* (5th ed.). New York, NY: John Wiley & Sons.

Lande, G. (2011). Fel de se: Soldiers suicides in America's Civil War. *Military Medicine, 176,* 531–536.

Lande, R. (1992). Suicides and the military justice system. *Suicide and Life-Threatening Behavior, 22,* 341–349.

Lee, J., Sudom, S., & Rammsayer, T. (2011). Higher-order model of resilience in the Canadian Forces. *Canadian Journal of Behavioural Science, 43,* 222–234.

Leenaars, A. (1988). *Suicide notes.* New York, NY: Human Sciences.

Leenaars, A. (1989a). Suicide across the adult life-span: An archival study. *Crisis, 10,* 132–151.

Leenaars, A. (1989b). Are young adults' suicides psychologically different from those of other adults? (The Shneidman lecture). *Suicide and Life-Threatening Behavior, 19,* 249–263.

Leenaars, A. (1991). Suicide in the young adult. In A. Leenaars (Ed.), *Life-span perspectives of suicide* (pp. 121–136). New York, NY: Plenum.
Leenaars, A. (1992). Suicide notes from Canada and the United States. *Perceptual and Motor Skills, 74*, 278.
Leenaars, A. (Ed.). (1993). *Suicidology: Essays in honor of Edwin S. Shneidman.* Northvale, NJ: Jason Aronson.
Leenaars, A. (1995a). Clinical evaluation of suicide risk. *Psychiatry and Clinical Neurosciences, 49*(Suppl. 1), 561–568.
Leenaars, A. (1995b). Suicide and the continental divide. *Archives of Suicide Research, 1*, 39–58.
Leenaars, A. (1996). Suicide: A multidimensional malaise. (The presidential address). *Suicide and Life-Threatening Behavior, 26*, 221–236.
Leenaars, A. (1997). Rick: A suicide of a young adult. *Suicide & Life-Threatening Behavior, 27*, 15–27.
Leenaars, A. (Ed.). (1999). *Lives and deaths: Selections from the works of Edwin S. Shneidman.* Philadelphia, PA: Brunner/Mazel.
Leenaars, A. (2004). *Psychotherapy with suicidal people: A person-centred approach.* Chichester, UK: John Wiley & Sons.
Leenaars, A. (2005). Effective public health strategies in suicide prevention are possible: A selective review of recent studies. *Clinical Neuropsychiatry, 2*, 21–31.
Leenaars, A. (2006). Psychotherapy with suicidal people: The commonalities. *Archives of Suicide Research, 10*, 305–322.
Leenaars, A. (2007). Suicide: A cross-cultural theory. In F. Leong & M. Leach (Eds.), *Suicide among racial and ethnic minority groups* (pp. 13–37). New York, NY: Routledge.
Leenaars, A. (2009a). Death systems and suicide around the world. In J. Morgan, P. Laungani, & S. Palmer (Eds.), *Death and bereavement around the world* (Vol. 5, pp. 103–138). Amityville, NY: Baywood.
Leenaars, A. (2009b). Controlling the availability of means of suicide—Gun availability and control. In D. Wasserman & C. Wasserman (Eds.), *The Oxford textbook of suicidology and suicide prevention: A source book* (pp. 577–581). Oxford, UK: Oxford University Press.
Leenaars, A. (2010a). *Suicide and homicide-suicide among police.* Amityville, NY: Baywood.
Leenaars, A. (2010b). Lives and deaths: Biographical notes and selections from the works of Edwin S. Shneidman. *Suicide and Life-Threatening Behavior, 40,* 476–491.
Leenaars, A., Cantor, C., Connolly, J., EchoHawk, M., Gailiene, D., He, Z. . . . Wenckstern, S. (2002). Controlling the environment to prevent suicide: International perspectives. *Canadian Journal of Psychiatry, 45,* 639–644.
Leenaars, A., De Wilde, E., Wenckstern, S., & Kral, M. (2001). Suicide notes of adolescents: A life span comparison. *Canadian Journal of Behavioural Science, 33,* 47–57.
Leenaars, A., Fekete, S., Wenckstern, S., & Osvath, P. (1998). Suicide notes from Hungary and the United States. *Psychiatrica Hungarica, 13,* 147–159. [In Hungarian]
Leenaars, A., Girdhar, S., Dogra, T., Wenckstern, S., & Leenaars, L. (2010). Suicide notes from India and the United States: A thematic comparison. *Death Studies, 34,* 426–440.
Leenaars, A., Haines, J., Wenckstern, S., Williams, C., & Lester, D. (2003). Suicide notes from Australia and the United States. *Perceptual and Motor Skills, 92,* 1281–1282.

Leenaars, A., & Lester, D. (1994). Suicide and homicide rates in Canada and the United States. *Suicide & Life-Threatening Behavior, 24,* 184–191.

Leenaars, A., & Lester, D. (Eds.). (1996). *Suicide & the unconscious.* Northvale, NJ: Aronson.

Leenaars, A., & Lester, D. (1998). Social factors and mortality from NASH in Canada. *Crisis, 19,* 73–77.

Leenaars, A., Lester, D., Lopatin, A., Schustov, D., & Wenckstern, S. (2002). Suicide notes from Russia and the United States. *Social and General Psychiatry, 12,* 22–28. [In Russian]

Leenaars, A., Lester, D., & Wenckstern, S. (1999). Suicide notes in alcoholism. *Psychological Reports, 85,* 363–364.

Leenaars, A., Lester, D., Wenckstern, S., & Heim, N. (1994). Suicide notes from Germany and the United States. *Suizidprophylaxe, 3,* 99–101. [In German]

Leenaars, A., & Maltsberger, J. (Eds.). (1994). The Inman diary: Some reflections on treatments. In A. Leenaars, J. Maltsberger, & R. Neimeyer (Eds.), *Treatment of suicidal people* (pp. 227–236). Washington, DC: Taylor & Francis.

Leenaars, A., Maris, R., & Takahashi, Y. (Eds.). (1997). *Suicide: Individual, cultural, international perspectives.* New York, NY: Guilford.

Leenaars, A., Park, B., Collins, P., Wenckstern, S., & Leenaars, L. (2010). Martyrs' last letters: Are they the same as suicide notes? *Journal of Forensic Science, 55,* 660–668.

Leenaars, A., Sayin, A., Candansayar, S., Akar, T., Demirel, B., & Leenaars, L. (2010). Different cultures, same reasons: A thematic comparison of suicide notes from Turkey and the United States. *Journal of Cross-Cultural Psychology, 41,* 253–263.

Leenaars, A., & Wenckstern, S. (1991). Posttraumatic stress disorder: A conceptual model for postvention. In A. Leenaars & S. Wenckstern (Eds.), *Suicide prevention in schools* (pp. 173–195). New York, NY: Hemisphere.

Leenaars, A., & Wenckstern, S. (1993). Trauma and suicide in our schools. *Death Studies, 17,* 151–171.

Leenaars, A., & Wenckstern, S. (1998). Principles of postvention: Applications to suicide and trauma in schools. *Death Studies, 22,* 357–391.

Leenaars, A., & Wenckstern, S. (Eds.). (2004). Altruistic suicide: From sainthood to terrorism. *Archives of Suicide Research, 8,* 1–136.

Leenaars, A., Wenckstern, S., Sakinofsky, I., Dyck, R., Kral, M., & Bland, R. (Eds.). (1998). *Suicide in Canada.* Toronto, Canada: University of Toronto Press.

Leong, F., & Leach, M. (Eds.). (2008). *Suicide among racial and ethnic minority groups.* New York, NY: Routledge.

Lester, D. (1987). *Suicide as a learned behavior.* Springfield, IL: Thomas.

Lester, D. (1994). A comparison of fifteen theories of suicide. *Suicide and Life-Threatening Behavior, 24,* 80–88.

Lester, D. (Ed.). (2004). *Katie's diary.* New York, NY: Routledge.

Lipset, S. (1990). *Continental divide.* New York, NY: Routledge.

Litman, R. (1984). Psychological autopsies in court. *Suicide & Life-Threatening Behavior, 14,* 188–195.

Litman, R. (1988). Psychological autopsies, mental illness and intention of suicide. In J. Nolan (Ed.), *The suicide case: Investigation and trial of insurance claims* (pp. 69–82). Chicago, IL: American Bar Association.

Litman, R. (1995, May). *Suicide without a clue*. Paper presented at the American Association of Suicidology Conference, Phoenix, AZ.

Litman, R., Curphey, T., Shneidman, E., Farberow, N., & Tabachnick, N. (1963). Investigations of equivocal suicides. *Journal of the American Medical Association, 184,* 924–929.

Lomas, D., & Lester, D. (Eds.). (2011). *Understanding and preventing college student suicide.* Springfield, IL: Charles C. Thomas.

Lyons, J. (2007). The returning warrior: Advice for families and friends. In C. Figley & W. Nash (Eds.), *Combat stress injury: Theory, research and management* (pp. 311–324). New York, NY: Routledge.

Macdonald, J. (1964). Suicide and homicide by automobile. *American Journal of Psychiatry, 121,* 366–370.

Mahon, M., Tobin, J., Cusack, D., Kelleher, C., & Malone, K. (2005). Suicide among regular-duty military personnel: A retrospective case-control study of occupation-specific risk factors for workplace suicide. *American Journal of Psychiatry, 162,* 1688–1696.

Mallon, T. (1984). *A book of one's own: People and their diaries.* New York, NY: Ticknor & Fields.

Maloney, L. (1988). Posttraumatic stresses of women partners of Vietnam veterans. *Smith College Studies in Social Work, 58,* 122–143.

Maltsberger, J. (1986). *Suicide risk: The formulation of clinical judgment.* New York, NY: New York University Press.

Maltsberger, J. (2001). Grandiose fury. *Crisis, 22,* 144–145.

Managing Suicidal Behavior Working Group. (n.d.). *Air Force guide for managing suicidal behavior—Strategies, resources and tools.* Brooks AFB TX: Air Force Medical Operations Agency (AFMOA) Population Health Support Division.

Mann, J., Apter, A., Bertolote, J., Beautrais, A., Currier, D., Haas, A., Hegerl, U. et al. (2005). Suicide prevention strategies: A systematic review. *The Journal of the American Medical Association, 294,* 2064–2074.

Mann, J., Waternaux, C., Haas, G., & Malone, K. (1999). Toward a clinical model of suicidal behavior in psychiatric patients. *The American Journal of Psychiatry, 156,* 181–189.

Mansfield, A., Bender, R., Hourani, L., & Larson, G. (2011). Suicidal or self-harming ideation in military personnel transitioning to civilian life. *Suicide and Life-Threatening Behavior, 41,* 392–405.

Maris, R. (1981). *Pathways to suicide.* Baltimore, MD: Johns Hopkins University Press.

Maris, R., Berman, A., Maltsberger, J., & Yufit, R. (Eds.). (1992). *Assessment and prediction of suicide.* New York, NY: Guilford.

Maris, R., Berman, A., & Silverman, M. (2000). *Textbook of suicidology.* New York, NY: Guilford.

Martin, J., & McClure, P. (2000). Today's active duty military family: The evolving challenges of military family life. In J. Martin, L. Rosen, & L. Sparacino (Eds.), *The military family: A practice guide for human service providers* (pp. 3–24). Westport, CT: Praeger.

Maslow, A. (1966). *The psychology of science.* New York, NY: Harper & Row.

McDowell, C., Rothberg, J., & Lande, R. (1994). Homicide and suicide in the military. In R. Zajtchuk & R. Bellamy (Eds.), *Textbook of military medicine: Military psychiatry: Preparing in peace for war* (pp. 91–113). Washington, DC: Office of the Army Surgeon General, Department of the Army.

McGurk, D., Cotting, D., Britt, T., & Adler, A. (2006). Joining the ranks: The role of indoctrination in transforming civilians to service members. In T. Britt, A. Adler, & C. Castro (Eds.), *Military life: The psychology of service in peace and combat. Operational stress* (Vol. 2, pp. 13–31). Westport, CT: Praeger.

McIntyre, J. (1996, May 16). Navy's top officer dies of gunshot, apparently self-inflicted. *CNN News*.

McKelley, R. (2007). Men's resistance to seeking help: Using individual psychology to understand counseling-reluctant men. *Journal of Individual Psychology, 63*, 48–58.

McKinley, J., Jr. (2009, November 12). Suspect in Fort Hood attack is charged on 13 murder counts. *New York Times*. Retrieved February 6, 2010 from http://www.nytimes.com/2009/11/12/US/suspect.html

Meehl, P. (1978). Theoretical risks and tabular asterisks: Sir Karl, Sir Richard, and the slow progress of soft psychology. *Journal of Consulting and Clinical Psychology, 46*, 806–834.

Meehl, P. (1986). Diagnostic taxa as open concepts. Metatheoretical and statistical questions about reliability and construct validity in the grand strategy of nosological revision. In T. Millon & G. Klerman (Eds.), *Contemporary directions in psychopathology* (pp. 215–231). New York, NY: Guilford.

Meehl, P. (1990). Appraising and amending theories: The strategy of Lakatosian defense and two principles that warrant it. *Psychological Inquiry, 1*, 108–141.

Meehl, P. (1993). Philosophy of science: Help or hindrance. *Psychological Reports, 72*, 707–733.

Meichenbaum, D. (2012). *Road map to resilience: A guide for military, trauma victims, and their families*. Clearwater, FL: Institute Press.

Menninger, K. (1938). *Man against himself*. New York, NY: Harcourt, Brace & Co.

Menninger, K. (1963). *The vital balance*. New York, NY: The Vicking Press.

Mental Health Advisory Team. (2003). *Operation Iraqi Freedom: Executive summary*. Washington, DC: U.S. Army Surgeon General, Department of the Army.

Michie, P. (1885). *The life and letters of Emory Upton, Colonel of the Fourth Regiment of Artillery, and Brevet Major-General, U.S. Army*. New York, NY: D. Appleton.

Milburn, D. (2009, October 7). Suicide prevention: You never fight alone. *The Maple Leaf*.

Mill, J. (1984). *Systems of logic*. London, UK: George Routledge. (Original work published 1892)

Miller, M., Barber, C., Azrael, D., Calle, E., Lawler, E., & Mukamal, K. (2009). Suicide among US veterans: A prospective study of 500,000 middle-aged and elderly men. *American Journal of Epidemiology, 170*, 494–500.

Millon, T. (2010). Classification considerations in psychopathology and personology. In T. Millon, R. Krueger, & E. Simonsen (Eds.), *Contemporary directions in psychopathology* (pp. 149–173). New York, NY: Guilford.

Millon, T., Krueger, R., & Simonson, E. (Eds.). (2010). *Contemporary directions in psychopathology*. New York, NY: Guilford.

Mills, P., Huber, S., Watts, B., & Bagian, J. (2011). Systemic vulnerabilities to suicide among veterans from the Iraq and Afghanistan conflicts: Review of case reports from a national veterans affairs database. *Suicide and Life-Threatening Behavior, 41*, 21–32.

Minsky, A. (2010a, October 29). MacKay "outraged over DND treatment of family." *The Windsor Star*, p. C1.

Minsky, A. (2010b, November 1). Military assaults "troubling." *The Windsor Star*, p. B1.

Montgomery, V. (2009). *Healing suicidal veterans*. Far Hills, NJ: New Horizon.

Morris, J. (2000). Emory Upton. In D. Heidler & J. Heidler (Eds.), *Encyclopedia of the American Civil War: A political, social, and military history* (pp. 2006–2008). New York, NY: Norton & Co.

Murphy, G. (1992). *Suicide in alcoholism*. New York, NY: Oxford University Press.

Murray, H. (1938). *Explorations in personality*. New York, NY: Oxford University Press.

Murray, H. (1967). Death to the world: The passions of Herman Melville. In E. Shneidman (Ed.), *Essays in self-destruction* (pp. 3–29). New York, NY: Science House.

Nash, W. (2007a). The stressors of war. In C. Figley & W. Nash (Eds.), *Combat stress injury* (pp. 11–31). New York, NY: Routledge.

Nash, W. (2007b). Operational stress adaptations and injuries. In C. Figley & W. Nash (Eds.), *Combat stress injury* (pp. 33–63). New York, NY: Routledge.

National Center for Health Statistics. (1967). *Eighth revision international classification of diseases, adapted for use in the United States*. PHS Pub. No. 1693. Public Health Service, Washington, DC: U.S. Government Printing Office.

Newman, M. (2009, November 5). 12 dead, 31 wounded in base shootings. *New York Times*. Retrieved February 6, 2010 from http://www.nytimes.com/2009/11/06texas.forthood.shootings/index.html

*New York Times*. (2009, December 2). Fort Hood suspect faces new charges. Cited in Wikipedia, Nidal Malik Hasan. Retrieved February 6, 2010 from http://en.wikipedia.org/wiki/Nidal_Malik_Hasan

Nolan, J. (1988). Suicide, sane or insane and suicidal intent. In J. Nolan (Ed.), *The suicide case* (pp. 51–65). Chicago, IL: American Bar Association.

North, S. (Ed.). (1899). *Biographies of Genessee County, New York*. New York, NY: Boston History.

O'Connor, R., & Leenaars, A. (2004). A thematic comparison of suicide notes drawn from Northern Ireland and the United States. *Current Psychology, 22*, 339–347.

O'Connor, R., Sheeby, N., & O'Connor, D. (1999). A thematic analysis of suicide notes. *Crisis, 20*, 106–114.

Orbach, I., Gilboa-Schechtman, E., Ofek, H., Lubin, G., Mark, M., Bodner, E., Cohen, D., & King, R. (2007). A chronological perspective on suicide—The last days of life. *Death Studies, 31*, 909–932.

Pace, J. (2010). Obama: More post-traumatic stress help for vets. *Associated Press*. Retrieved July 10, 2010 from http://news.yahoo.com/S/AP/US_obama.veteran

Palmer, B., Pankrantz, V., & Bostwick, J. (2005). The lifetime risk of suicide in schizophrenia: A re-examination. *Archives of General Psychiatry, 62*, 247–253.

Pap, A. (1953). Reduction sentences and open concepts. *Methodos, 5*, 3–30.

Park, B. (2004). Sociopolitical contexts of self-immolations in Vietnam and South Korea. *Archives of Suicide Research, 8*, 81–97.

Paulsen, G. (1998). *A soldier's heart*. New York, NY: Random House.

Pavese, C. (1961). *The burning brand: Diary 1935–1950*. (A. Murch, Trans.). New York, NY: Walker.

Payne, S., Hill, J., & Johnson, D. (2008). The use of unit watch or command interest profile in the management of suicide and homicide risk: Rationale and guidelines for the military mental health professional. *Military Medicine, 173*, 25–35.

Perreaux, L. (2011, December 23). Holiday season adds stress for ailing vets. *The Globe and Mail*, p. A4.

Pokorny, A. (1983). Prediction of suicide in psychiatric patients: Report of a prospective study. *Archives of General Psychiatry, 40*, 249–257.

Pompili, M., Girardi, P., Tatarelli, G., & Taterelli, R. (2006). Suicidal intent in single-car accident drivers. *Crisis, 27* 92–99.

Pompili, M., & Tatarelli, R. (Eds.). (2010). *Evidence-based practice in suicidology*. Gottingen, Germany: Hogrefe.

Psychological autopsies. (2001, June 4). *Health affairs policy letter*. Retrieved from http://www.ha.osd.mil/policies/2001/01_016.pdf. Cited in Ritchie, E., & Geller, M. (2002). Psychological autopsies: The current Department of Defense effort to stabilize training and quality assurance. *Journal of Forensic Sciences, 47*, 1370–1372.

Purcell vs. U.S. (2011). *No. 10-3734. United States Court of Appeals, Seventh Circuit.* Retrieved August 28, 2011 from http://www.thetruthhaschanged.com/2011/08/24/navy-can-not-be-sued-over-purcell-suicide/

Reay, D., & Hazelwood, R. (1970). Deaths in military police custody and confinement. *Military Medicine, 135*, 765–771.

Regan, J., Outlaw, F., Hamer, G., & Wright, A. (2005, July). Mental health series: Suicide in the military. *Tennessee Medicine*, 400–401.

Reger, M., Etherage, J., Reger, G., & Gahm, G. (2008). Civilian psychologists in an army culture: The ethical challenge of cultural competence. *Military Psychology, 20*, 21–35.

Risen, J. (1996, November, 25). Details of top admiral's suicide note disclosed. *Los Angeles Times*. Retrieved May 1, 2011 from http://articles.latimes.com/1996-11-25/news/mn-2849_1_suicide-note

Ritchie, E. (2010, April). *Suicide prevention: Valuable information learned from Army surveillance and research*. Plenary presented at the American Association of Suicidology Conference, Orlando, FL.

Ritchie, E., Benedek, D., Malone, R., & Carr-Malone, R. (2006). Psychiatry and the military: An update. *Psychiatric Clinics of North America, 29*, 695–707.

Ritchie, E., & Geller, M. (2002). Psychological autopsies: The current Department of Defense effort to stabilize training and quality assurance. *Journal of Forensic Sciences, 47*, 1370–1372.

Ritchie, E., Keppler, W., & Rothberg, J. (2003). Suicidal admissions in the United States military. *Military Medicine, 168*, 177–181.

Rosenblatt, P. (1983). *Bitter, bitter tears*. Minneapolis, MN: University of Minnesota Press.

Rothberg, J., Bartone, P., Holloway, H., & Marlowe, D. (1990). Life and death in the US Army. *Journal of the American Medical Association, 264*, 2241-2244.

Rothberg, J., & Jones, F. (1987). Suicide in the U.S. Army: Epidemiological and periodic aspects. *Suicide and Life-Threatening Behavior, 17*, 119–132.

Rothberg, J., Ursano, R., & Holloway, H. (1987). Suicide in the United States military. *Psychiatric Annals, 17,* 63–75.

Rourke, R., Young, G., & Leenaars, A. (1989). A childhood learning disability that predisposes those afflicted to adolescent and adult depression and suicide risk. *Journal of Learning Disabilities, 22,* 169–175.

Royal Commission on Aboriginal Peoples. (1995). *Choosing life: Special report on suicide among aboriginal people.* Ottawa, Canada: Ministry of Supply and Services.

Rudd, M. (2000). Integrating science into the practice of clinical suicidology: A review of the psychotherapy literature and a research agenda for the future. In R. Maris, S. Canetto, J. McIntosh, & M. Silverman (Eds.), *Review of suicidology 2000* (pp. 47–83). New York, NY: Guilford.

Rudofossi, D. (2006). *Working with traumatized police officer-patients: A clinician's guide to complex PTSD syndromes in public-safety professionals.* Amityville, NY: Baywood.

Runyan, W. (1982). *Life histories and psychobiology.* New York, NY: Oxford University Press.

Sageman, M. (2004). *Understanding terrorist networks.* Philadelphia, PA: University of Pennsylvania.

Sakinofsky, I., Lesage, A., Escobar, M., Wong, A., Loyer, M., & Vanier, C. (1996). *Suicide in the Canadian armed forces with special reference to peacekeeping.* Ottawa, Canada: Directorate of Health Protection and Promotion, National Defence Headquarters.

Sareen, J., Belik, S-L., Afifi, T., Asmundson, G., Cox, B., & Stein, M. (2008). Canadian military personnel's population attributable fractions of mental disorders and mental health service use associated with combat and peacekeeping operations. *American Journal of Public Health, 98,* 2191–2198.

Sareen, J., Cox, B., Afifi, T., Stein, M., Belik, S-L., Meadows, G., & Asmundson, G. (2007). Combat and peacekeeping operations in relation to prevalence of mental disorders and perceived need for mental health care: Findings from a large representative sample of military personnel. *Archives of General Psychiatry, 64,* 843–852.

Satcher, D. (1998). Bringing the public health approach to the problem of suicide. *Suicide and Life-Threatening Behavior, 28,* 325–327.

Scoville, S., Gubata, M., Potter, R., White, M., & Pearse, L. (2007). Deaths attributed to suicide among enlisted U.S. armed forces recruits, 1980–2004. *Military Medicine, 172,* 1024–1031.

Scurfield, R. (1985). Post-trauma stress assessment and treatment. Overview and formulation. In C. Figley (Ed.), *Trauma and its wake* (pp. 219–256). New York, NY: Mazel.

Selaković-Bursić, S. (2001). Suicidal behavior during NATO bombing of Yugoslavia. *Revista Internacional de Tanatologia Y Suididio, 4,* 40–44.

Selaković-Bursić, S., Haramic, E., & Leenaars, A. (2006). The Balkan Piedmont: Male suicide rates pre-war, wartime and post-war in Serbia and Montenegro. *Archives of Suicide Research, 10,* 225–238.

Shenon, P. (1996, May 18). Admiral, in suicide note, apologizes to "My Sailors." *The New York Times,* pp. A1, A9.

Sher, J. (2011, March 7). Soldiers suffer high rate of brain trauma. *The Globe and Mail,* p. A6.

Sherrow, V. (2007). *Women in the military.* New York, NY: Chelsea House.

Shineski, E. (2000, Spring). Suicide prevention "Could I have done more?" *Hot Topics: Current Issues for Army Leaders,* 3–16.

Shneidman E. (1963). Orientations toward death. In R. White (Ed.), *The study of lives* (pp. 200–227). New York, NY: Atherton.

Shneidman, E. (1967). Sleep and self-destruction: A phenomenological approach. In E. Shneidman (Ed.), *Essays in self-destruction* (pp. 510–539). New York, NY: Science House.

Shneidman, E. (1969). Suicide, lethality, and the psychological autopsy. In E. Shneidman & M. Ortega (Eds.), *Aspects of depression* (pp. 225–250). Boston, MA: Little, Brown.

Shneidman, E. (1971). Perturbation and lethality as precursors of suicide in a gifted group. *Suicide & Life-Threatening Behavior, 1,* 23–45.

Shneidman, E. (1973a). *Deaths of man.* New York, NY: Quadrangle.

Shneidman, E. (1973b). Suicide. In *Encyclopedia Britannica, 21,* 383–385. Chicago, IL: Williams Benton.

Shneidman, E. (1975). Postvention: Care of the bereaved. In R. Pasnau (Ed.), *Consultation-liaison psychiatry* (pp. 245–256). New York, NY: Grune & Stratton.

Shneidman, E. (1977). The psychological autopsy. In L. Gottschalk, F. McGuire, E. Dinovo, H. Birch, & J. Heiser (Eds.), *Guide to the investigation and reporting of drug-abuse deaths* (pp. 42–56). Washington, DC: U.S. Department of Health, Education and Welfare.

Shneidman, E. (1980a). *Voices of death.* New York, NY: Harper & Row.

Shneidman, E. (1980b). The reliability of suicide statistics: A bomb-burst. *Suicide and Life-Threatening Behavior, 10,* 67–69.

Shneidman, E. (1982). The suicidal logic of Cesare Pavese. *Journal of the American Academy of Psychoanalysis 10,* 547–563.

Shneidman, E. (1985). *Definition of suicide.* New York, NY: Wiley.

Shneidman, E. (1991). The commonalities of suicide across the life span. In A. Leenaars (Ed.), *Life span perspectives of suicide* (pp. 39–52). New York, NY: Plenum.

Shneidman, E. (1993). *Suicide as psychache.* Northvale, NJ: Aronson.

Shneidman, E. (1994). Clues to suicide reconsidered. *Suicide & Life-Threatening Behavior, 24,* 395–397.

Shneidman, E. (1996). *The suicidal mind.* New York, NY: Oxford University Press.

Shneidman, E. (1999). On "Therefore I must kill myself." In A. Leenaars (Ed.), *Lives and deaths: Selections from the works of Edwin S. Shneidman* (pp. 72–76). Philadelphia, PA: Brunner-Mazel.

Shneidman, E. (2001). *Comprehending suicide: Landmarks in 20th century suicidology.* Washington, DC: American Psychological Press.

Shneidman, E., & Farberow, N. (Eds.). (1957). *Clues to suicide.* New York, NY: McGraw-Hill.

Simpson, M. (1980). Self-mutilation as indirect self-destructive behavior: "Nothing to get so cut up about. . . ." In N. Farberow (Ed.), *The many faces of suicide* (pp. 257–283). New York, NY: McGraw-Hill.

Smith, K., Conroy, M., & Ehler, P. (1984). Lethality of suicide attempt rating scale. *Suicide & Life-Threatening Behavior, 14,* 215–242.

Snarr, J., Heyman, R., & Smith Slep, A. (2010). Recent suicidal ideation and suicide attempts in a large-scale survey of the U.S. Air Force: Prevalences and demographic risk factors. *Suicide and Life-Threatening Behavior, 40,* 544–552.

Stack, S. (1997). Homicide followed by suicide: An analysis of Chicago data. *Criminology, 35,* 95–99.
Stack, S. (2000). Suicide: A 15-year review of the sociological literature. *Suicide & Life-Threatening Behavior, 30,* 163–176.
Stack, S., & Bowman, B. (2011). *Suicide movies.* Gottingen, Germany: Hogrefe.
Stahl, S. (2000). *Essential psychopharmacology.* Cambridge, MA: Cambridge University Press.
Stander, V., Hilton, S., Kennedy, K., & Robbins, D. (2004). Surveillance of completed suicide in the Department of the Navy. *Military Medicine, 169,* 301–306.
Statistics Canada. (2005). *Canadian Persian Gulf cohort study: Summary report* (Catalogue no. 82-580-XIE). Ottawa, Canada: Minister of Industry.
Stengel, E. (1964). *Suicide and attempted suicide.* Baltimore, MD: Penguin Books.
Stoff, D., & Mann, J. (Eds.). (1997). *The neurobiology of suicide: From the bench to the clinic.* New York, NY: New York Academy of Sciences.
Strom, T., Leskela, J., James, L., Thuras, P., Voller, E., Weigel, R., et al. (2012). An exploratory examination of risk-taking behavior and PTSD symptom severity in a veteran sample. *Military Medicine, 177,* 390–396.
Styron, W. (1990). *Darkness visible.* New York, NY: Random House.
Sullivan, H. (1962). Schizophrenia as a human process. In H. Perry, N. Gorvell, & M. Gibbens (Eds.), *The collected works of Harry Stack Sullivan, Vol. II.* New York, NY: W. W. Norton.
Sullivan, H. (1964). The fusion of psychiatry and social sciences. In H. Perry, N. Gorvell, & M. Gibbens (Eds.), *The collected works of Harry Stack Sullivan.* New York, NY: W. W. Norton.
Taber, J. (2010, September 20). Ottawa makes new deal for injured veterans. *The Globe and Mail,* p. A7.
Tartaro, C., & Lester, D. (2009). *Suicide and self-harm in prisons and jails.* Lanham, MD: Lexington.
Teasdale, T., & Engberg, A. (2001). Suicide after traumatic brain injury: A population study. *Journal of Neurology, Neurosurgery, and Psychiatry, 71,* 436–440.
Thompson, M. (2012, June 8). U.S. Military suicides in 2012:155 days, 154 dead. *Time.* Retrieved October, 6, 2012 from http://nation.time.com/2012/06/08/lagging-indicator/
Thoresen, S., & Mehlum, L. (2004). Risk factors for fatal accidents and suicides in peacekeepers: Is there an overlap? *Military Medicine, 169,* 988–993.
Thoresen, S., & Mehlum, L. (2006). Suicide in peacekeepers: Risk factors for suicide versus accidental death. *Suicide & Life-Threatening Behavior, 36,* 432–442.
Thoresen, S., & Mehlum, L. (2008). Traumatic stress and suicidal ideation in Norwegian male peacekeepers. *The Journal of Nervous and Mental Disease, 196,* 814–821.
Thoresen, S., Mehlum, L., & Moller, B. (2003). Suicide in peacekeepers—A cohort study of mortality from suicide in 22,275 Norwegian veterans from international peacekeeping operations. *Social Psychiatry and Psychiatric Epidemiology, 38,* 605–610.
Thoresen, S., Mehlum, L., Røysamb, E., & Tønnessen, A. (2006). Risk factors for completed suicide in veterans of peacekeeping: Repatriation, negative life events and marital status. *Archives of Suicide Research, 10,* 353–363.
Tien, H., Acharya, S., Donald, A., & Redelmeier, D. A. (2010). Preventing deaths in the Canadian military. *American Journal of Preventive Medicine, 38,* 331–339.

Toynbee, A. (1968). *Man's concern with death*. New York, NY: McGraw-Hill. Quoted in Shneidman, E. (Ed.). (1973). *Death: Current perspectives*. Palo Alto, CA: Mayfield.
Triandis, H. (1995). *Individualism versus collectivism*. Boulder, CO: Westview.
U.S. Air Forces. (2004). A landmark program. *Preventing Suicide, 3*, 2–8.
U.S. Department of Health and Human Services, National Center for Health Statistics. (1996). *National Health Interview Survey, 1986–1994* [computer file]. 2nd ICPSR release. Washington, DC: U.S. Department of Health and Human Services, National Center for Health Statistics (producer), 1986–1995. Ann Arbor, MI: Inter-university Consortium for Political and Social Research (distributor), 1989–1996.
Unnithan, C., Corzine, J., Huff-Corzine, L., & Whitt, H. (1994). *The currents of lethal violence*. Albany, NY: State University of New York Press.
Vasterling, J., & Vergaellie, M. (2009). Posttraumatic stress disorder: A neurocognitive perspective. *Journal of the International Neuropsychological Society, 15*, 826–829.
Virginia Board of Medicine. Practitioner Information. (2010). Nidal Malik Hasan, MD. *Wikipedia*. Retrieved February 6, 2010 from http://www.vahealthprovider.com/results_generalinfo.asp?License_No=101238630
von Bertalanffy, L. (1968). *General systems theory* (Rev. ed.). New York, NY: George Braziller.
von Clausewitz, C. (1982). *On war*. (J. Graham, Trans.). New York, NY: Penguin Classics. (Original published in 1832)
Van Praag, H. (1997). Some biological and psychological aspects of suicidal behavior: An attempt to bridge the gap. In A. Botsis, C. Soldatos, & C. Stefanis (Eds.), *Suicide: Biopsychosocial approaches* (pp. 73–92). Amsterdam: Elsevier.
Warden, D. (2006). Military TBI during the Iraq and Afghanistan wars. *Journal of Head Trauma Rehabilitation, 21*, 398–402.
Warner, C., Appenzeller, G., Greger, T., Belenkiy, S., Breitbach, J. Parker, J., et al. (2011). Importance of anonymity to encourage honest reporting in mental health screening after combat deployment. *Archives of General Psychiatry, 68*, 1065–1071.
*Washington Post*. (2009a, November 6). Major Nidal M. Hasan. Retrieved February 6, 2010 from http://www.washingtonpost.com/wp_dyn/content/article/2009/11/06/AR2009110601978.html
*Washington Post*. (2009b, November 10). The Koranic world view as it relates to Muslims in the U.S. military. Retrieved February 6, 2010 from http://www.washingtonpost.com/wp_dyn/content/gallery/2009/11/10GA20091110000920.htp
Wasserman, D., & Varnik, A. (1998). Reliability of statistics on violent deaths and suicide in the former USSR, 1970–1990. *Acta Psychiatrica Scandinavica, 394*(Suppl.), 34–41.
Weber, M. (1949). *The methodology of social sciences*. Glencoe, IL: Free Press. (Original work published in various years, early 20th century)
Wertsch, M. (1991). *Military brats: Legacies of childhood inside the fortress*. New York, NY: Harmony.
West, P. (1966). *Murder followed by suicide*. Cambridge, MA: Harvard University Press.
Wikipedia. (2010). *Nidal Malik Hasan*. Retrieved February 6, 2010 from http://en.wikipedia.org/wiki/Nidal_Malik_Hasan
Wikipedia. (2011). *Jeremy Michael Boorda*. Retrieved March 6, 2011 from http://en.wikipedia.org/wiki/Jeremy_Michael_Boorda
Wilhelm, T. (2011, July 14). Family blames military after ex-soldier's death. *The Windsor Star*, pp. A1, A4.

Wilson, J., Smith, W., & Johnson, S. (1985). A comparative analysis of PTSD among various survivor groups. In C. Figley (Ed.), *Trauma and its wake* (pp. 142–172). New York, NY: Brunner/Mazel.

Windelband, W. (1904). *Geschichte und Naturwissenchaft* (3rd ed). Strassburg, Germany: Heitz.

Wolfgang, M. (1958). An analysis of homicide-suicide. *Journal of Clinical and Experimental Psychopathology, 19,* 208–218.

Wong, A., Escobar, M., Lesage, A., Loyer, M., Vanier, C., & Sakinofsky, I. (2001). Are UN peacekeepers at risk for suicide? *Suicide & Life-Threatening Behavior, 31,* 103–112.

Wong, D., Le Gras, L., & Mains, G. (1988). *Suicide behaviour trends in Canadian forces training system 1980–1986* (CFTS COSP Report 2/88). Ontario, Canada.

Woodward, S., Kaloupek, D., Grande, L., Stegman, W. K., Kutter, C. J., Leskin, L., et al. (2009). Hippocampal volume and declarative memory function in combat-related PTSD. *Journal of the International Neuropsychological Society, 15,* 830–839.

World Health Organization (WHO). (2002). *World report on violence and health.* Geneva, Switzerland: Author.

World Health Organization (WHO). (2006). *Preventing disease through healthy environments.* Geneva, Switzerland: Author.

Wortzel, H., Binswanger, I., Anderson, C., & Adler, L. (2009). Suicide among incarcerated veterans. *The Journal of the American Academy of Psychiatry and the Law, 37,* 82–91.

Yufit, R., & Lester, D. (Eds.). (2005). *Assessment, treatment and prevention of suicidal behavior.* New York, NY: John Wiley & Sons.

Zachar, P., & Kendler, K. (2010). Philosophical issues in the classification of psychopathology. In T. Millon, R. Krueger, & E. Simonson (Eds.), *Contemporary directions in psychopathology* (pp. 127–148). New York, NY: Guilford.

Zilboorg, G. (1936). Suicide among civilized and primitive races. *American Journal of Psychiatry, 92,* 1347–1369.

Zilboorg, G. (1937). Considerations on suicide, with particular reference to that of the young. *American Journal of Orthopsychiatry, 7,* 15–31.

Zoroya, G. (2010, April 23). Military's health care costs booming. *USA Today,* p. 1A.

# Index

AAS. *See* American Association of Suicidology
accepting the unacceptable, 244, 327
accidental deaths, 194–195
    accidents and the military, 196–197
Ad Hoc Committee for the Preparation of a Memorial Service for the Nation's Martyrs and Victims of Democratization Movement, 113
Adams, D., 227–228
Adler, A., 31
Afghanistan War, 26, 158
    PTSD and suicide during/after, 229–230
*Air Force Guide for Managing Suicidal Behavior—Strategies, Resources and Tools,* 246
alcohol abuse, 166–167, 194–197, 199
alcoholism, 235–236
Allen, N., 185
Allport, Gordon, 91–92, 259
Al-Qaeda, 116
altruistic suicide, 107–109
    common suicide *vs.,* 115–116, 118–119
    current study of, 111–112
    murder and, 117–118
    notes, methods, results, 113
    protocol sentences, frequency of endorsement, 113, 114–115*t*
    social bonds and, 116–117
Alvarez, Alfred, 139
American Association of Suicidology (AAS), 76, 161

*American Journal of Preventative Medicine,* 168
American Psychiatric Association, 51
*American Psychologist,* 86
Apter, A., 166
*Are Suicides Considered Less Honorable* (Gould), 307–308
*Armies of Europe and Asia, The* (Upton), 124
*Army & Navy Journal,* 129
Asberg study, 34
*Ascent of Man, The* (Bronowski), 327
attempted suicide, 33–34, 38, 158–160, 205
Atwood, Margaret, 324
Ayer, A., 48

Bagley, S., 254–255
Baker, Joshua, 190
*Baltimore Sun,* 143
Barraclough, B., 37–38
barriers, 251–253. *See also* green walls
Barry, Richard, 126
Basham, C., 245
Beck, A., 31, 33, 44, 49
behavioral clues
    cognitive clues, 38
    emotional clues, 39
    life-threatening behavior, 39
    previous attempts, 38
    sudden behavioral changes, 39
    suicide notes, 40–42, 61–62
    verbal statements, 38

Belik, S., 230–231
Berg-Cross, L., 76–77
Binswanger, I., 31
Boorda, Michael, 75–76, 121
    Arlington National Cemetery Website on, 144–146
    background, early years, 143
    biography, 144–145
    CNN report, 147
    death, 145–147
    harassment and, 145
    marriage, family, 144
    Navy career, 143
    opinions reached, bias therefore, 148–149
    suicide notes, 147–148
    Tailhook scandal, 145
Boscarino, J., 213, 224
brain dysfunction, traumatic brain injuries (TBIs), 35–37
Brenner, L., 35–36, 231–233
Breshears, R., 35–36
Bronowski, Jacob, 327
Broz (Tito), Joseph, 21
Bruyea, Sean, 250–251, 257
Bryan, C., 233–234, 251–253
*Buddy-to-Buddy, a Citizen Soldier Peer Support Program to Counteract Stigma, PTSD, Depression, and Suicide* (Greden, Valenstein, Spinner, Blow, Gorman, Dalak, Marcus, Kees), 248–249
Buteau, J. 187

call to action, 323
Camp, T., 237
Campbell, Joseph. *See* Scott-Campbell case
Canadian Forces (CAF), 244
    academic studies, 166–167
    alcohol abuse in, 166–167
    causes of death, 168–169
    healthy soldier effect, 166
    position of, 165–168
    professional studies, 166
    selection, screening, 166
    suicide rates, 168, 171–172
    war, suicide in, 168–169

Canadian Community Health Survey: Mental Health and Well-Being Canadian Forces Supplement (CCHS-CFS), 231
*Canadian Forces Expert Panel on Suicide Prevention, The* (Jetly, Cooper), 170–171
*Canadian Persian Gulf Cohort Study: Summary Report,* 168
cancer, 213
Caplan, G., 77
cardio-vascular disease, 213
Carnap, R., 48
Carr, J., 175
Carr, R., 25
Carr-Malone, R., 189
Carter, Jimmy, 16
case studies, 81–82
    advantages, problems, 91
*Castle, The* (Kafka), 88
casualties, 234–235
catch 22 (double bind, no-win cycle), 14, 26
CCHS-CFS. *See* Canadian Community Health Survey: Mental Health and Well-Being Canadian Forces Supplement
Centers for Disease Control (CDC), 224
Centre for Suicide Prevention, SIEC database, 19–20
challenges (in war), 251–253, 313
*Challenges and Considerations for Managing Suicide Risk in Combat Zones* (Bryan, Kanzler, Durham, West, Green), 251
Christian, J., 4, 5–6, 15
*Chronological Perspective on Suicide—The Last Days of Life, A* (Orbach), 208–209
Church, William, 129
Civil War, vi, 111, 123–125, 140, 243, 256, 312, 314, 325
Clark, S., 324
Clinton, Bill, 146
Cobain, Kurt, 40–42, 76
cognitive style, faulty syllogisms, 72–75
Cohen, J., 238–239

collectivism *vs.* individualism, 4–6, 20–21
combat exposure, 233–234
combat stress, 7–8
*Combat Stress Injury* (Figley, Nash), 7, 217, 235
combat-related experiences, 223
confidentiality, 249–251
contextual observations
  biological roots, 34–35
  brain dysfunction, traumatic brain injuries (TBIs), 35–37
  physical disability, illness, 37–38
*Continental Divide* (Lipset), 323
Cooper, K., 170–172
Corzine, J., 183, 184
courage, 327
crisis intervention, 255
Cukrowicz, K., 233–234
culture, military culture, 10–11
Curphey, Theodore, 84

Dahlberg, L., 8, 18
*Darkness Visible* (Styron), 90
Datel, William, 174–175
*Death Studies,* 208
*Deaths of Man* (Shneidman), 24
*Definition of Suicide* (Shneidman), 30, 34, 108–109
dehumanization, 6
Dekel, R., 234–238
Department of Defense Medical Mortality Registry, 175
deployment stress, 223
depression measurement, 48
Dieserud, G., 326
disenfranchisement, 26
disgrace, 142, 149, 197, 305
dissembling (mask[ing]), 14, 75–77, 79, 168, 205, 209, 221, 325
*Donnie Darko,* 277
draft lottery, 22
Drapeau, Michael, 250–251
Durkheim, Emile, 20–23, 30–31, 103, 106–111, 116, 325–326
Dyregrov, K., 326

Eaton, K., 176
ecological model, 8–9, 9f, 18, 182
Einstein, A., 44
"Emory Upton, Military Genius" (Barry), 126
environmental control (gun control), 256
Erikson, E., 299
estranged, 252
*Everlasting Lives,* 113
Everson, R., 237
evidenced-based practice, 19–20
evidenced-based theory, suicide definition, 51
*Example of an Equivocal Death Clarified in a Court of Law, An* (Shneidman), 93–94
*Explorations in Personality* (Murray), 57
*External-Cause Mortality After Psychologic Trauma: The Effects of Stress Exposure and Predisposition* (Boscarino), 213

Fadiman, J., 71
family system, parent absence, 15
Farberow, Norman, 179, 197–198, 201, 225–226, 235
Feig, J., 254
*Felo De Se: Soldiers Suicides in America's Civil War* (Lande), 111
Ferri, E., 183
fight, 327
Figley, Charles, 3, 7, 217, 235, 238
firearms, 157
5-H1 AA biochemical marker, 34
Flaherty, Anne, 206–207
forensic investigation, 87
Foster, Vincent, 121
Fragala, M., 212
Frager, R., 71
Freeman, A., 222, 235
Freeman, S., 222, 235
French Armed Forces, military suicide, 170
Freud, Sigmund, 31, 45, 183, 194, 219–220

Gaines, T., 205–206
Gates, Robert, 256
Geller, Michael, 101
gender differences, 70–71
Gilligan, James, 14–15
Goffman, E., 78
good soldier, 210
Gould, Joe, 307–308
Greden, J., 248–249
green culture, 10–11
green walls, 11, 243, 253, 312, 314. *See also* barriers; dissembling; secrets; taboo
Gulf War, 154, 165

Hagmann, Douglas, 192
Hall, Lynn, 10–15
hardiness, resilience, 6, 253, 254
Harper, Stephen, 193
Hasan, Nidal, 33, 190–194
Hasbrouck, Henry, 128, 134
Hawton, K., 34, 198–199
*Healing Suicidal Veterans* (Montgomery), 16
Hempel, C., 50–51
Hendin, H., 255
Henry, A., 183
Hitler, Adolf, 40
Hoge, C., 8, 175, 214, 246–247
holidays, 225
Homer, 327
homicide
　ecological view, 182
　rates, 182
　suicide *vs.*, 182–183
　types, definition, 181–182
　World Health Organization (WHO) on, 181–182
*Homicide and Suicide in the Military* (McDowell, Rothberg, Lande), 188
homicide-suicide, 185–188
　in military, 188–190
honor, 6, 110, 142, 146, 296, 326
hospitalization, 248, 256
*Hot Topics: Current Issues for Army Leaders,* 246

*Hundreds of PTSD Soldiers Likely Misdiagnosed* (Flaherty), 206–207
Hutchinson, Kay Bailey, 192
Huxley, A., 327
Hyatt, Dan, 100–101
Hyer, L., 226–227
Hyson, John, 127–129

identity, 12, 71
*In Service of Two Masters: The Ethical-Legal Dilemma Faced by Military Psychologists* (Jeffrey, Rankin, Jeffrey), 250
incarceration, 199–201
individualism, 221
indoctrination, 6–7
Inman, Arthur, 89
intentionality, 3, 32–33
internalization, 6
interpersonal stage
　identification-egression, 59–60
　interpersonal relations, 56–57
　Murray's psychological needs, 57, 58$t$
　rejection, aggression, 57–59, 58$t$
interpersonal stressors, 223
intrapsychic drama
　cognitive constriction, 53–54, 73–74
　ego, vulnerable ego, 56
　inability to adjust, psychopathology, 54–56
　indirect expressions, 54
　unbearable psychological pain, 52–53
intrapsychic, open concepts, 46–47, 51
Iraq War, 158
　PTSD and suicide during/after, 229–230
Islamic extremists, 69–70

*Jacob's Ladder,* 277
"Jan," 16–18
Jankowski, Stefan, 240–241
Janoff-Bulman, R., 219–220, 238

Jeffrey, T., 250
Jenkins, Rachel, 311
Jetly, R., 170–172
job demands, 222
*Journal of Forensic Science,* 101
*Journal of Personality Assessment,* 35
*Journal of the American Medical Association,* 160
Jung, C., 31, 119

Kafka, Franz, 88
Kamerman, J., 25
Kang, H., 158
Kant, Emmanuel, 118–119
Kanzler, K., 251–253
Kaplan, Mark, 154–156, 157
*Kate's Diary* (Lester), 89
"Katie," 89
Keen, Sam, 13–14
Kelly, G., 31
Kelly, Megan, 6, 222–223
Kennedy, Carrie, 16
King, L., 6
Knox, K., 254
Koren, D., 37, 217–218, 230
Kraepelin, Emil, 105–106
Krug, E., 8, 18
Kuhn, T., 44, 74

Lambert, M., 214, 243
Langridge, Stuart, 257
last straw, 213
Le Sage, A., 187
Lester, David, 89, 184–185, 199–200
lethal violence rate (LVR), 184
lethal violence, stream analogy, 183–184
lethality, 33, 43, 52
Lieberman, Joseph, 193
life, 327
*Life and Letters of Emory Upton, The* (Michie), 125–127
Linnaeus, Carolus, 105–106
Lipset, Seymour, 323
literature review, 19–20
Litman, R., 31–32
Litts, D., 254

*Lives and Deaths* (Shneidman), 83
*Living and Surviving in Harm's Way* (Freeman, Moore, Freeman), 222, 235
living, working conditions, 223
Los Angeles Suicide Prevention Center, 84
*Los Angeles Times,* 148
LVR. *See* lethal violence rate

MacKay, Peter, 256
Malone, Rosemary, 16, 189
Maltsberger, J., 35, 70
Mann, J., 19, 34
*Many Faces of Suicide, The* (Farberow), 179, 201, 235
Maris, R., 43, 82
martyr suicide. *See* altruistic suicide
Maslow, A., 92
McCain, John, 145
McDowell, C., 188–189, 253
McGurk, D., 6
medical-physical evaluation boards, 212
medication, 256
Meehl, Paul, 8, 48
Mehlum, L., 196–197, 228
Meichenbaum, D., 253
Melville, Herman, 79
Mendeleyev, Dimitri, 105–106
Menninger, Karl, 31, 243
Michie, Peter, 123, 125, 143
Mid-Atlantic Mental Illness Research, Education and Clinical Center Work group, 229
military and war, unique system, 9–10
*Military Medicine,* 111, 127, 174
*Military Policy of the United States, The* (Upton), 123, 124, 132–133
military suicide. *See also* psychology of military suicide
 classic population study, 105
 Department of Defense (DoD) on, 162
 firearms and, 156–157
 historical rates, 153–154
 Iraq, Afghanistan wars, 158
 personal stories, vignettes, 162–163

[military suicide]
  physical injury and, 156
  predictors, 156
  recent rates, 153–156
  study of, vii–viii
  war as cause, 153
military suicide, psychology of, 205
  stigma, the military, tactics, 206–207
  suicide soldier as good soldier, 210
  why soldiers die by suicide, 207–210
military suicide statistics, surveillance and reliability, 173–177
  current military reporting reliability, 175
  Department of Defense (DoD) on, 176
  epidemiological data, historical perspective, 173–175
  surveillance, investigations, research, 176–177
military system, psychological traits, 13
*Military Times*, 307
military training, indoctrination, 6
Mill, John Stuart, 44, 91, 259
Miller, M., 157
Millon, Theodore, 46, 50–51
Mills, P., 234–235
*Moby-Dick* (Melville), 79
Montgomery, Victor, 16–17
Moore, B., 222
Morselli, E., 183
Mosberg, William, 127, 128, 129
Motto, Jerome, 96–99
Munjas, B., 254–255
Murphy, G., 236
Murray, Henry, 31, 34, 51, 57, 58*t*

narcissistic injury, 15, 57, 137–138, 208, 313
NASH, 181
Nash, William, 7–8, 217, 218, 220, 235
National Defence (Canada) (2010), 320–321
natural, accident, suicide, homicide (NASH) modes of death, 30, 190, 193–195
needs, 58

*New York Times*, 126, 147
*Newsweek*, 147
Newton, I., 44
Novi Sad (Vojvodina), 23

Obama, Barack, 257
Olds, Perry, 95–96
*On War* (Clausewitz), 7
open concepts, 46–47
Orbach, I., 208–209
Oswald, Lee Harvey, 89

Pavese, Cesare, 90
Payne, S., 247–248
perceived threat, 223
persevere, 327
personal documents, 88–92
perturbation, lethality, 33–34
physical disability, illness, 37–38
poems, poets, 90
police officers suicides, New York, 25
policies and prevention
  challenges, considerations for managing suicide risk in harm's way, 313
  historical speculation on continental divide, 324–325
  methods of difference, U.S. and Canada, 323–324
  military suicide, 314–322
  policies and procedures, recommendations, 323
  policies, procedures from military, 313–314
  prevention, 318–320
  prevention strategies, 320–322
  recommendations from U.S., Canada, 315
  suicidal soldiers, call to action, 325–327
  suicide prevention in military, 311–313
Pompili, M., 43, 195
posttraumatic stress disorder (PTSD), 207–208, 210–211, 213, 217
  adjustments to trauma, 219–220

[posttraumatic stress disorder (PTSD)]
  casualties allowed, 234–235
  combat exposure, increased risk, 233–234
  combat stress injury, research, 222–223
  DSM-IV criteria for, 219
  individual/cultural differences, stigmatization, 220–222
  Kelly and Vogt stresses list, 222–223
  military stress types, 222–223
  posttraumatic stress reactions, 218–219
  secondary traumatization, families, spouses, children, 236–239
  suicide and, 223–225
  suicide and TBIs, 231–233
  suicide, during/after Iraq and Afghanistan Wars, 229–230
  suicide, during/after Vietnam war, 225–228
  suicide, during/after Yugoslavian War, 228–229
  system gaps, 234, 241
  trauma, alcoholism, suicide, 235–236
  traumatic conclusion, 239–241
*Posttraumatic Stress Disorder, Traumatic Brain Injury, and Suicide Attempt History among Veterans Receiving Mental Health Services,* (Brenner, Betthauser, Homaifer, Villarreal, Staves, Huggins), 231–232
Powell, Colin, 3, 243
*Predicting Suicidal Behavior in Veterans with Traumatic Brain Injury: The Utility of the Personality Assessment Inventory,* 36–37
*Preventing Deaths in the Canadian Military* (Tien, Acharya, Donald, Redelmeier), 168
prevention/intervention/postvention, 77–78. *See also* policies and prevention
*Profile of Major Nidal Malik Hasan* (Hagmann), 192
protective factor(s), 253–254, 326–327
*Psychiatry* (Kraepelin), 105
Psychoanalytic Society, 57

psychological autopsy (PA), 1, 23–24, 32
  conducting, 84–86
  death-follow-up procedure, 86–87
  Foster, 122
  military court and, 93
  military perspective on, 101–102
  procedure overview, 83–84
  purpose, 84
  Shneidman categories, 85
psychology of military suicide. *See also* posttraumatic stress disorder
  costs of military suicide, 214–215
  faces of military suicide, 212–214
  shame, disgrace, 211
  transitions, repatriation, incarceration, and suicide, 210–211
psychopathology (mental disorder, inability to adjust)
  adjustment disorder, 54–56, 244
  antisocial personality disorder, 54–56
  anxiety disorder, 54–56, 227
  attention deficit/hyperactivity disorder, 35
  bipolar disorders (manic depressive disorder), 54–55, 229
  borderline personality disorder, 54–56, 206–207
  depressive disorders, 54–56, 227, 229
  learning disability (disorder), 35
  posttraumatic stress disorder, 54–56. *See also* posttraumatic stress disorder
  schizophrenic disorder, 54–56
psychotherapy, 255–256
public health, systems approach, 18–19
Purcell, Chris, 257, 261
  alcohol abuse, 271, 278–281, 293, 302
  autopsy report, 263–265
  cause of death, 304
  cognitive constriction, 306
  cognitive style of, 300
  death, court account, 262–263
  developmental view, young adults' suicides, 300
  Google searches by, 276
  inability to adjust, psychopathology, 298, 306

[Purcell, Chris]
  identification-egression, 299–300, 307
  indirect expressions, 298, 306
  interpersonal relations, 299, 307
  interpersonal stage, 299–300
  interviews, psychological autopsy, 282–295
  intrapsychic drama, 298, 305–306
  last chat, 273–276
  last chat, analysis, 273–276
  medical record, 278–282
  military family, culture and, 295–297
  personal documents, 265–268
  psychological autopsy opinions reached, basis thereof, 304–305
  psychological report, 268–271
  psychological theory of his suicide, 305–307
  psychopathology (mental disorder, inability to adjust)
    alcohol abuse, 278, 301
    anxiety, 285, 302
    bipolar disorder (manic-depressive disorder), 205, 302
    borderline personality disorder, 303
    depression, 285, 302
    global assessment of functioning (GAF), 278, 280, 301, 303
    mood disorder, NOS, 280, 301
    posttraumatic stress disorder, 301
    psychosocial and environmental problems, 301, 303
  psychopathology assessment, 300–303
  reason for suicide, 295
  rejection-aggression, 307
  suicide movies, 276–277
  suicide notes, 271–272
  suicide notes #2, analysis, 272
  unbearable psychological pain, 305–306
  vulnerable ego, 298, 306
  younger vs. older soldiers and, 297
Purcell, Helene, 257, 261–262
Purcell, Mike, 257, 261, 307–308

rape, sexual assault, 16–17, 248, 326
relatedness vs. separateness culture, 4
*Reliability of Mortality Count and Suicide Count in the United States Army, The* (Datel), 174
*Reliability of Suicide Statistics: A Bomb-Burst, The* (Shneidman), 173
repatriation (transitions), 210–211
research on theory, empirically based understanding, 60–63, 64–69*t*, 69–70
responsibility, 246
"Richard Cory" (Robinson), 75
Ritchie, Elspeth, 101, 161, 189–190, 212
Robinson, Edwin, 75
Root, Elihu, 125
Rothberg, J., 188–189, 253
Royal Commission on Aboriginal Peoples, 10–11
Rudd, D., 255
Rudofossi, D., 26, 325


Sakinofsky, Issac, 165
Sanborn, George, 127, 128, 129
Santayana, George, 103
Sareen, J., 167, 207–208, 230–231
Satcher, David, 18–19, 202
Scott-Campbell case
  autopsy, 96
  character consistent with homicide question, 98
  charge, 94–95
  crime scene, 96
  defense psychiatric consultation report, 96–98
  emotional stress, 97–98
  Hyatt-Shneidman questions, responses, 100–101
  issues, 95
  local newspaper account, 94–95
  military investigator pretrial report, 95–96
  Motto report, 96–99
  newspaper account of verdict, 101
  Olds report, 95–96

[Scott-Campbell case]
   Olds report, comment, 99
   precipitating event possibility, 98
   Shneidman testimony excerpts, 100–101
   summary, 96
   summary, conclusions, 99
   time of death, deceased's nudity, 96
   vulnerability to suicidal impulses, behavior, 97
Scoville, S., 154
second leading cause of death, 245
secondary traumatization, 234, 236–239
   children, 236–239
   families, 236–239
   spouse, 236–239
secrecy, stoicism, denial, 13, 76–77
secrets, 13, 76–77
Selaković-Bursić, S., 21–24
self-harm, 197–199
self-immolators, 116
Seneca, 119, 123
sexual assault, 223
sexual harassment, assault, 222
shame, 14, 142, 149, 197, 305
Shenon, Philip, 147
Sherman, William, 325
Shineski, Eric, 246
Shneidman, Edwin, 24–25, 29–31, 34, 45, 75–77, 81, 83, 85–87, 108–109, 122, 139, 160, 173–175, 313, 326
Short, J., 183
Simpson, M., 198
Sisyphus, 327
Snarr, J., 159–160
Snow, John, 18, 151, 173, 176
Socrates, 27, 326
Soldier Readiness Center, Fort Hood, 33
South Korean self-immolators, 113, 116
Stack, Steven, 22, 185–187, 276
Stahl, S., 256
Stander, V., 156
Statistical Bureau of the Republic of Serbia, 21
Statistical, demographic reports, 87
Stengel, Erwin, 33, 93, 205

stigma, 8, 206–207, 220, 234, 245–246
Stoff, D., 34
Strength, 327
Styron, William, 52, 90
suicide. *See also* theory and suicide
   by battle, 30
   best understood, 45–47
   clues, 75–77
   cross-cultural studies, 4
   defined, 30–34
   Durkheim's typology of, 106–111
   external causes, 29
   homicide *vs.*, 182–183
   intentional, subintentional, unintentional, 31
   intrapsychic, interpersonal, 46–47, 51–52
   multidimensional event of, 29
   multidisciplinary perspectives, 82
   neurobiology of, 34–35
   psychological theories, 31, 45
   risk evaluation, 42–44
   study of, 82–83
*Suicide: A Study in Sociology* (Durkheim), 30–31, 106, 109, 116
*Suicide Among Incarcerated Veterans* (Wortzel, Binswanger, Anderson, Adler), 200
*Suicide Among Male Veterans: A Prospective Population-Based Study* (Kaplan, Bentson, McFarland, Huguet), 154–155
*Suicide and Attempted Suicide* (Stengel), 93
*Suicide and Life-Threatening Behavior*, 173
*Suicide and Self-harm in Prisons and Jails* (Tartaro, Lester), 199–200
*Suicide Behaviour Trends in Canadian Forces Training System 1980–1986* (Wong, Le Gras, Mains), 168
suicide diary, 88
suicide ideation, attempts, 158
*Suicide in the Canadian Armed Forces with Special Reference to Peacekeeping* (Sakinofsky), 165

*Suicide Movies* (Stack, Bowman), 276
suicide notes, 40–42, 61–62, 92–93, 113, 121–122. *See also* personal documents; suicide diary
*Suicide of General Emory Upton: A Case Report, The* (Hyson, Mosberg, Sanborn, Whitehorne), 127
suicide prevention in military, 257
   challenges, considerations for managing suicide risk in harm's way, 251–253
   confidentiality and military, 249–251
   Department of Defense (DoD) actions, 244–245
   everyone's order, 247–249
   priorities for military, 245–247
   protective factors, resilience, 253–254
   suicide prevention programs, systematic review, 254–255
   treatment, common factors, 255–256
   what armed forces can do, 234–244
*Suicide Surveillance in the U.S. Military—Reporting and Classification: Biases in Rate Calculations* (Carr, Hoge, Gardner, Potter), 175
Sullivan, H., 31
surviving suicide in armed forces, 24–25
Survivor Outreach Program (SOS), 322
survivors, 24–25, 326–327
   dyadic event, 25
Süssmilch, Johan Peter, 105
system, culture, and military, 12–15
*System of Logic* (Mill), 44
systems theory, 9

taboo, 15
Tartaro, C., 199–200
Tepper, Martin, 166
terrorist, 70, 111, 116–118, 190–194
TGSP. *See* Thematic Guide for Suicide Prediction

Thematic Guide for Suicide Prediction (TGSP), 62–63, 64–69*t*
theory and suicide, 44
   best understood, 45–47
   cluster analysis, 50
   construct validation, 48–49
   empirical approaches, computers and, 50
   psychological theories, 45, 49–50
   validation approaches, 48–51
Thoresen, S., 196–197, 211, 228
Tien, H., 168
Tito (Joseph Broz), 21
Toynbee, Arnold, 24
Tragedy Assistance Program for Survivors (TAPS), 322
trauma, loss, 26–27
traumatic brain injuries (TBIs), 35–37
   PTSD, suicide and, 231–233
*Trial, The* (Kafka), 88
Triandis, H., 4

unbearable pain, 46, 48
unit watch, 247
United Kingdom (UK), military suicide, 169–170
Unnithan, C., 183, 184
Upton, Emory
   analysis of letters, 135
   author, 123
   battles, tactics, 124
   cognitive constriction, 140
   Coroner's Verdict, 128
   early years, 122
   funeral, 142–143
   Hasbrouck letters, 134
   identification-egression, 141–142
   inability to adjust, psychopathology, 140
   indirect expressions, 140
   interpersonal relations, 141
   intrapsychic drama, 135, 140–141
   letters of, 129–135
   letters on near approach of death, 133
   letters to parents, sister, 132–134
   life context, 135–136

[Upton, Emory]
  marriage, 130–131
  Michie biography, 122
  military background, 123
  *Military Policy of United States, The,*
    letters, 132–133
  military stage, opinions reached, 135
  opinions reached, bias therefore,
    135–138
  posts after war, 124
  psychological theory of Upton's
    suicide, 139–142
  psychopathology (mental disorder,
    inability to adjust)
    adjustment disorder, 135, 137, 140
    depression, 128–129, 137
    posttraumatic stress disorder, 137, 140
  rejection-aggression, 141
  shame, disgrace, 142
  suicide notes, 135–136
  suicide of, 127–129
  theories on his suicide, 129
  unbearable psychological pain, 140
  vulnerable ego, 141
U.S. Department of Defense (2010), 313,
  315–318
U.S. Department of the Army (2009),
  313, 318–320, 322

Vietnam War, 26, 160, 213, 313, 325
  suicide during/after, 225–228
violence
  rate, cost of, 8
  WHO definition, 3
violence and war, systems theory for, 3–8
violence, suicide, ecological model, 8–9,
  9f
Vogt, Dawne, 6, 222–223
*Voices of Death* (Shneidman), 139
von Bertalanffy, Ludwig, 9
von Clausewitz, Carl, 7, 217, 325

war and suicide, 20–24, 153
warrior identity, 12
*Washingtonian,* 148
Waugh, Evelyn, 88
Weber, Max, 324
Wenckstern, S., 218, 220–221
white walls, 249
Whitehorne, Joseph, 127, 128,
  129
Williams, Russell, 193–194
Wilson, J., 220
Wilson, James, 125–127
*Windsor Star, The,* 240
Windelband, W., 81, 91
women, integration into modern military,
  16–18
Wong, D. H., 168
Woods, M., 226–227
Woolf, Virginia, 88
work-family conflict, 222
World Health Organization (WHO), vi, 3,
  8, 18
*World Report on Violence and Health*
  (WHO report), vii
World War I, 214
World War II, 1, 81, 173–175, 214
Worldwide Casualty System, 157
Wortzel, H., 200

young adulthood, 71–72
Yugoslavia (former), 21–23
  military suicide, 170
Yugoslavian War, 20–24, 165, 170
  PTSD and suicide during/after,
    228–229

Zachar, P., 48–49
Zilboorg, G., 29, 31, 59